Functions in Biological and Artificial Worlds

Vienna Series in Theoretical Biology
Gerd B. Müller, Günter P. Wager, and Werner Callebaut, editors

Functions in Biological and Artificial Worlds
Comparative Philosophical Perspectives

edited by
Ulrich Krohs and Peter Kroes

The MIT Press
Cambridge, Massachusetts
London, England

MIT Press books may be purchased at special quantity discounts for business or sales promotional use. For information, please email special_sales@mitpress.mit.edu or write to Special Sales Department, The MIT Press, 55 Hayward Street, Cambridge, MA 02142.

This book was set in 10 on 13 pt Times Roman by SNP Best-set Typesetter Ltd., Hong Kong. Printed and bound in the United States of America.

Library of Congress Cataloging-in-Publication Data

Functions in biological and artificial worlds : comparative philosophical perspectives / edited by Ulrich Krohs and Peter Kroes.
    p.   cm. — (Vienna series in theoretical biology)
    Includes bibliographical references and index.
    ISBN 978-0-262-11321-2 (hard cover : alk. paper)   7014510
    1. Biology—Philosophy.   2. Technology—Philosophy.   I. Krohs, Ulrich.   II. Kroes, Peter, 1950–.
    QH331.F897   2009
    570.1—dc22

10 9 8 7 6 5 4 3 2 1

2008031061

Contents

Series Foreword

Biology is becoming the leading science in this century. As in all other sciences, progress in biology depends on interactions between empirical research, theory building, and modeling. But whereas the techniques and methods of descriptive and experimental biology have evolved dramatically in recent years, generating a flood of highly detailed empirical data, the integration of these results into useful theoretical frameworks has lagged behind. Driven largely by pragmatic and technical considerations, research in biology continues to be less guided by theory than seems indicated. By promoting the formulation and discussion of new theoretical concepts in the bio-sciences, this series intends to help fill the gaps in our understanding of some of the major open questions of biology, such as the origin and organization of organismal form, the relationship between development and evolution, and the biological bases of cognition and mind.

Theoretical biology has important roots in the experimental biology movement of early-twentieth-century Vienna. Paul Weiss and Ludwig von Bertalanffy were among the first to use the term *theoretical biology* in a modern scientific context. In their understanding the subject was not limited to mathematical formalization, as is often the case today, but extended to the conceptual problems and foundations of biology. It is this commitment to a comprehensive, cross-disciplinary integration of theoretical concepts that the present series intends to emphasize. Today theoretical biology has genetic, developmental, and evolutionary components, the central connective themes in modern biology, but also includes relevant aspects of computational biology, semiotics, and cognition research, and extends to the naturalistic philosophy of sciences.

The "Vienna Series" grew out of theory-oriented workshops, organized by the Konrad Lorenz Institute for Evolution and Cognition Research (KLI), an international center for advanced study closely associated with the University of Vienna. The KLI fosters research projects, workshops, archives, book projects, and the journal Biological Theory, all devoted to aspects of theoretical biology, with an emphasis on integrating the developmental, evolutionary, and cognitive sciences. The series editors welcome suggestions for book projects in these fields.

Gerd B. Müller, University of Vienna and KLI
Günter P. Wagner, Yale University and KLI
Werner Callebaut, Hasselt University and KLI

Preface

The notion of function is an integral part of the way of thinking in biology as well as in technology. Traits and organs of organisms as well as technical artifacts and their components have or are attributed functions. The concept of function, however, is notoriously obscure. The same holds for the relationship between biological organisms and technical artifacts. This relationship is obscure because there are, on the one hand, many parallels that never hold completely—evolvability, wholeness, hierarchical and modular organization—and, on the other hand, many important differences that may nevertheless have analogies in the other class—natural selection versus intentionality, propagation versus (series) production, fitness versus usefulness. The concept of "function" is obscure because it seems to imply reference to goals or norms even in cases where intentionality is absent (such as with biology), to effects where the effect is absent (in the case of dysfunction) or it is even missing for principle reasons (in the case of a misinformed design of, e.g., a perpetual motion machine), and because it even may be regarded as unclear whether it is not merely used metaphorically in its biological sense.

Throwing more light on the sketched topics is a highly challenging task for philosophers of biology and technology. Scholars have tried for decades to save the notion of function from obscurantism. This has yielded some highly elaborate explications but not as yet a consensus about which one is acceptable in which case. Scholars have less often tried to clarify the relations between artifacts and organisms, though it is quite common to use one as a model for the other—in both directions—again without coming up with results that go beyond stating common principles like those already mentioned. We decided to combine both issues and to investigate the relationship between organisms and artifacts exactly with respect to the obscure matter of functionality. The reason is that we believe that this very issue is the root of many of the difficulties linked to a proper understanding of biological organisms, technical artifacts, and the relations between the two. Consequently the problems should be treated in an integrative way rather than separated when one aims at a new perspective that sheds light on each of the problems.

The 15th Altenberg Workshop in Theoretical Biology, "Comparative Philosophy of Technical Artifacts and Biological Organisms," held in the Konrad Lorenz Institute for Evolution and Cognition Research (KLI) in Altenberg, Austria, in September 2006, fostered an integrative view on the two topics. The participants traveled to the Danube from all over the world. Discussions in the library at Altenberg were extraordinarily lively and fruitful. Ultimately the positions of the participants did not converge, but that was not our intention. The workshop was held to juxtapose opposing positions and thus broaden the scope for future work and highlight relevant observations and results from the different perspectives requiring consideration. However, all participants significantly rewrote their papers for submission to the present volume. So the reader has in hand the results of the workshop discussions rather than the workshop contributions. On this subject, we want to acknowledge the efforts of those who have supplied the content of this volume. We wish to thank all contributors for their engaged participation in the workshop and for the effort put into writing their chapters after the workshop. Thank you all for your contributions and for your patience and collaboration during the editing process.

The editors also wish to thank the board of the KLI for its generous support of the workshop. This official support was financial and even included permanent Lucullan pleasures—this at least is our recollection. But we also enjoyed immaterial support of various kinds. There was much encouragement and help in the preparation phase for which one of the editors (Ulrich Krohs), then fellow of the KLI, wishes to thank Gerd Müller and Werner Callebaut. During the workshop we took advantage of the perfect logistics, courtesy of the KLI staff. The workshop ran so smoothly that the organizers were able to fully concentrate on scientific content, discussions, and participants. We wish to thank Eva Karner, the secretary, and Astrid Jütte, the executive manager, for their great support. We owe thanks to Maarten Ottens for help with the index. One of the editors, Peter Kroes, would like to thank the Netherlands Institute for Advanced Study (NIAS) for providing him with the opportunity, as a Fellow-in-Residence, to work on the preparation of this volume.

The workshop on functionality had a forerunner in the form of the conference "Artifacts in Philosophy," held at Delft University of Technology, the Netherlands, in 2004. At this conference, the fruitfulness of a comparative approach became visible in many contributions and the basis for the 2006 workshop was laid. We hope that these two meetings will mark the beginning of a fruitful discourse on the philosophy of biology and technology in an integrative and comparative perspective.

I INTRODUCTION

1 Philosophical Perspectives on Organismic and Artifactual Functions

Ulrich Krohs and Peter Kroes

The nature of functionality is one of the big and difficult questions shared by the philosophies of biology and of technology. The ascription of a function to a biological trait goes beyond a mere description of what the trait does. Mammalian hearts move blood and, like most but not all other hearts, they move it through the animal's blood vessels. This biological finding is descriptive in the same way that the geological finding that magma chambers below the base of volcanoes extrude lava out of a crater through a conduit in the volcano or the physical finding that two masses attract each other are descriptive. However, biologists ascribe to the heart not only the action but also the function of pumping blood. In contrast to the mere description, the function ascription allows one also to talk about malfunction or dysfunction, a situation in which a function is impaired or not performed at all (Neander 1995; Davies 2000). This contrasts sharply with the situation in physics. Physicists do not talk about malfunction if some expected physical interaction does not occur but rather about a new phenomenon that requires explanation (though they may refer to malfunctioning technical equipment). Even an inactive volcano is not said to have a malfunctioning magma chamber. In such cases the descriptors "dormant" and "extinct" are used in a metaphorical way to refer to a volcano that is no longer active. In contrast to the findings of these other natural sciences, biological function ascriptions do involve reference to a norm (in a weak, nonmoral sense), which delineates dysfunction from function.

Functionality is not restricted to biological entities. The most obvious domain of function ascription is technology. Again, the ascription of a function to, for instance, an Archimedean screw, goes beyond being a mere description of what this technical artifact does when it moves water upward. It is the function of an Archimedean screw to move water upward, that is, it is supposed to or ought to move water upward. As a consequence, just like biological organs, technical artifacts may dysfunction. In the domain of technology, functionality is even more familiar than in biology and it is often claimed that the concept stems from the former field and that its proper use may be primarily in relation to technical artifacts. Here the intentionality of designers, makers, and users comes into the picture and may well be the source of the normativity of technical functions.

It turns out, however, that also with respect to technical functions, things are not that easy.

Entities of a third kind, namely social institutions, are also often described in a functional way. We are convinced that functional approaches to sociology may also profit from a comparative perspective when analyzing functions, but that is not included in the present comparative approach. The reasons are as follows. First, there is much less consensus among sociologists than among biologists or technicians on whether or not the systems they are dealing with should really be conceived as functional systems. Sociological structuralism manages without function ascriptions, so the problem of functionality seems to depend much more on the general approach adopted within the sociological field than in the fields dealt with in this volume (Krohs 2008a). Second, social institutions, if conceived functionally, combine aspects that are found to be relevant to function ascriptions in biology (evolution, development, and organization) and to technology (designing, use, and, again, organization), probably blended in many different proportions. Insofar as these aspects exhaust the notion of "social functions," the latter do not add a new perspective to the ones included in this book.

It should be mentioned that there are more than the clear-cut cases of functional systems so far mentioned. Between each of these poles, all kinds of hybrid systems are to be found (figure 1.1). There are biotechnological hybrids, such as genetically engineered organisms. There are systems that are described as intermediates between biological organisms and

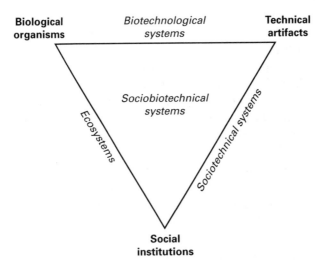

Figure 1.1
A map of the kinds of functionally organized entities. While the entities shown in the corners of the triangle are typically the subject of inquiry within the respective disciplines of biology, technology, and sociology, the intermediate and hybrid systems often, but not always, belong to transdisciplinary fields of research.

social institutions, such as insect states and possibly also ecosystems as conceived in syn-ecology. Intermediates of a third kind are sociotechnical systems such as a company running a factory or—paradigmatically—a coal mine. These can be located on the map of functionally organized entities between technical artifacts and social institutions. Finally, there are sociobiotechnical systems, which involve all three kinds of aspects. Among these systems are farms together with companies that are active in the biotechnological produc-tion sector.

For the reasons mentioned, we confine the scope of the present volume to the cases of biological organisms and technical artifacts. Considerations pertaining to hybrids thus are included only where they relate to intermediates between these two kinds of systems.

Given that organisms and artifacts have both been described in functional terms since antiquity, it is not surprising to find in the history of philosophy and also in biology and technology many attempts to use entities of one kind as a model or explanation for another. The transfer goes in both directions—compare the machine analogy for biological organ-isms (Descartes 1985a [1637], 1985b [posth. 1664]; La Mettrie 1960 [1747]) and see the evolutionary account of technological development (Basalla 1988; Ziman 2000; Lewens 2004). With respect to functions, it was taken for granted that the concept of a technical function is the better-understood concept in a seminal contribution to the debate: "Of the two, natural functions are philosophically the more problematic" (Wright 1973). Artifact functions became a kind of implicit point of reference in the discussion on biological functions (Millikan 1984; Kitcher 1993). Unfortunately it turned out that philosophy of technology was far behind its twin discipline at those times: the concept of a "technical function" was not at all well explicated. When this was taken up as a philosophical chal-lenge, it turned out to be all but an easy task to resolve. In fact philosophy of technology often relied upon theories of function from the biological domain. This had two strange and undesired consequences. First, as far as theories of biological function refer to specific biological processes such as evolution by natural selection or to features characteristic of living organisms—not all, but some of the most prominent ones do—it seems rather arti-ficial to transfer them to technology. Second, and worse still, the definitions are at risk of becoming circular insofar as the technological concepts themselves are explicated in terms of biological concepts.

In the early days of philosophizing about functions, the divide between the biological and the artificial world was not yet an issue. Aristotle conceived of all entities and pro-cesses in the world as being subjected to four causes or origins (*aitiai*), one of which was the teleological cause, inherent in the answer to "what for" questions (Aristotle, *Physica*). He even maintained that objects like a stone have a goal that causes it to fall: its natural place is on the ground or in the center of the universe. However, teleology—goal-directed-ness—is in fact a much stronger concept than functionality. But his concept of "*energeia*," the actuality of an entity, in its application to living entities may indeed be seen as a fore-runner of present-day concepts of function: according to Aristotle, this is the way living entities are "working." Aristotle often used the alternative term *entelecheia,* which translates "having the end within itself," when writing in particular on the *energeia* of living entities. Though this is not a clear terminological distinction in his writings, the use of the term *entelecheia* may show how close he conceived the connection between the *energeia* and ends or goals to be. Notably he often explains the *energeia* of living entities by means of technical analogies. Already his talk about "organs" involves this analogy, since *organon* is Greek for "tool."

During and after the Renaissance, teleology was banned from the physical world, and causality within physical processes became restricted to effective causality. Kant then noticed that biological organisms cannot be conceived of without some kind of reference to teleology, but he banned the goals of goal-directedness and gave teleology the status of a regulative idea—something we need to assume in an "as if" mode to understand living nature but not something that constitutes nature. What remains is teleological judgment instead of teleological explanation: we cannot do without this kind of judgment but we cannot know whether teleology is indeed present (Kant 2007 [1791]). For Kant the reason for needing teleological judgment at all was that there is "cyclic causality" in living organisms. What is well known to us, for instance, from feedback control, was unimaginable to him—an explanation of a causal chain being closed to a loop in strictly physical terms. For him only linear or branched causal chains were imaginable. He had of course no concept of a system far from equilibrium or of a dissipative system, which makes cyclic causal processes easily understandable. Nevertheless he anticipated modern concepts of regulation in his notion of cyclic causality, which he related to teleological judgment. So in Kant's writing we encounter the idea that biological functionality is related to a particular organization of a living entity rather than to goals.

1.1   The Challenge of Dysfunction

We have already seen that wherever a function is ascribed, dysfunction immediately comes into play. A function may be performed well or poorly or even not at all. One refers to or poses a norm when ascribing a function, a norm that may not be met. However familiar expressions like "a bad heart" or a "good coffee machine" may sound, explicating the normative aspects of functions is not a trivial undertaking at all. Several major problems arise. In the first place, function talk sits uneasy within a naturalistic approach to the world, an approach that roughly takes the outcomes of modern science and its descriptive methodology as its point of departure. Within a "naturalistic" perspective, the ascription of functions to objects, in particular natural objects, is rather problematic. In contrast to the Aristotelian approach, the idea that physical objects or chemical substances have functions has lost its validity in modern physical sciences. Only within the biological sciences has the attribution of functions to organs, traits, or the behavioral patterns of organisms stayed alive and it is, according to many biologists, an indispensable facet of their conceptual toolkit. They claim that an adequate description and explanation of biological phenomena requires recourse to the notion of function. If that is indeed the case, then the problem of how normative statements with regard to functions may be reconciled with the underlying descriptive methodology arises. One way to avoid this conflict is by assuming that there are, after all, norms in at least biological nature. In that case the statement "this is a bad heart" is simply the objective description of a normative state of affairs in the biological world. This leads to a form of normative realism—that is, the idea that there are normative states of affairs in the world—with all its problematic aspects. Another way to avoid the conflict is by denying that function talk is necessary in the biological sciences. In his contribution to this volume, Davies argues that biologists who cling to function talk are suffering from "conceptual conservatism" and that function talk should be given up. If that is done, it becomes possible to remain faithful to the descriptive methodology but again at a considerable price: as far as statements like "this is a bad heart" or "this heart ought to behave like this or that" have any meaning at all, they describe nonnormative states of affairs. Whether this is an adequate interpretation of the meaning of prima facie normative statements with regard to functions remains controversial.

In the second place, it is far from clear how normative statements about technical artifacts are to be interpreted. One of the first attempts to interpret the "goodness" of artifacts stems from von Wright (1963). In contrast to normative statements about biological entities, normative statements about technical artifacts appear to be intimately related to human action. Humans make use of technical artifacts and it is quite common to (partially) ground the functions of technical artifacts in intentional human action (Kroes and Meijers 2006). This grounding of technical functions in intentional human action opens the possibility to explicate normative statements about technical artifacts in terms of normative statements about human action. The contribution by Franssen to this volume contains one of the rare attempts to spell out the details of how this might be done. It shows that the interpretation of normative statements about technical artifacts is far from self-evident.

1.2   Disanalogies Between Biology and Technology

The main disanalogy between functions in biology and technology that immediately springs to mind is indeed functions' relation to intentional human action. Theories of biological functions make no reference to human intentionality (Searle [1995] being a notable exception). By contrast, in most theories of technical functions, human intention plays a constitutive role in the sense that without human intention (of designers, producers, users, etc.) it does not make sense to claim that technical artifacts have or may be attributed functions. Within the technological domain, functions may be interpreted in terms of means-ends relations and the ends involved may be simply interpreted as the ends of human beings. Within the biological domain, an interpretation of functions in terms of means-ends relations is much more problematic because the status of ends within the biological world is problematic.

This difference in the role of intentionality with regard to the notion of function in biology and technology may prove to be a major obstacle to attempts to develop a unified account of normative aspects of biological and artificial functions. This is simply a special aspect of the general problem of whether it is possible to develop a general theory of functions applicable to the biological and the artifactual domains. If indeed the normative aspects of technical artifacts are derivative of human action, then the prospects for such a general theory of functions appear dim. By analogy, the normativity of biological functions of organs, for instance, would have to be grounded in the use organisms make of such organs. However, generally speaking, it hardly makes sense to say that organisms make use of their organs. Moreover, grounding the normativity of biological functions in the use that humans make of organisms seems out of the question, since their organs have functions independent of any human use.

The search for a unified account of normativity may be in vain for different reasons. To start off with, there may well be different sources of normativity in biology and technology. Another possibility is that with respect to biology, talk about normative functions may not be justified. The problem is that there is a difference between regarding the reference state of a particular function as brought into being, for example, through natural selection, and viewing its very status as a reference as a product of evolution. Some contributions to this volume deal with the difficulty of establishing a naturalized, nonnormative account of biological function (McLaughlin; Davies).

Another disanalogy between the fields is that a technical artifact is usually ascribed a function as a whole, while organisms as wholes are not considered to have functions. The function of a car is to enable rapid movement on streets, the function of a lathe is to turn wood or metal workpieces, the function of a molding press is to form plates. But killing mice or looking majestic is not the function of an eagle, nor is it the function of a dormouse to sleep for a considerable part of the year. So while in biology functions are only ascribed to components of organisms, artifacts-as-wholes do have functions. This need not mean that the difference holds from any perspective. It may well be that the functions of artifacts-as-wholes are relational with respect to the system in which they are used, for example, to a functionally organized sociotechnical system (Krohs 2008a). On the other hand, moving up one level in biology we have to consider ecosystems as conceived in synecology. If these can be described as functionally organized, then organisms may well have functions-as-wholes insofar as they are components of an ecosystem. What remains to be seen, however, is whether the concept of an ecological function is normative in the same sense as functions in sociotechnical systems.

1.3   A Brief Survey of the Parallels Between the Fields

Despite the differences discussed in the past two sections, there are many parallels between organisms and artifacts. If there are reasons to apply the concept of "function" to both kinds of entities, it seems plausible to look for them in such shared or at least similar features. An obvious parallel that holds between organisms and the more complex of the technical artifacts is to be found in hierarchically organized systemic structure. Consequently one of the basic notions of function refers to function as the role of a component within a system (Cummins 1975; his concept is more elaborate than is apparent from this sketch; cf. the contribution of Mark Perlman in this volume). However, such a notion is also applicable to many physical systems, such as solar systems, atoms, or the hydrological cycle, precisely because it lacks normativity. One has thus to look at more peculiar parallels when explicating a notion of normative functionality. We list several in this section, pointing each time to the differences between both fields with regard to each aspect and clarifying the different terminology used in both fields to refer to comparable features.

Another important parallel, apart from organization, is to be found in evolution, which occurs in the biological and technical realms. Just as mammals and birds evolved from reptiles, so jet planes are said to have evolved from less sophisticated airplanes. However, the underlying processes of variation and the retention of variants may follow largely different mechanisms in both cases. Variation is mostly considered to be blind in the organismic case and directed within the technical domain. As an aside—looking for the origin of this parallel, it should be observed that Darwin (1988 [1859]) describes the process of natural selection as parallel to the breeding process, that is, to a process that belongs at least partly to the artificial domain.

Biological development finds its equivalent in technical construction. In both cases, deviations from what may be regarded as developmental pathways fixed in the genome or as instructions laid down in a construction plan may occur. So development and construction really do have a modifying influence on the resulting entity and to that extent on its functionality or functional organization. Again the influences in the technical domain, but not in biological cases, are at least in part intentional.

In a way biological reproduction may be paralleled with technical series production. However, in the biological case propagation and multiplication are the sources of variation, while in engineering there is usually avoidance of variation in the multiplication process. Instead variation is sought in separate steps.

A final parallel we want to mention is the way in which biological and technical entities retain their integrity. Biological recovery, regulation, and self-repair can be seen as counterparts to technical maintenance and repair. The big difference is that usually these processes are internal in biological organisms, being performed autonomously to the degree laid down in or allowed for by the internal structure of the organism. There are often strict limits to biological regeneration. Mammals cannot regenerate lost limbs, though many amphibians can and do. In technical cases maintenance and repair are brought about by external agents, partly on a regular basis and partly ad hoc, in accordance with requirements and feasibility.

Various theories of function refer to one or more of the aspects mentioned. Insofar as these aspects are shared between the fields—though differently termed—it seems promising to apply theories of function that refer to one or more of the topics in both fields (Krohs 2008b). However, the already-mentioned differences show that any unified approach also has to face the fundamental differences between the biological and the artificial realms. On the other hand, in the growing class of biotechnical hybrid systems these differences may either fade out or else the hybrids might at least prove that the occurrence of biological functions is not excluded from the technical world and vice versa.

1.4   The Aim of This Volume

Up until now contributions to the debate on the concept of "function" usually have been biased in that they are oriented to one of the fields. The other field was used just as a reference—without acknowledging that the problems on the other side are as big as those an author sees in his or her own field. Due to this habit, the authors forfeited the chance to profit from a view that takes both fields in question into account. This book aims at doing justice to both sides, to the functionality of organisms and of artifacts, and it aims to present proper philosophical analyses of the concept of function from a perspective that embraces both fields of function ascription. In this way it aims at a better understanding not only of the concept of "function" itself but also much more generally of the similarities and differences between organisms and artifacts insofar as they are related to functionality. The contributions to this volume fulfill this aim by presenting ontological, epistemological, and phenomenological comparisons. This helps clarify problems that are at the very center of the philosophies of biology and technology. The results are also valuable to the philosophy of social science.

This volume also seeks to contribute to the emancipation of the philosophy of technical artifacts. Within philosophy, artifacts in general, but technical artifacts in particular, have been neglected for a long time. It is only during the past decades that artifacts have become a topic of philosophical analysis in their own right (Dipert 1993; Preston 1998; Thomasson 2003, 2007; Hilpinen 2004). Even within this emerging field of the philosophy of artifacts, *technical* artifacts often play only a marginal role and if they are taken into consideration it is usually in the form of technical artifacts that are produced by craftsmanship and not by modern engineering. However, it is one thing to compare a beaver dam to a stone ax but another thing to compare it to a modern Airbus 380. If we take into account the complexity in the structure of the technical artifacts involved as well as the complexities of the production processes, then we may question on valid grounds whether stone axes are representative for modern engineered technical artifacts. Whether the constitutive role of human intentions and of physical structures in realizing the functions of these different kinds of technical artifacts may be treated in the same way is, for instance, debatable. Such problems belong to the philosophy of engineering design, a field that is virtually nonexistent (Kroes and Meijers 2001; Krohs 2004).

Philosophers of biology and of technical artifacts may learn a lot from a comparative analysis of functions in both domains, even where such an analysis leads to the conclusion that we are dealing here with two fundamentally different kinds of functions. In such a case we have to develop, for instance, different explications of the normativity associated with these different kinds of functions and stop using misleading analogies between function talk in both domains. However, it would surely be premature to draw such a conclusion. It is precisely the recent development of functional theories for technical artifacts that offers a unique opportunity for such a comparative approach. The overall question about the extent to which it will be possible to arrive at a common interpretation for the notion of "function" that is viable for both the domains of biology and technology remains an open issue. This volume presents a number of significant results that must be taken into account in future discussions about the possibility of a unified function theory for biology and technology.

References

Aristotle. Physica. *The Works of Aristotle, Vol. II.* (Ross, D., ed.). Oxford: Clarendon Press, 1970.

Basalla, G. (1988). *The Evolution of Technology.* New York: Cambridge University Press.

Cummins, R. (1975). Functional analysis. *The Journal of Philosophy, 72:* 741–765.

Darwin, C. (1988 [1859]). On the origin of species by means of natural selection. In: *The Works of Charles Darwin, Vol. 15,* (Barrett, P. H., Freeman, R. B., eds.). London: Pickering.

Davies, P. S. (2000). Malfunctions. *Biology and Philosophy, 15:* 19–38.

Descartes, R. (1985a [1637]). Discours de la méthode/Discourse on the method. In: *The Philosophical Writings of Descartes, Vol. 1* (Cottingham, J., Stoothoff, R., Murdoch, D., trans.). Cambridge: Cambridge University Press.

Descartes, R. (1985b [posth. 1664]). "La Description du corps humain/Description of the Human Body and All of Its Functions." In: *The Philosophical Writings of Descartes, Vol. 1* (Cottingham, J., Stoothoff, R., Murdoch, D., trans.). Cambridge: Cambridge University Press.

Dipert, R. R. (1993). *Artifacts, Art Works, and Agency.* Philadelphia: Temple University Press.

Hilpinen, R. (2004). Artifact. In: *The Stanford Encyclopedia of Philosophy* (Zalta, E. N., ed.). http://plato.stanford.edu/archives/fall2004/entries/artifact/.

Kant, I. (2007 [1791]). *Critique of Judgement.* Oxford: Oxford University Press.

Kitcher, P. (1993). Function and design. *Midwest Studies in Philosophy, 18:* 379–397.

Kroes, P., Meijers, A. (2001). *The Empirical Turn in the Philosophy of Technology.* Amsterdam, New York: JAI Press.

Kroes, P., Meijers, A. (2006). The dual nature of technical artefacts. *Studies in History and Philosophy of Science, 37:* 1–4.

Krohs, U. (2004). *Eine Theorie Biologischer Theorien.* Berlin: Springer.

Krohs, U. (2008a). Co-designing social systems by designing technical artifacts: A conceptual approach. In: *Philosophy and Design: From Engineering to Architecture* (Vermaas, P. E., Kroes, P., Light, A., Moore, S. A., eds.), 233–245. Dordrecht: Springer.

Krohs, U. (2008b). *Functions as Based on a Boncept of General Design.* Synthese, in press.

La Mettrie, J. O. de. (1960 [1747]). *L'Homme Machine.* Princeton: Princeton University Press.

Lewens, T. (2004). *Organisms and Artifacts: Design in Nature and Elsewhere.* Cambridge, Mass.: The MIT Press.

Millikan, R. G. (1984). *Language, Thought, and Other Biological Categories: New Foundations for Realism.* Cambridge, Mass.: The MIT Press.

Neander, K. (1995). Misrepresenting and malfunctioning. *Philosophical Studies, 79:* 109–141.

Preston, B. (1998). Why is a wing like a spoon? A pluralist theory of function. *Journal of Philosophy, 95:* 215–254.

Searle, J. (1995). *The Construction of Social Reality.* London: Penguin Books.

Thomasson, A. L. (2003). Realism and human kinds. *Philosophy and Phenomenological Research, 67:* 580–609.

Thomasson, A. L. (2007). Artifacts and human concepts. In: *Creations of the Mind: Theories of Artifacts and Their Representation* (Margolis, E., Laurence, S., eds.), 52–73. Oxford: Oxford University Press.

von Wright, G. H. (1963). *The Varieties of Goodness.* London: Routledge & Kegan Paul.

Wright, L. (1973). Functions. *Philosophical Review, 82:* 139–168.

Ziman, J. (ed.) (2000). *Technological Innovation as an Evolutionary Process.* Cambridge: Cambridge University Press.

II BRIDGING FUNCTIONS OF ORGANISMS AND ARTIFACTS

One of the primary topics in a comparative perspective on biological and technical functions must be the question of whether there is a gap between both kinds of functions and, if so, whether it can be bridged. The gap is visible in particular when considering the difference in the most commonly accepted bases of the normativity of function, namely the selection history of the evolutionary adaptation of biological traits on the one hand and the intentionality of a human designer or user on the other hand. An easy way to bridge the gap would be by giving up the normativity of functions and thus any allusion to normativity altogether in order to understand functional talk merely in terms of a description of systemic roles of components of a system. This may seem to be an innocent move that allows one to understand functional talk as being purely descriptive and therefore acceptable in any naturalistic account without need for further justification. However, the potential to distinguish function from dysfunction would in this way be renounced, which would go against the biological and the technological use of the concept of functionality.

Perlman votes to not refrain from teleology but rather to refocus the debate on teleological functions. He first recommends giving up positions in the debate that may seem natural to many participants but that he regards as too restrictive. His recommendations are 1) Don't draw a hard line between natural functions and functions of artifacts, 2) Don't let teleofunctional theories neglect contexts, 3) Don't make designer's intentions essential to artifact function, and 4) Don't let theories of teleology spiral out of control into definitional oblivion. Recommendations 2 and 3 are, in part, specifications of 1; they specify the basic steps that are required to avoid the hard dividing line between natural and biological functions. Recommendation 4 is designed to protect us from definitorial sophistry that distracts our attention from the fundamental issue we are tackling. Perlman regards such sophistry as inevitable in approaches to functionality that aim at a unificatory approach. To avoid it he proposes allowing for different facets of functionality in both fields, biological and artificial functionality. Emphasis should be shifted to explanatory use of function ascriptions of different kinds—an approach he calls "Pragmatic Teleo-Pluralism."

Preston considers the problem of bridging organismic and artifactual functions from another perspective. She deals with the problem of making sense of the concept of a proper function—that is, a function that "sticks" to its bearer like a proper name to a person—in both considered domains. Proper functions were initially defined by referring to the evolutionary etiology of a trait in terms of a history of selection and reproduction (Wright 1973; Millikan 1984). The idea that natural selection picks out biological proper functions in that way has been, according to Preston, progressively eroded, which now makes it doubtful that natural selection has any such relationship to proper function. She shows that the same doubts hold with respect to the parallel case where cultural selection is regarded as the basis of the occurrence of proper functions of technical artifacts and of material culture in general. She then assesses the idea of transferring more recent modifications of the etiological account of biological functions to functions of material culture. She concludes that neither reference to only the recent selection history (instead of to the whole evolutionary past), nor a definition based on the contribution made by a function bearer to the biological fitness of an organism can be transferred to artifactual functions. Fitness in particular does not have a good analogue in material culture. She concludes that definitions of proper function that are suitable for biology are unlikely to be of much use as models for a definition of proper function in material culture. Preston proposes that proper functions of material culture are rather based in use and reproduction.

Longy chooses a completely different approach to discuss the gap between biological and artifactual functions. She questions whether there is a conceptually clear distinction between both domains that needs to be bridged at all. She takes it that biological functions currently are described as selected effects and artifactual or cultural functions as intended effects. Pointing to functions in domesticated animals and cultivated plants that are at once biological and artifactual, she argues that no ontological basis for the mentioned distinction can be found. From this observation she concludes that we need to adopt a different perspective on functions than the selected-versus-intentional dichotomy can offer. This dichotomy refers to the origins of functions, but precisely the origins are often of a mixed nature, comprising both selection and intention. Consequently Longy proposes placing less weight on these origins. An acceptable account of functionality should be more abstract and allow for different (selective) mechanisms rather than make the concept dependent on a particular mechanism. The approach to functionality she exposes is closely related to Wright's (1973) account but avoids some shortcomings that Wright's approach is often accused of.

Vermaas concentrates on another aspect that divides biological from artifactual functions. With respect to biology he refers to the main stream of approaches clustered around Wright's and Millikan's proper function accounts and takes it that biological functions are usually considered to be features the function bearers objectively possess. In contrast, functions of technical artifacts are viewed as subjective features of the artifacts that depend on the beliefs of agents. He argues that though theories of biological function often fit in well with the objectivity presupposition and theories of artifactual function with the subjectivity view, the difference between the theories is not a categorical one. First, he introduces sufficient conditions to transpose function-ascription theories into theories about functions-as-properties. He then shows how theories that analyze functions of technical artifacts in terms of beliefs of agents can be transposed into theories in which artifacts have functions as properties. He shows how this transformation allows a theory of (subjective) artifact functions to also be applied to the realm of (objective) biological functions. The resulting unified theory of function is an epistemic one, so the objectivity of ascribed functions is given only within the epistemic framework of the ascribing theory. This will still not satisfy those who are looking for an objective account of biological functions that fits within a unified approach. But the demonstration of how theories of objective and subjective function may be unified sheds new light on the question of the gap between biological and artifactual functions.

References

Millikan, R. G. (1984). *Language, Thought, and Other Biological Categories: New Foundations for Realism.* Cambridge, Mass.: The MIT Press.

Wright, L. (1973). Functions. *The Philosophical Review, 82:* 139–168.

2 Changing the Mission of Theories of Teleology: DOs and DON'Ts for Thinking About Function

Mark Perlman

Teleology has a long history. Religious doctrines often give teleology a central role—the gods infuse Nature with their own goals and purposes, and rains and droughts and earthquakes occur to bring about divine goals, or a single God is said to instill functions by design throughout Creation. These views seek to explain natural events on the model of something we think we understand—intelligent human creation of objects for a purpose. Aristotle made generous use of teleology in describing objects, organisms, and their interactions, and even as the basis of ethics and metaphysics. This cornerstone of his philosophy remained influential for more than eighteen hundred years. But with the Scientific Revolution and the Enlightenment, talk of the function of natural objects, teleological function, began to be viewed with suspicion as the mechanical model of the world replaced the old Aristotelian model. To a large degree nonnatural explanations based on religion have given way to science. Yet there are still areas of nature that seem at their core to involve teleological functions—the parts of organisms. Biology seems to be unable to do without functions, and even after science rejected functions so forcefully, it became clear that even a dedication to naturalistic explanation seems to require retaining teleological functions. So philosophers grudgingly allowed functions, albeit on a Deductive-Nomological model of explanation with cumbersome and unwieldy formulations that really didn't put any explanatory weight on functions being an important part of scientific explanation.

The modern philosophical movement to legitimize teleology began in the early 1970s, with Wright's 1973 paper "Functions," which proposed a definition of function that was naturalistic, historical (or etiological) in nature, simple, and elegant:

The function of X is Z *means*:

(a) X is there because it does Z, and

(b) Z is a consequence (or result) of X's being there.

This was joined by a 1976 book by Wright and important papers by Boorse (1976, 1977), Wimsatt (1972), Mayr (1974), Woodfield (1976), and others. In 1975, Robert Cummins's paper "Functional Analysis" broke new ground in giving a naturalistic and ahistorical analysis of functions based on causal role in a system, an approach that had little trace of the metaphysical, and proposed an account acceptable to science. The historical, or etiological, side to naturalizing teleology was brought to prominence with Ruth Garrett Millikan's landmark 1984 book, *Language, Thought, and Other Biological Categories* (hereafter abbreviated as "*LTOBC*"), and subsequent papers (1986, 1989a, 1989b, 1989c, 1990, 1993, 2002). Following her book, there was a flood of different accounts of functions, how they would or would not serve as the basis of this and that, and the functions literature took off.

We now have various streams or families of views on teleology, with reductionist theories dividing into systematic (causal-role, Cummins-style) functions on one end, and various historical/etiological views on the other. The historical versions further divide into those that focus on evolution by natural selection in the distant past (Millikan [ibid.]; Neander 1991a, 1991b; Papineau 1987; Griffiths 1993; Allen and Bekoff 1995) and those looking only to goal-contribution in the more recent past (Nagel 1977; Woodfield 1976; Nissen 1997; Boorse 1976, 1977, 2002; Godfrey-Smith 1994). We can also look to the future for a basis of functions—functions as propensities (Bigelow and Pargetter 1987). Then there are the nonreductionist views of functions, from Emergentism (Bedau 1990, 1991, 1992a, 1992b; Cameron 2003) to Conventionalism (Searle 1998). In my 2004 article "The Modern Philosophical Resurrection of Teleology," I present the following taxonomy diagram of theories of function (see figure 2.1).

The terrain is getting thick with competing views of teleological function. In one sense this is a good thing—we're reexamining an area that seemed virtually closed off from philosophical consideration only half a century ago. From the viewpoint of theory of science, it is unclear whether we are in what Kuhn called the chaotic "Pre-Paradigm" phase, or whether we have competing paradigms. Or perhaps the paradigm is naturalistic reductionism, and the proliferation of views is just the philosophical side of "normal science".

Things become even more complicated when we look to the issue of functions of artifacts. It may have appeared that the functions of artifacts were easily explainable by reference to the intentions of the makers of the artifacts. But we now see difficulties with such a simplistic account, and philosophers are making use of the developments in the teleological view of organisms to answer questions about artifact functions.

However, rather than pledge my allegiance to one or the other camp in the debate, I want to look at the debate itself, and see if perhaps some of the conflict between theories is misplaced or misguided. Among these various views, the counterexample game is now going strong—clever exceptions lead to revisions, extensions, exceptions, or rejections of the functional account. Perhaps the problem lies in the mission of seeking "*the* function" in the first place. I argue that many of these theories have something right about them—the mistake is thinking that we can use only one. By pursuing a more multifaceted approach

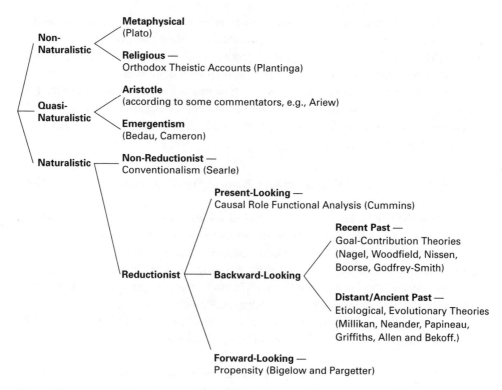

Figure 2.1
Categories of philosophical theories of teleological functions.

to identifying and using functions, and focusing on explanation, we can move to a richer and more successful use of teleological functions. I use an analogy with epistemological theories to shed some light on the shape of the teleology debate.

In what follows, I suggest some factors to consider in the further development of theories of teleofunctions. They amount to a set of DOs and DON'Ts, mainly four DON'Ts. Perhaps they can be called "guidelines," though they may be merely pieces of advice. I can't say I provide deductive proofs that one should avoid these things, or that they could not possibly yield results. But I do provide arguments against them. First, I argue that there is no principled difference between natural functions and functions of artifacts. There are many cases of supposedly "natural" objects that are actually so altered by human action as to be as much artifact as fully natural. So my first recommendation in thinking about function is 1) DON'T draw a hard line between natural functions and functions of artifacts. Then I examine the role of context in teleofunctional explanations, and urge more attention to it. Thus, 2) DON'T let teleofunctional theories neglect contexts. Then I move to the artifacts side, and urge caution in appealing to intentions of the designers of objects as the basis of assigning function. Although it is tempting to say that the function of an artifact comes from what its maker meant for it to do, there are many instances where this is not the case. Thus, 3) DON'T make designer's intentions essential to artifact function. Finally, I use an analogy with theories of knowledge and epistemological justification to put the teleofunction theories in perspective. The result of this is first a warning, 4) DON'T let theories of teleology spiral out of control into definitional oblivion, and then a more substantive positive recommendation of where we should go in thinking of teleofunctions—Pragmatic Teleo-Pluralism.

2.1 DON'T Make a Hard Line between Natural Functions and Artifact Functions

It was often thought to be relatively easy to identify the functions of human-produced artifacts—just look at the intentions of the designers of the objects. The designers built the objects for some purpose, so the objects' purpose rests in what they were intended to do (even if they do not often or ever do it). It was thought that the natural objects, for which naturalism deprives us of the Creator's intentions, are the ones for which it is difficult to clearly identify teleofunctions. However, this dichotomy assumes a hard line between natural objects and artifacts, and this line has been eroded. It might be thought that modern technology is the culprit, but in fact humans have been responsible for creating "natural" objects for thousands of years.

Consider the shape of the landscape. Fly over the Midwest of the United States and you'll see geometrical shapes everywhere—farm fields cut the plains into squares, rectangles, and circles. The whole state of Iowa is almost one giant cornfield. This is obviously not natural—humans have manipulated the environment for their own purposes. Consider also the elaborately manicured gardens of most of the royal palaces of Europe. From Windsor Castle to Versailles to the gardens of Hannover to the gardens of Schloss Schönbrunn in Vienna, plants are meticulously molded to fit human designs (see figure 2.2).

Not only gardens but individual plants can be shaped by human design—Japanese bonsai trees are a prime example. These plants have parts with functions, but given the degree of human intervention in their development, they seem as much artifacts as natural. Consider also the vast varieties of roses—www.everyrose.com lists more than 7,250 varieties of roses, with names from Aafje Heynis and Admired Miranda to Zephirine Drouhin and Zwemania in colors from red, yellow, pink, and white to coral, peach, deep burgundy, and black. Doubtless some of these are close to what could be considered "natural," but the vast majority has been bred by humans for specific color or other characteristics.

Humans have also had a hand in steering the evolution of many animal species. In fact all domesticated animals are designed by humans to a large degree. Dairy cows are bred to produce vastly more milk than they would need in any natural setting—so much so their health can be endangered without regular milking by people, for no calf could drink

Figure 2.2
Gardens of Schloss Schönbrunn in Vienna.

Figure 2.3
A dairy cow—nature did not design it like this!

enough. The udders were designed by nature to deliver milk, but their size and productivity are designed by human intentions (see figure 2.3). These domesticated animals, while designed by human breeding to be good for our purposes, are no longer well suited to living in the wild. Thus they now depend on human care for their existence.

The amazing variety of breeds of dogs is virtually entirely due to human breeding (see figure 2.4). We have bred them for more than two thousand years for various purposes. We created retrievers and other gun dogs (or bird dogs) for retrieving birds in hunting, various kinds of shepherd dogs for herding sheep, pit bulls and other dogs for attack, huskies for pulling sleds through snow and tundra, Saint Bernards for snow rescue,

Figure 2.4
Varieties of dog breeds—various sizes and shapes and talents all designed by human intervention.

greyhounds for running fast, and pugs seemingly for looking silly and cute and breathing loudly. (I am still wondering about the Chihuahua.) Surely, for all these dogs, their eyes have the function of seeing and their hearts have the function of pumping blood. But their size and shape and certain specific characteristics are bred into them for human purposes. There is a reason a golden retriever will be relentless if it senses even the vaguest possibility of a game of catch-the-tennis-ball—we bred them to have instincts to chase down moving objects. Though all dogs have excellent senses of smell (far better than a human's), some, like bloodhounds, have this sense far in advance of even other dogs. Why? Because we bred them for that purpose. Good shepherding ability was responsible for people continuing to breed shepherd dogs, and for their selecting for reproduction the shepherd dogs with such good shepherding abilities. Many people today (lacking flocks of sheep) like to keep such dogs as pets, but on the etiological view, the dogs' function may still be shepherding even if they almost never do that job.

The functions of dogs and their parts become even more complicated when we consider modern breeding for pedigree and shows. Dog varieties are now bred to fit a specific set of characteristics that have been determined by the dog-show authorities to be ideal for that breed. Thus the German shepherd show dog is not now bred for good shepherding ability so much as to look like the stipulated ideal of that breed. This has taken a toll on them—they now are often victims of hip displacement, no doubt due to the physical shapes they have been designed to have. Many domestic breeds of animals have physical characteristics that diminish their abilities to hunt. Pugs and bulldogs, with their pushed-in snouts and loud breathing, could never sneak up on prey, so they are totally dependent on their human owners for food. Some long-haired dogs (and cats) have so much fur that they often cannot keep it groomed themselves—requiring special grooming. Left alone their fur would be hopelessly tangled and knotted and catch on branches. They depend on human grooming, and can't live healthily without it. These results of breeding are evolutionarily disadvantageous, except that as pets these animals get to live cushy comfortable lives. Many breeds are designed, among other things, to be friendly and good pets. Given that their lovability is what gets them fed by their owners, it may turn out that being cute, friendly, and lovable is what these organisms are *for* (on an etiological model going back hundreds of years). Their human-instilled teleofunction may well be to look pleasing and be cute. This sounds odd against a background of biological stories of natural selection, but doesn't sound odd if we focus on the artificial selection that has shaped these diverse breeds. The etiologists may want to object and restrict the historical forces to those in the distant (predomestication) past so as to retain the functions that natural selection yielded. But this would be begging the question. Selection can be done by accident in nature or according to human purposes.

Domesticated plants and animals, as well as human-designed environments, might then be thought to create a special problem for theories of teleological function, crossing the line between natural objects and artifacts. But I would argue that once we focus on specific structures to be explained, there is no special puzzle. If breeders breed dogs that have smooshed-up faces (like bulldogs and pugs), then the prevalence of smooshed-up faces is explained by selection. If people breed smooshed-faced dogs because smooshed-up faces are actually pleasing to some people, then smooshed-up faces are *for* pleasing people. This is a straightforward answer to the question of function, and in the end the fact that humans had a part in the selection doesn't make the domestic animal case any different in kind than the wild animal case. So here the artifactual function of domesticated animals is determined in the same way as the natural proper function was. Why do dogs have hearts? Evolution will tell us. Why are there so many pug dogs around? The answer rests with people's desire for these animals, with their peculiar physical traits. So despite the urge to see artifact function as vastly different from natural function, this class of artifacts— living, biological human-designed artifacts—requires nothing special from a theory of the functions.

Furthermore, the blurring of the line between "natural" and "artifact" has occurred at an environmental scale. Phoenix, Arizona, is in the middle of the desert in the American Southwest, yet it has more than two million residents. This is possible only because of the damning of rivers and providing air-conditioning in every building. But the entire city has become even hotter due to the large amount of concrete and asphalt, and that famous "dry heat" has become increasingly humid due to swimming pools and lawn sprinklers. The once clear desert air is often thick and brown with dust and car exhaust. For another

Figure 2.5
The "American Falls" of Niagara Falls, seen currently (left) and in 1969 (right) when water was blocked off for "restoration."

example, it is not well known that the shape of Niagara Falls (on the U.S.–Canada border) is not what the water eroded naturally—its appearance has been partly engineered by humans. For one thing, 50 to 75 percent of the river's water is diverted upstream by tunnels for hydroelectric power generation, so the falls have much less water than they would naturally. Furthermore, the very shape of the falls has been carved by human intervention, dynamited to blast off loose rock and bolted together (while the water source was blocked to leave the falls dry) to secure the rock and reduce the natural erosion (see figure 2.5). So while the presence of a large waterfall in that location is natural, the exact appearance of Niagara Falls has been significantly affected by intentional human intervention.[1]

For very large-scale human alteration of nature, consider global warming. It has become clear now that human activity is increasing the average temperature of the entire planet, such that nothing can be said to be truly "natural" if by that we mean untouched by human tampering.

One might attempt to retain the distinction between natural and artificial functions by admitting that there is no hard distinction between natural and artificial *objects* but insisting that the *functions* of these objects can be divided into two distinct kinds—natural and artificial. The functions may remain separable even if many objects have a hybrid status. Such a view would be tempting if the kind of human action on natural objects was a recent phenomenon, an innovation of the technological age of the past century or two. But human intervention in nature is not really new—we have been breeding domesticated animals for thousands of years. Moreover, we are ourselves produced by nature, so one wonders how our behavior can be so fully deemed unnatural. It may well be that recent technology has greatly increased the magnitude of human alteration of nature, but the existence of human changes to the natural world with biological and geological artifacts predate recorded history. So what exactly separates the supposed functions into two dis-

Figure 2.6
A chimpanzee using a rock as a tool.

tinct kinds? These "natural" objects have been modified by human activity longer than we can remember.

Further blurring the natural-artificial distinction, some artifacts have biological characteristics. Consider computer viruses—they engage in self-replication, supposedly an exclusive biological trait, so much so that we use biological terms like *virus* and *spread* to describe them. Yet they are clearly designed by human beings, and have teleological functions, though not very admirable ones. Add to that the instances of animals creating tools: apes modifying branches, ants building anthills, bees making beehives and honeycombs, otters smashing shells against rocks they put on their bellies as they float, and many others (see figure 2.6). Then there are the various cases of cats, apes, and elephants putting brush to canvas (or walls) and producing paintings—artworks (allegedly) that are "artifacts," but not human-made. The old view of man as the only animal that uses tools has been decisively debunked by more than thirty years of new discoveries of animal behavior. Of course animal tools are not as sophisticated as human artifacts, but that is a matter of degree, not kind. So the hard-line distinction between natural objects and human artifacts has become blurred, and the hard distinction between natural and artifact functions blurs with it.[2]

2.2   DON'T Let Teleofunctional Theories Neglect Contexts

One aspect of teleological functions that seems often to be neglected is context. Objects develop or are given functions in a context. Cummins (1975) very clearly makes his notion of function dependent on context. It was also a central feature of Millikan's *LTOBC*—she even gives us technical terms—"Normal conditions" and "Normal explanations." But other theories spend too much time focused on trying to define *function,* and pay too little attention to context. This can lead to an over-reliance on function, when considering the context would do most of the explaining.

A well-known account of function comes in Dretske (1986), who invokes biological functions in an attempt to solve the problem of misrepresentation. His teleological account of misrepresentation rests on a relatively simple claim: a representation M means that p if and only if it is M's function to indicate the condition of p. In his view, functions distinguish events that, for adaptational reasons, are supposed to occur (and thus determine content) from those that merely happen to occur. When a token instance of M occurs because of a condition it is its function to represent, M correctly represents that condition. When a token of M occurs because of a condition other than one it is M's function to indicate, M fails to perform its function and is a misrepresentation. Dretske (1986: 26) illustrates his point with an oft-cited biological example:

Some marine bacteria have internal magnets (called magnetosomes) that function like compass needles, aligning themselves (and, as a result, the bacteria) parallel to the earth's magnetic field. Since these magnetic lines incline downwards (towards geomagnetic north) in the northern hemisphere (upwards in the southern hemisphere), bacteria in the northern hemisphere, oriented by their magnetosomes, propel themselves towards geomagnetic north. The survival value of magnetotaxis (as the sensory mechanism is called) is not obvious, but it is reasonable to suppose that it functions so as to enable the bacteria to avoid surface water. Since these organisms are capable of living only in the absence of oxygen, movement towards geomagnetic north will take the bacteria away from oxygen-rich surface water and towards the comparatively oxygen-free sediment at the bottom. Southern hemispheric bacteria have their magnetosomes reversed, allowing them to swim towards geomagnetic south with the same beneficial results. Transplant a southern bacterium in the North Atlantic and it will destroy itself—swimming upwards (towards magnetic south) into the toxic, oxygen-rich surface water.

Dretske invokes the complexity of the organism to deal with this problem: with more complex organisms, organisms capable of learning, we are entitled to the more liberal function of oxygen-indication, whereas in less complex organisms without the resources for expanding their information gathering resources, the more conservative magnetic-field-indication is the function.

Millikan's (1989a, 1989b) answer to this indeterminacy problem is that, from the standpoint of the "consumer" of the representation, the content is clear. The consumer needs only to have a representation that will lead it to behave in the right way. By that point in the organism's physiology, the history of the representation does not make a difference to the effect that consuming the representation will have. Thus Millikan's position is that, from the consumer end, the representational content is univocal and determinate—the direction of oxygen-free water (not magnetic north). This content assignment has the transplanted organisms misrepresenting their surroundings, as their environment does not have oxygen-free water in that direction, and this kills them.

Many others have weighed in on the broad or narrow scope of the content of the bacteria's representations. But my point is that the question isn't really about function and content, it is about context. If we want to explain why the bacteria die, focusing so much on their representational content and its function can make us miss the most important factor. They seem to be functioning correctly, but in a foreign context. The lesson is not that an organism can misrepresent its environment in a foreign context—it is rather that even if you do everything right, it can get you killed. The bacteria function properly, but unfortunately for them, doing it the way they do is fatal in this particular foreign context. The change in context or environment is what explains their demise much more obviously than obsessing about the function and content of their mental representations. Their problem is that their systems cannot really perceive or comprehend their larger environmental context, and thus cannot adjust to certain kinds of changes in it. For limited creatures like that, change in context is more than they can adapt to, and they die. Focusing on functions muddies the water, whereas focusing on context easily explains why switching hemispheres would give us a bunch of dead bacteria.

2.3 DON'T Make Designer's Intentions Essential to Artifact Function

Human beings are (many of them) intelligent designers and creators, and often build or produce objects *for* certain purposes, thus instilling in them *proper functions*. As long as it is safe to talk of human intentions for producing such objects, we can simply put the intentions into the objects as functions. While there seem to be long-standing questions about ascribing teleology to natural objects, it seems easy to place teleology into human artifacts. It seems much easier to have a purpose and a function in a lawnmower or can opener than in a heart or a bee dance.

Karen Neander (1991a: 462) gives us a clear example of such an intentionalist view of artifact functions, where the function "is the purpose or end for which it was designed, made, or (minimally) put in place or retained by the agent." McLaughlin (2001: 52) similarly has it rest in "the actual intentions of the designer, manufacturer, user, etc." Millikan (in *LTOBC*), not surprisingly, has artifact function based on the reproduction of the artifact, in this case by the designers or users. Thus for Millikan an artifact's functioning the way it does causes its reproduction and retention—survival of the technically fittest—but it still rests indirectly (or in part) with the intentions of the designers and/or users who make the judgments about whether or not an artifact is doing what they want it to do.

Vermaas and Houkes (2003) make one of the key taxonomical distinctions among teleological theories whether or not a theory of function is "intentionalist" or "nonintentionalist." Virtually all the theories of biological function are nonintentionalist, having no creator with intentions. The only two exceptions are the religious (theistic) approach, which bases function on the conscious intentions of a Divine Creator, and Searle, who sees functions as merely intentional human ascriptions. But even if we were attracted to the nonintentionalist accounts of biological functions, Vermaas and Houkes argue that an adequate theory of the functions of artifacts must be intentionalist, based on the intentions of the designers. In my taxonomy of theories of functions (figure 2.1), the reason I do not separate these categories of "intentionalist" and "nonintentionalist" is that there is already a category for them—the Recent Past Backward-looking Reductionist category. Intentions of a designer do reach back in time, even if they don't reach back very far in history, especially when we consider the great time durations involved in evolutionary history. But they do precede the artifact and its function. Even the name of many of these Recent Past Backward-looking theories seems taken from the artifact side: "Goal-Contribution" theories.

However, though the intentionalist approach to artifact function is attractive, I would like to urge caution, because this approach overlooks crucial insights in psychology, sociology, and anthropology. In studying people, their behavior, and their creation and use of artifacts, it is tempting, and easy, to rely on their intentions in so behaving and using objects (insofar as we can discover their intentions). But this is not the end of the story. Psychologists, anthropologists, and sociologists are (or should be) very wary of letting such intentional ascriptions cloud over the *real* function of behavior and artifacts and their uses. Good methodology in social and cognitive science has us study all of what *is* being done, including what people *say* about what they do, and then use *all* the evidence in diagnosing the functions of behaviors and artifacts. Good methodology makes room for the possibility that the best analysis of human practices (including artifact use) may deviate from what the subjects themselves think of their own practices. Human behavior results from many different factors, including tradition and habit and routine. People often do not themselves know exactly why they do the things they do. When they do provide explanations of their behavior and use of objects, an "objective" outsider might ascribe different functions to behaviors than the practitioners do. Thus some religious ritual might be seen by those who practice it (i.e., from the intentional side) as involving tribute to a deity, and the artifactual objects used in the ritual might be said by the adherents of that religion to have various supernatural functions. But an anthropologist might analyze the situation as one in which the behaviors serve to reinforce kinship relations, and the object functions as status indicator and economic vehicle, even if no one in the group has that intention, and even if no one in the past ever had that intention. So a ceremonial artifact might have the function of, say, assisting in reinforcement of kinship ties, and that function might not be based on any intentions at all. Of course enlightened scientists realize the dangers of imposing outside judgments on cultural practices, and seek to avoid ethnocentrism. It is a delicate business to balance the descriptions that the practitioners give with such "outside" analyses, and this balance is extensively examined in the social sciences. But it is clear that the best explanations of behavior and artifacts may include ecological, environmental, social, economic, physiological, physical, and perhaps also genetic factors that have little or nothing to do with intentions.[3]

We see a similar kind of balancing on a more individual basis in psychology and psychoanalysis. The best psychological explanation of certain behaviors may include assigning functions to behaviors (including the design and creation of artifacts) that are far different from those the subject would describe, and assigning intentions vastly different from what the subject would *say* his or her intentions were.

So when we seek to explain behavior and function of artifacts, we should of course give the designers' (and manufacturers' and users') intentions significant weight. But to require a theory of artifactual functions to be intentionalist would ignore important methodological considerations in psychology, sociology, and anthropology. We must value the reports of the people we're studying, yet also be willing to acknowledge that the best explanation of their behavior, and the functions of their artifacts, is sometimes not given by their intentions but by other factors.

2.4   Lessons from Epistemology: DON'T Let Theories of Teleological Function Spiral Out of Control into Definitional Oblivion

It seems to me that the development of the functions debate since Millikan's *LTOBC* is similar to the way theories of knowledge and justification developed in the 1960s and 70s after the famous "Gettier Problem" became their focal point in 1963. In a short space I can give only the briefest overview of this theme in the development of epistemology, but I hope I can show how analytic approaches to knowledge and justification met and overcame a serious roadblock in their development.

In his *Theaetetus*, Plato was the one who gave us the standard equation: Knowledge Is Justified True Belief. This formula stood fairly well for quite a long time, through the disputes between rationalists and empiricists that raged during the Enlightenment. Then in 1963 Edmund Gettier wrote a short paper that presented a case in which one allegedly has a true belief that is justified, yet intuitively does not seem like something we should consider knowledge.[4] The trick is that part of the belief is true, and another part justified, and minor logical manipulation and conjunction yields a belief that is both true and justified, but not knowledge.[5] Gettier's paper caused an uproar, and the game was afoot—find an additional factor to add to truth and justification to yield knowledge and escape the Gettier problem.

The dust cleared somewhat to show that what seemed to be needed was a way of designating justification as "defeasible" or "indefeasible"—having true, indefeasibly justified beliefs equals having knowledge. Many competing definitions of *defeasibility* were proposed, some bizarrely complex. One solution memorable for its length and extravagance is Swain's (1974)[6]:

D6   *S* knows that *h* iff (i) *h* is true, (ii) *S* is justified in believing that *h* (that is, there is a true body of evidence *e* such that *S* is justified in believing *e* and *e* justified *h*), (iii) *S* believes that *h* on the basis of his justification and (iv) *S*'s justification for *h* is indefeasible (that is, it is not logically possible that there is a body of evidence *e*' such that the conjunction of *e* and *e*' fails to justify *h*).

This is followed by various revisions of clause (iv) culminating in the following:

(ivg)   *S*'s justification for *h* is indefeasible (that is, there is an evidence-restricted alternative Fs* to *S*'s epistemic framework Fs such that (i) "S is justified in believing that *h*" is epistemically derivable from the other members of the evidence component of Fs* and (ii) there is some subset of members of the evidence component of Fs* such that (a) the members of this subset are also members of the evidence component of Fs and (b) "S is justified in believing that *h*" is epistemically derivable from the members of this subset).

I quote this definition at length as a warning—let us resist the urge to produce similarly out-of-control overblown definitions for teleological functions. Yet that is exactly the direction we have been headed since 1984.

This is especially ironic because Wright's influential 1973 paper on functions was a huge step in simplifying the philosophical definitions of functions. In the 1950s and 60s, the Deductive-Nomological model of explanation was still dominant, and highly complex definitions of function from Hempel (1959), Beckner (1959), Nagel (1961), and Canfield (1966) were just what Wright was reacting against in devising his elegant and concise view of function. The point is not that complexity is bad in itself or that explanations are never complicated. But the basic principles of nature do indeed tend to be simple (i.e., $E = mc^2$), though they are instantiated in complex ways. Thus one could hope that a few simple notions of function should be the basis for multifaceted explanations of phenomena.

To continue our brief tour through epistemology, what emerged were three competing approaches—Foundationalism, Coherentism, and Reliabilism.[7] (Reliabilism is related to the additional camp known as Naturalized Epistemology, which seeks to step out from armchair theorizing and base epistemology on empirical results in cognitive science.) For some period these three big camps battled it out over the correct approach to the definition of knowledge and the nature of justification. Then in the late 1980s, the smoke cleared, and people stopped worrying about the Gettier problem, and the three camps quieted the battles. It wasn't that the Gettier problem had been solved but rather that people recognized that it was not productive to continue to make it the central focus of epistemology. Moreover, it began to look as if each of these three big camps of Anglo-American epistemology had something to offer, and perhaps they might be arguing past one another. Many philosophers began to think that defining the necessary and sufficient conditions as to when *S* knows that *p* isn't so vital after all, and shouldn't be the exclusive focus of epistemology. More important are the conditions of justification. Even there, it seems to have become clear that there are simply different senses of justification, and each has a role to play in explaining certain puzzling problems, and they need not be seen as competing mutually exclusive options but rather as alternative parts of a larger picture. (Almost) nobody writes papers defining when "*S* knows that *p*" anymore. Even papers declaring the nature of justification have waned—in favor of examinations of the various aspects to justification, be they biological, ecological, psychological, social, political, and so forth. No one factor is the magic single answer about justification—they each have interesting things to tell us, and which factor is the deciding one depends on the specific circumstances of a particular instance.

This is the kind of result I envision, and would recommend, for modern theories of teleological functions. To some extent we see these same kind of battle lines now being drawn between present-looking systematic (Cummins-style) functions, backward-looking views, both distant-past historical/evolutionary functions (Millikan 1984, 1986, 1989a, 1989b, 1989c, 1990, 1993, 2002; Neander 1991a, 1991b; Papineau 1987; Griffiths 1993; Allen and Bekoff 1995) and recent-past/goal contribution/design views (Boorse 1976, 1977, 2002; Godfrey-Smith 1994; Kitcher 1993), and forward-looking propensity functions (Bigelow and Pargetter 1987). What I would recommend is a truce, and a realization that all of these notions have their virtues, and each has the potential to explain things we would like to explain. Let us stop the functions debate from imitating the kinds of chaos and splintering that epistemologists have wisely moved away from.

2.5   DO Focus on Explanation: Pragmatic Teleo-Pluralism

I have now described four DON'Ts—things I think we should avoid in our theories of teleofunctions. The upshot of the last one is that we should stop pursuing conceptual analysis of some single unified concept of "function" that will cover every case, defuse every counterexample, and explain everything. There is no one intuitive notion of "function" (or "teleofunction") for which we will find the magic set of necessary and sufficient conditions.[8] There are many different kinds of questions to be answered, and different phenomena to be explained, and various accounts of function can explain some and not others (or perhaps different aspects of the same phenomenon). The urge to find one account that explains everything is bound to lead to complex, convoluted, gerrymandered theories that may get some things right, but rarely sound like basic principles of nature.[9] So what *should* we do? We should acknowledge the different advantages of the various views of function, and use them all as they help explain things. For lack of a better term, I call this attitude Pragmatic Teleo-Pluralism.

I am not alone in urging this cessation of hostilities. The maven of teleology, Millikan, herself tried to avoid it in *LTOBC* by explicitly stating that her notion of "proper function" was a technical term, not an analysis of what we all mean by "function." More recently (1989c, 2002) she recognized what she termed an "ambiguity" in the notion of function, acknowledging the validity and usefulness of Cummins-style systematic functions, and argued for a kind of pluralism about functions. Others have advocated pluralism as well, such as Peter Godfrey-Smith (1993), Beth Preston (1998), and Peter Schwartz (2004). But I think we should also not view the various kinds of functions theories as merely expounding technical notions (as Millikan urges in *LTOBC*). If any of these varieties of theories of function is to help us explain events, objects, and structures in the world around us, then I would think the functions of which they speak must be real parts or aspects of the world. To play instrumentalist about these notions of functions would deprive them of real application to the world. But this is not to say that any or all of them are *the* meaning behind a single intuitive notion of "function."

A pluralist approach does not mean we must beg off of all the debates and adopt a kind of tentative relativism. For instance, Cummins (2002) gives powerful and compelling arguments against Millikan's view of about function and natural selection. As he points out, natural selection doesn't select between winged and wingless sparrows, but between sparrows with better and worse wing designs, where both the good and the bad wings have the same proper function—enabling flight. So Cummins concludes that selection is not sensitive to function, and does not pick out proper functions. Nothing about teleo-pluralism prevents us from endorsing some conclusions about what certain kinds of functions can and cannot explain. We should recognize that each of the various notions of "function" has different strengths and weaknesses, and will be appropriate in explaining different things. I argue elsewhere (Perlman 2000, 2002) against teleosemantics, the view that teleofunction can explain the mental representation and misrepresentation. The problems of indeterminacy of mental content reappear in the teleofunctions that are supposed to explain content, and thus functions can't hold the weight of explaining content.[10] So to be a pluralist about functions is not to be a noncommittal relativist. And yet there is still room for those who (like Millikan) think functions can be determinate enough to ground mental content to argue their case.

While various people have tried to unify the notions of "function" into one (Griffiths 1993; Buller 1998; Kitcher 1993, with his notion of "design"; Walsh and Ariew 1996, with the relational theory), I think we shouldn't focus the argument on what the "real" definition of *function* is, but rather on which conception of "function" will adequately explain which phenomena. That being said, we should also not say that the different notions of "function" are restricted to different fields within biology. The difference has to do with kinds of explanation, and what is being explained, not the field of investigation itself. In some cases a focus on Cummins-style systematic functions will give us the best explanation of the phenomenon in question (many of the issues in biochemistry, neuroscience, or developmental biology). In other cases, an evolutionary account such as Millikan's account will be preferable. In still others, a focus on recent-past (goal-contribution) will do better than focus on distant evolutionary past (especially for talking about the designer's intention as a source of the function of a human artifact). In still other cases, future-looking propensity and disposition will tell us more and explain more. Let's give ourselves an arsenal of all of these useful concepts, and not insist that one of them is the ultimate and exclusive account of function. The efforts to battle it out over which notion of "function" is the right one has led many people to lose sight of the goal of the whole endeavor—to explain things. Let us use whatever works, whatever conception of "function" is most useful, in explaining whatever puzzles us today.

Acknowledgments

I wish to thank all the participants of the September 21–24, 2006, Workshop on Comparative Philosophy of Technical Artifacts and Biological Organisms, held in Altenberg, Austria, and especially thank its organizers Ulrich Krohs and Peter Kroes, as well as everyone at the Konrad Lorenz Institute for sponsoring the workshop. Thanks to Wikipedia.com, my source for all the photographs (which were public domain).

The Table of Theories of Functions in this chapter is from my earlier paper in *The Monist,* January 2004, vol. 87, no. 1, titled "The Modern Philosophical Resurrection of Teleology," and is copyright © 2004, *The Monist: An International Journal of General Philosophical Inquiry*, Peru, Illinois, 61354, USA. Reprinted by permission.

Notes

1. See http://www.niagaraparks.com/nfgg/geology.php and http://en.wikipedia.org/wiki/Niagra_Falls.

2. In her chapter in this volume, Françoise Longy also argues that the biological often mixes with the artifactual. Where she and I differ is that she holds out hope for a unified concept of "function," whereas I go for the pluralist position.

3. The chapters in this volume by Beth Preston and Françoise Longy make arguments along similar lines.

4. Some claim that Bertrand Russell proposed the same problem early in the twentieth century. I do not wade into that dispute—what matters here is that it was Gettier's paper that led to the debate in question.

5. In one Gettier example, suppose I believe that Jones owns a Ford automobile, because I have seen Jones and his Ford (i.e., a good reason). We can pick any random statement for which there is no good reason—"Smith is in Barcelona"—and link it with "Jones owns a Ford" in a disjunction. I can believe that "Jones owns a Ford or Smith is in Barcelona" and be justified. But suppose unbeknownst to me, Jones has sold his Ford, but also Smith just happens to actually be in Barcelona. The disjunctive belief is still true, and justified. But it does not seem like knowledge, because the disjunctive belief is true because of Smith's location (unjustified) but justified because of Jones's car (no longer true). Thus a justified true belief may not be knowledge.

6. When I was a student of his at the Ohio State University, Marshall Swain conceded to me in conversation (around 1988) that even he thought this kind of formulation had probably gone too far and was too complicated, elaborate, and cumbersome.

7. Foundationalism defended by, among others, Robert Audi (1988), John Pollock (1986), and William Alston (1989). Coherentism lead by Laurence BonJour (1985) and Keith Lehrer (1974, 1990, 1997). Externalism (reliabilism) developed by David Armstrong (1973) and Alvin Goldman (1976, 1986).

8. Contra Neander's (1991b) defense of conceptual analysis of teleofunction. Peter Schwartz (2004) makes a similar point against conceptual analysis.

9. In a different field, that of mental content, I see the same kind of problem. Fodor's (1987) notion of "asymmetric dependence" as a theory of content may well get all the cases right, and skirt the counterexamples, but it just seems too convoluted to be a basic rule of nature, that is, too arcane and messy to be what content or meaning *is*.

10. See also Fodor (1990), Neander (1995, 1996), and Enç (2002).

References

Allen, C., and Bekoff, M. (1995). Function, natural design, and animal behavior: Philosophical and ethological considerations. *Perspectives in Ethology, 11:* 1–47.

Alston, W. (1989). *Epistemic Justification.* Ithaca: Cornell University Press.

Ariew, A. (2002). Platonic and Aristotelian roots. In: *Functions: New Essays in the Philosophy of Psychology and Biology* (Ariew, A., Cummins, R., Perlman, M., eds.), 7–32. Oxford and New York: Oxford University Press.

Armstrong, D. (1973). *Belief, Truth, and Knowledge.* Cambridge: Cambridge University Press.

Audi, R. (1988). *Belief, Justification, and Knowledge.* Belmont, Calif.: Wadsworth Publishing Company.

Beckner, M. (1959). *The Biological Way of Thought.* New York: Columbia University Press.

Bedau, M. A. (1990). Against mentalism in teleology. *American Philosophical Quarterly, 27:* 61–70.

Bedau, M. A. (1991). Can biological teleology be naturalized? *The Journal of Philosophy, 88:* 647–655.

Bedau, M. A. (1992a). Goal-directed systems and the good. *The Monist, 75:* 34–49.

Bedau, M. A. (1992b). Where's the good in teleology? *Philosophy and Phenomenological Research, 52:* 781–806.

Bedau, M. A. (1993). Naturalism and teleology. In: *Naturalism: A Critical Appraisal* (Wagner, S., and Warner, R., eds.), 23–51. Notre Dame, Ind.: University of Notre Dame Press.

Bigelow, J., and Pargetter, R. (1987). Functions. *The Journal of Philosophy, 84:* 181–196.

BonJour, L. (1985). *The Structure of Empirical Knowledge.* Cambridge, Mass.: Harvard University Press.

Boorse, C. (1976). Wright on functions. *Philosophical Review, 85:* 70–86.

Boorse, C. (1977). Health as a theoretical concept. *Philosophy of Science, 44:* 542–573.

Boorse, C. (2002). A rebuttal on functions. In: *Functions: New Essays in the Philosophy of Psychology and Biology* (Ariew, A., Cummins, R., Perlman, M., eds.), 63–112. Oxford and New York: Oxford University Press.

Buller, D. (1998). Etiological theories of function: A geographical survey. *Biology and Philosophy, 13:* 505–527.

Cameron, R. (2003). How to be a realist about *sui generis* teleology—yet feel at home in the 21st century. *The Monist, 87:* 72–95.

Canfield, J. (1966). Introduction. In: *Purpose in Nature* (Canfield, J., ed.), 1–7. Englewood Cliffs, N.J.: Prentice-Hall.

Cummins, R. (1975). Functional analysis. *The Journal of Philosophy, 72:* 741–765.

Cummins, R. (2002). Neo-teleology. In: *Functions: New Essays in the Philosophy of Psychology and Biology* (Ariew, A., Cummins, R., Perlman, M., eds.), 157–172. Oxford and New York: Oxford University Press.

Dipert, R. R. (1993). *Artifacts, Art Works, and Agency.* Philadelphia: Temple University Press.

Dretske, F. I. (1986). Misrepresentation. In: *Belief: Form, Content and Function* (Bogdan, R. J., ed.), 17–36. Oxford: Oxford University Press.

Enç, B. (2002). Indeterminacy of function attributions. In: *Functions: New Essays in the Philosophy of Psychology and Biology* (Ariew, A., Cummins, R., Perlman, M., eds.), 291–313. Oxford and New York: Oxford University Press.

Fodor, J. (1987). *Psychosemantics.* Cambridge, Mass.: Bradford/The MIT Press.

Fodor, J. (1990). A theory of content I. In: *A Theory of Content and Other Essays.* Cambridge, Mass.: Bradford/The MIT Press.

Gettier, E. (1963). Is justified true belief knowledge? *Analysis, 23:* 121–123.

Godfrey-Smith, P. (1993). Functions: Consensus without unity. *Pacific Philosophical Quarterly, 74:* 196–208.

Godfrey-Smith, P. (1994). A modern history theory of functions. *Noûs, 28:* 344–362.

Goldman, A. (1976). Discrimination and perceptual knowledge. *Journal of Philosophy, 73:* 771–791.

Goldman, A. (1986). *Epistemology and Cognition.* Cambridge, Mass.: Harvard University Press.

Griffiths, P. E. (1993). Functional analysis and proper functions. *The British Journal for the Philosophy of Science, 44:* 409–422.

Hempel, C. G. (1959). The logic of functional analysis. Reprinted in: *Aspects of Scientific Explanation,* 297–330. New York: Free Press, 1965.

Kitcher, P. (1993). Function and design. *Midwest Studies in Philosophy, 18:* 379–397.

Lehrer, K. (1974). *Knowledge.* Oxford: Oxford University Press.

Lehrer, K. (1990). *Theory of Knowledge.* Boulder, Colo.:Westview Press.

Lehrer, K. (1997). *Self Trust: A Study of Reason, Knowledge and Autonomy.* Oxford: Oxford University Press.

Mayr, E. (1974). The multiple meanings of teleological. Reprinted with a new postscript in: Mayr, E. (1988). *Towards a New Philosophy of Biology.* Cambridge, Mass: The MIT Press.

McLaughlin, P. (2001). *What Functions Explain.* Cambridge: Cambridge University Press.

Millikan, R. G. (1984). *Language, Thought, and Other Biological Categories: New Foundations for Realism.* Cambridge, Mass.: Bradford/The MIT Press.

Millikan, R. G. (1986). Thoughts without laws: Cognitive science without content." *Philosophical Review, 95:* 47–80.

Millikan, R. G. (1989a). In defense of proper functions. *Philosophy of Science, 56:* 288–302 (also in Millikan 1993).

Millikan, R. G. (1989b). Biosemantics. *The Journal of Philosophy, 86:* 281–297 (also in Millikan 1993).

Millikan, R. G. (1989c). An ambiguity in the notion of "function." *Biology and Philosophy, 4:* 172–176.

Millikan, R. G. (1990). Truth rules, hoverflies, and the Kripke-Wittgenstein paradox. In: *White Queen Psychology and Other Essays for Alice.* Cambridge, Mass.: The MIT Press, 1993.

Millikan, R. G. (1993). *White Queen Psychology and Other Essays for Alice.* Cambridge, Mass.: The MIT Press.

Millikan, R. G. (2002). Biofunctions: Two paradigms. In: *Functions: New Essays in the Philosophy of Psychology and Biology* (Ariew, A., Cummins, R., Perlman, M., eds.), 113–143. Oxford and New York: Oxford University Press.

Nagel, E. (1961). *The Structure of Science: Problems in the Logic of Scientific Explanation.* New York: Harcourt, Brace & World.

Nagel, E. (1977). Teleology revisited. *The Journal of Philosophy, 74:* 261–301.

Neander, K. (1991a). The teleological notion of "function." *Australasian Journal of Philosophy, 69:* 454–468.

Neander, K. (1991b). Functions as selected effects: The conceptual analyst's defense. *Philosophy of Science, 58:* 168–184.

Neander, K. (1995). Misrepresenting and malfunctioning. *Philosophical Studies, 79:* 109–41.

Neander, K. (1996). Dretske's innate modesty. *Australasian Journal of Philosophy, 74:* 258–74.

Nissen, L. (1997). *Teleological Language in the Life Sciences.* Lanham, Md.: Rowman & Littlefield.

Papineau, D. (1987). *Reality and Representation.* Oxford: Blackwell.

Perlman, M. (2000). *Conceptual Flux: Mental Representation, Misrepresentation, and Concept Change.* Dordrecht: Kluwer Academic Publishers.

Perlman, M. (2002). Pagan teleology: Adaptational role and the philosophy of mind. In: *Functions: New Essays in the Philosophy of Psychology and Biology* (Ariew, A., Cummins, R., Perlman, M., eds.), 263–290. Oxford and New York: Oxford University Press.

Perlman, M. (2004). The modern philosophical resurrection of teleology. *The Monist, 87:* 3–51.

Plantinga, A. (1993a). *Warrant: The Current Debate.* New York: Oxford University Press.

Plantinga, A. (1993b). *Warrant and Proper Function.* New York: Oxford University Press.

Plantinga, A. (2000). *Warranted Christian Belief.* New York: Oxford University Press.

Pollock, J. (1986). *Contemporary Theories of Knowledge.* Lanham, Md.: Rowman and Littlefield.

Preston, B. (1998). Why is a wing like a spoon? A pluralist theory of functions. *Journal of Philosophy, 95:* 5, 215–254.

Schiffer, M. (1992). *Technological Perspectives on Behavioral Change.* Tucson: University of Arizona Press.

Schwartz, P. (2004). An alternative to conceptual analysis in the function debate. *The Monist, 87:* 136–154.

Searle, J. (1998). *Mind, Language and Society, Philosophy in the Real World.* New York: Basic Books, 1998; London: Weidenfeld and Nicholson, 1999.

Swain, M. (1974). Epistemic defeasibility. *American Philosophical Quarterly II:* 15–25.

Vermaas, P. E., Houkes, W. (2003). Ascribing functions to technical artifacts: A challenge to etiological accounts of functions. *British Journal of the Philosophy of Science, 54:* 261–289.

Walsh, D. M., and Ariew, A. (1996). A taxonomy of functions. *Canadian Journal of Philosophy, 26:* 493–514.

Wimsatt, W. C. (1972). Teleology and the logical structure of function statements. *Studies in the History and Philosophy of Science, 3:* 1–80.

Woodfield, A. (1976). *Teleology.* Cambridge: Cambridge University Press.

Wright, L. (1973). Functions. *Philosophical Review, 82:* 139–168.

Wright, L. (1976). *Teleological Explanations.* Berkeley: University of California Press.

3 Biological and Cultural Proper Functions in Comparative Perspective

Beth Preston

3.1 Proper Function

It is widely acknowledged that both biological traits and artifacts have proper functions. That is, there are performances in which they are "supposed" to engage, and failure to engage in these performances counts as malfunctioning. The literature on proper function has been devoted almost exclusively to proper function in biology, but it is widely assumed that some version of this account will work for proper function in material culture. But due to this initial biological focus, accounts of proper function have typically made liberal use of biological concepts such as "selection," "fitness," "reproduction," and so on. So adapting this account will require finding cultural analogues of these concepts. This creates an opportunity for things to go wrong in two ways. The biological concepts may not succeed in picking out either biological or cultural proper functions; or, alternatively, they may succeed for biological proper functions but fail for cultural proper functions because they have no good cultural analogues. I argue that things have gone wrong in the first way with regard to selection, and in the second way with regard to fitness. Finally, I argue that the only way forward is to closely examine the phenomena of reproduction and use in material culture.

3.2 Selection

The problem for theorists of proper function is to pick out proper functional performances from all the performances in which a thing engages. So, for example, we want to be able to say that the proper functions of leaves are photosynthesis and transpiration, even though leaves do a lot of other things as well, such as shade roots, conceal fruit, harbor insects, provide nutritious mulch, and so on. It is thought that natural selection picks out proper functions by indicating which performances have historically accounted for the reproduction of the trait in question. Importantly, a trait has been selected only if it has coexisted with variant traits, and if it has been so much better at contributing to the survival and reproductive success of its possessors that it has persisted and proliferated while the variants have slowly but surely disappeared. As Ruth Millikan puts it:

> The [proper] functional trait must be one that is there in *contrast* to others that are *not* there, because of historic difference in the results of these alternative traits. It must be tied to genetic materials that were selected from among a larger pool of such materials because of their *relative* advantagiousness. . . . Graphically, whether my shoulders have as a biological [proper] function to hold up my clothes depends not on what proportion of my ancestors used their shoulders that way to advantage but on whether there were once shoulderless people who died out because they had nothing to hang their clothes on. (Millikan 1993: 38)

A difficulty noted immediately concerned change or loss of proper function. For example, feathers are thought to have been originally selected for thermoregulation, and only more recently, in the case of the wing feathers of birds, for their aerodynamic capacities. Moreover, wing feathers may have lost their capacity for thermoregulation in the process—they may now be vestigial with respect to that earlier proper function. To solve this problem, the account of proper function was adjusted to appeal to recent selection (Godfrey-Smith 1994) or to a recent evolutionarily significant time period (Griffiths 1993). This adjustment is known as the *modern history* account of proper function.

However, Peter Schwartz (1999, 2002) suggests this adjustment itself needs adjustment. The problem is that for many traits there has been no selection in recent evolutionary history, even though there was selection in the distant past. Schwartz identifies two reasons for this: 1) lack of recent variation, and 2) the possibility that of two currently maintained performances, the one that is now undergoing selection is not the one originally selected. The wing feathers of birds might be an example of (1) if there has been no significant variation in the recent evolutionary past. And they might be an example of (2) as well. Suppose there is a variation in the wing feathers that renders them simultaneously better for flying and better for thermoregulation. But suppose (what is plausible, given global warming!) that there is current selection pressure only for airworthiness. Then the variation might be selected for flying efficiency alone, even though it continues to perform its original, thermoregulatory function, and does so better than ever. We may call this problem ambiguous variation. In both cases—lack of recent variation and ambiguous variation—we want to say that the proper function is retained, but because it is not currently being selected for we cannot do so. To solve this problem, Schwartz recommends an account with two conditions. The proper function of a trait is a performance that was a) selected for in the distant past, and that has b) continued to contribute to the fitness of the organism in the recent past. This continued contribution to fitness may be in the form of enabling survival or reproductive success in some way without variation or selection. Schwartz calls this the *continuing usefulness* account of proper function. It weakens the previous total reliance on selection to pick out proper functions by requiring selection only at some point in the past and relying on continuing usefulness to pick out proper functions in the present.

David Buller (1998) weakens this reliance on natural selection still more. He gives three reasons why selection may never have operated in the case of some erstwhile proper functional traits: 1) genetic drift rather than natural selection may have caused their proliferation, 2) there may never have been significant variations, and 3) in addition to assigning proper functions to whole traits we also want to assign them to the component parts of those traits, but a given component might be invariant in all variations of a trait, and thus not under selection in its own right. Reasons (2) and (3) are clearly the same problem—lack of variation. This is a problem already identified by Schwartz. And genetic drift is of course a favorite example of those who wish to point out that evolution at times proceeds by means other than natural selection. Buller's criticism is more radical than Schwartz's because Schwartz only identifies cases where natural selection is *no longer* operating, whereas Buller identifies cases where it *never* operated but where there is a performance contributing to survival and reproductive effectiveness that seems otherwise to warrant the proper function label. To solve this problem, Buller proposes a *weak etiological theory*. On this view a trait has a specific proper function if a) the performance associated with that function contributed to the fitness of the ancestors of present organisms with the trait, and if b) the trait is hereditary. "Strong" etiological theories such as Millikan's appeal exclusively to natural selection to pick out proper functions. Buller's "weak" theory, in contrast, appeals to fitness of inherited traits.

Let us now pause to ask whether there is an analogous process of cultural selection, and whether it is subject to the difficulties Schwartz and Buller have identified. To be analogous, cultural selection must involve competing variants of items of material culture, one of which proliferates while the others disappear. This phenomenon occurs in at least two contexts. First, design often involves the building and testing of a number of prototypes, one of which is then selected for reproduction while the others are consigned to the dustbin. Second, the economic processes of marketing and distribution often involve the appearance of a number of competing variants of a type of item, one of which proliferates while the others disappear. For example, quill pens disappeared with the advent of fountain pens; and Microsoft Word has arguably outcompeted other word processing systems, which are rapidly disappearing. So selection among alternative variants is a feature of material culture just as it is of biology; and it may be hypothesized to pick out proper functional performances of artifacts just as natural selection is hypothesized to pick out proper functional performances of biological traits.

But does this hypothesis run afoul of cultural phenomena analogous to those identified by Schwartz and Buller in the biological realm? Both of them pointed to lack of variation as a major problem. And examples of this in material culture are not far to seek. Simple, everyday implements like baskets, spoons, and brooms often remain virtually unchanged in material and design for centuries or millennia. Lack of variation in components also occurs in material culture—windows have varied a lot historically in design and construction, but they standardly have had panes of glass as components. Moreover, much of the variation that does occur in material culture does not substantially affect its utilitarian proper function. For example, artifacts often are varied for purely aesthetic reasons (e.g., decoration on the handles of tableware) or for reasons of social status (e.g., silver tableware instead of stainless steel). How much the proper function is affected is a relative matter, of course. In the case of silver tableware, for instance, there is more upkeep (polishing in addition to washing) and somewhat greater liability to damage (bent handles, dented bowls), so what is affected is not the function itself so much as its maintenance. Plastic tableware, on the other hand, does not work quite as well as metal tableware, especially for some foods (e.g., steak). Function is clearly somewhat impaired in this case, but not so much as to make plastic tableware useless, since it works just fine for most foods. In such cases the relative lack of impact on the exercise of the utilitarian proper function of the artifact typically results in a number of variations persisting indefinitely alongside one another rather than in one variation winning out and the others disappearing. Thus in material culture, even when there is considerable variation, there is often a lack of *sufficient* variation relative to the function under consideration, and the requisite process of selection over alternatives is stymied.

Sometimes even when variations do affect function, persistence or disappearance of variants is conditional upon extrafunctional features. Plastic tableware persists alongside metal tableware not only because it still functions relatively well for eating but because it has other features (cheapness, disposability) that endear it to consumers. This points to the operation of a cultural analogue of genetic drift. When a variation persists or disappears, the reasons do not necessarily have anything to do with how well it fulfilled its erstwhile function. Perhaps it actually worked better than all the competing variations, but the company was poorly managed and went under, or it ran afoul of Microsoft, or the advertising or distribution were not adequate, or the colors were unfashionable, or whatever. Or perhaps, like plastic tableware, it does not work as well, but is favored by certain human populations for other reasons, and thus persists alongside the more efficiently functioning variants. This phenomenon bears some resemblance to genetic drift in biology, where the frequency of genes in a population can change for reasons extraneous to natural selection, like natural disasters that randomly wipe out members of a population with no regard to their fitness, or locally favorable conditions in which otherwise nonadaptive features can persist.

Finally, Schwartz's second problem—ambiguous variation—is certainly imaginable. Suppose a basket originally made by a Native American group for winnowing grain becomes popular in the tourist trade. And suppose some variation makes it both better for winnowing and more attractive to tourists. But suppose that the need for disposable cash is what is driving the reproduction of these baskets at the moment. Then we want to say they still have winnowing as a proper function, even though that is not what they are currently being selected for.

So it looks like cultural selection will have all the same problems as natural selection with regard to picking out proper functions. So far the problem has been that although natural selection does succeed in picking out most proper functions, there are some functions we want to call proper functions where natural selection is not operative. But Robert Cummins (2002) has reentered the function theory fray with a much more fundamental critique. On his view, natural selection does not pick out proper functions even in the favorable cases in which it *does* occur. The target of Cummins's argument is a widely held view he calls neo-teleology. It claims, first, that biological organisms have the traits they have because of the functions those traits fulfill (e.g., vertebrates have hearts because hearts circulate blood), and second, that natural selection supplies the connection between function and the existence of traits by selecting traits for their functions (hearts have been selected because they circulate blood, and this accounts for their existence and ubiquity). On this view, natural selection accounts, first, for the very existence of traits because it "builds" them incrementally, and second, for their spread through populations of individual organisms. With regard to the "building" of traits, Cummins agrees that natural selection is largely responsible for this, but points out that it works by means of a piecemeal, long-term process that is entirely insensitive to the ultimate proper function of the trait (2002: 168–169). With regard to the spread of traits, Cummins again agrees that this is often accomplished by natural selection, which in this case *is* sensitive to function—but only to the relative success with which a function is performed by the current variants (2002: 164–165). For example, natural selection does not select between winged pileated woodpeckers and wingless pileated woodpeckers, but between pileated woodpeckers with better and worse wing designs. And the crucial point here is that both better and worse wings already have the proper function of enabling flight. In short, in the case of both the building and the spread of traits, what natural selection selects for does not correspond to the proper functions of these traits. So natural selection does not pick out proper functional performances.

The question then is whether an analogous argument goes through for material culture. Let us first consider spread. Like natural selection, cultural selection also appears to spread types and traits of artifacts only by selecting among better and worse variants with the same proper function. For example, quill pens and fountain pens both have the proper function of writing, but fountain pens are more efficient because they have a larger ink reservoir and a more durable writing point. So they proliferated while quill pens fell into disuse. Nor does the ink reservoir—which, as a component part, is more nearly analogous to the wings of birds in our biological example—represent a new proper function. The hollow shaft of a quill pen is also an ink reservoir but an inefficiently small one in comparison to that of the fountain pen. While there may be some rare cases in which a completely novel artifact capable of a novel performance is introduced, this is the vanishingly rare exception rather than the rule in material culture, as in biology. And as Cummins (2002: 166) remarks, if neo-teleology is applicable only to the spread of rare, radical novelties, then it is not a significant theory.

Now what about building items of material culture? Cummins calls this a Paley question, with reference to the watchmaker analogy made famous by William Paley. Paley's point was that if you find a complex artifact like a watch on a deserted beach, you naturally assume it must have had an intelligent designer in order to exist at all; likewise, if you find an eye or a stomach, you should assume an intelligent designer of these complex biological items. But what evolutionary theory shows is that you do not have to answer Paley questions about biological traits by appeal to intelligent design, because the observed results can be achieved by long-term, mechanical, incremental processes—including natural selection—that are insensitive to the eventual complex structural "design" as well as the ultimate proper function. It is widely assumed that Paley was right about material culture, though, and that cultural selection is necessarily sensitive to function and design while natural selection is not. But Cummins's argument will go through for material culture only if Paley was wrong—and wrong in the same way—about material culture. In short, the existence of a watch on a deserted beach, like the existence of a stomach, must be accounted for by a long history of incremental variations that was not from the beginning aimed at the creation of watches.

We have already been oriented in this direction by the preceding point about the rarity of radical novelty. In fact, the nature of human inventiveness is overwhelmingly a matter of making small changes in existing material culture rather than producing radical novelty out of nowhere. As Henry Petroski (1992, especially ch. 3) demonstrates at length and in detail, inventors are in the first instance critics of current technology, but constructive critics with ideas for incremental improvements. The resulting variations in artifact traits provide ongoing incremental changes on which cultural selection acts, just as mutation and various other evolutionary mechanisms provide incremental changes on which natural selection acts. In light of this observation, let us consider the history of mechanical watches of the sort Paley had in mind, the early forms of which appeared in the sixteenth century. First, such watches depend on the development of two basic technologies, glassmaking and metallurgy, both of which have histories stretching back many millennia. Second, they depend on the development of machining techniques for producing very small parts capable of precision operations, which also predate the advent of watches. Finally, watches depend on the prior history of mechanical clocks, which first appeared in the fourteenth century in the form of large, weight-driven tower clocks in public buildings. It is certain that early glassmakers and metallurgists were not aiming at watches. Neither were early machinists. And arguably, neither were the early clockmakers, who were working on a vertical mechanism driven by large weights that was not even conceivably portable. It was only with the invention of a spring-driven mechanism and early portable clocks (e.g., for use onboard ships) that the sort of personal, portable clock we now know as a watch could be realistically designed, let alone made.

The point is this: Positing an intelligent agent with an understanding of the structural design and proper function of a watch and its parts barely *begins* to explain the watch found on the deserted beach, because the existence of such an agent itself requires an explanation. An agent capable of designing or making a watch is not even remotely conceivable without the long history of incrementally built-up technologies sketched here, since no human agent, however intelligent, could possibly have invented such a thing utterly from scratch. In other words, a watch may imply a watchmaker, but a watchmaker in turn implies a cultural history during which the requisite technological resources and techniques are incrementally built up through the work of many intelligent agents who did *not* have in mind the structural design or proper functions of watches or their parts. Thus the existence of watches is not explained by appeal to a watchmaker but rather by appeal to the history of technologies and techniques on which watchmakers are utterly dependent and without which their production of watches is inconceivable. So artifacts are the result of long-term incremental processes that are insensitive to their ultimate proper functions, just as in the case of biological traits. The only difference between biology and culture is that the increments are implemented by intelligent agents in the latter case. But there is nothing to be made of this for the purposes of a counterargument, because those intelligent agents—early metallurgists, for instance—are sensitive only to the features of the increment they are implementing and not to the whole process that will result in a more complex material culture at some far future point. So an analogue of Cummins's argument goes through for the building of items of material culture just as it did for the spread of such items.

Let us summarize our results. Standard theories of biological proper function appeal to natural selection to pick out proper functional performances. But this appeal to natural selection has encountered a number of serious problems, all of which have analogues in cultural selection. In particular, neither natural selection nor cultural selection actually does pick out the proper functional performances. The most reasonable response is to abandon selection as criterial for proper function in favor of a weaker standard. Buller's weak etiological theory of biological proper function, which appeals to fitness instead of natural selection, is thus a logical alternative. Can it be adapted to proper function in material culture?

3.3 Fitness

Buller formulates the weak etiological theory thus:

A current token of a trait T in an organism O has the [proper] function of producing an effect of type E just in case past tokens of T contributed to the fitness of O's ancestors by producing E, and thereby causally contributed to the reproduction of Ts in O's lineage. (Buller 1998: 507)

Like the strong etiological theory, the weak theory appeals to the history of reproduction of a trait. But it grounds that history in *contributions* to fitness, not in the stricter condition of selection over alternative variations because of *superior* fitness. "Fitness" has been used in a number of different senses in biology and philosophy of biology (Endler 1986: 33–50). In addition, critics say it lacks explanatory power because organisms that survive and reproduce are by definition more fit (Sober 1984: ch. 2). So it will not do to explicate fitness as merely a propensity to survive and reproduce; the grounds of such a propensity must be spelled out. Buller (1998: 509) appeals to a widely accepted conception of "fitness" with four components: viability, fertility, fecundity, and ability to find mates. Are there analogues of these biological phenomena in material culture?

   Viability is the fundamental component of the fitness of individual biological organisms because they have to survive to sexual maturity (or to an appropriate size, in the case of asexual reproduction) in order to achieve fertility, fecundity, and the procurement of mates. But here a problem is apparent, for individual artifacts do not have a life cycle in this sense. It is true that many of them have a life cycle of sorts—they are made, last for a while, and then break or wear out. It is also true that some artifacts go through a maturation process of sorts. Cheese, whiskey, and firewood, for instance, have to spend a period of time in controlled storage before they are suitable for consumption; and some artifacts, such as shoes and clarinet reeds, must be broken in to work well. But these processes in material culture are not true analogues of growing to sexual maturity in biology because they are not connected with reproduction but with performance of other functions. You have to wait for your whiskey to mature for it to be drinkable, but not in order to make another batch; and you do not have to break in one clarinet reed before you can make another. The only possible exceptions are the rare cases where you need a "starter" and you get it by saving some of the current batch, as with yogurt or bread. But these exceptions occur because the artifacts in question incorporate biological organisms, and *they* have to mature to reproductive size or age. And there is a further disanalogy. If no individuals of a type of organism survive to reproductive maturity, that is the end of the lineage. Not so for artifacts. You can eat up all the brownies in existence, and so long as someone remembers how to make them or has the recipe, more brownies can be reproduced. In short, reproduction in material culture is not absolutely dependent on the survival or maturation of individual items of the same type in the way that reproduction in biology is. The exception is when the techniques and/or technology required to make a type of artifact have been lost, and then reproduction may depend on reverse engineering of surviving exemplars.

   What about the other three components? The ability to find mates is not applicable, because reproduction among artifacts is not sexual. But then this component is not applicable for all biological organisms, either, so we can safely ignore it. On the other hand, fertility and fecundity—the capacities to produce offspring and lots of them, respectively—do seem to have analogues in material culture. With regard to fertility, some prototype artifacts are reproduced while others are discarded or are used only by their maker. There are a variety of reasons for this. Some prototypes do not work as expected; others work well enough but are intended for some idiosyncratic purpose not shared by others; and reproduction in other cases depends on factors extraneous to proper function, such as aesthetic considerations, legal restrictions, or marketing constraints. With regard to fecundity, there are indeed differential rates of reproduction among artifacts. For example, there always seem to be a lot more chocolate cakes than red velvet cakes, and in that sense chocolate cakes are more "fecund." Here again, a variety of reasons may be operative, not all of which have to do directly with function. Red velvet cakes and chocolate cakes both serve the dessert function equally well, but chocolate is a preferred flavor in contemporary Western culture, and chocolate cakes are consequently more "fecund."

But these analogies are vague. Fertility and fecundity in biology are properties of individual organisms that singly or in pairs directly give rise to offspring like themselves. But the chocolate cake you are now baking is the offspring of the previous one (or two) you baked only indirectly, and mediated by your baking activity, know-how, and available raw materials. As Aristotle remarked, " . . . man is born from man, but not bed from bed" (*Physics* 193b8–9). On the other hand, as Aristotle also remarked:

Therefore it follows that in a sense health comes from health and house from house, that with matter from that without matter; for the medical art and the building art are the form of health and of the house, and when I speak of substance without matter I mean the essence. (*Metaphysics* 1032b11–14)

So the disanalogy is that the reproductive cycle of artifacts has an intermediate stage that is lacking in biological organisms. It is, in an etymologically correct sense, a larval stage in which the artifact exists in a distributed and partially mental, linguistic, or behavioral form. (The Latin root *lar* refers to tutelary household deities, often identified with the spirits of dead ancestors.) For Aristotle, the larval form is the essence—or more precisely, the formula of the essence—embodied in the art of producing the type of artifact in question. This account may need adjustment—it is not just the building art that is required but the existence and availability in the culture of suitable raw materials, for instance. But whatever the larval stage involves, the important point for us is that factors present in it affect the reproduction and reproductive rate of artifacts. This robs the analogy to biological fertility and fecundity of cogency, because biological organisms do not have a larval stage in this sense.

3.4   Use and Reproduction

Perhaps because of such disanalogies, Paul Griffiths (1993: 419–420) says that fitness in material culture is a vaguer notion than in biology, and suggests that an artifact's " . . . ability to fulfill its intended use gives it a propensity to be reproduced" (1993: 420). This is more promising. Artifacts are made for specific uses, and whether or not they are reproduced and at what rate plausibly depends on their actually fulfilling these uses. So rather than trying to find cultural analogues of biological phenomena like the four components of fitness discussed earlier, we can perhaps settle on just one factor as constituting the fitness of an artifact—its performing as intended by its makers and/or users. We may then pick out proper functions in accordance with this revised formula:

A current token of an artifact type has the proper function of producing an effect of a given type just in case producing this effect contributed to the intended use of past tokens of this type of artifact, and thereby contributed to the reproduction of such artifacts.

This formula has the added virtue of implicitly recognizing the role of the larval stage in the reproductive cycle of material culture by referring proper function in part to the intentions and activities of human agents *via* the notion of intended use.

But now we face two further difficulties. The first has to do with the qualification of use as intended. The problem is that there are established uses we want to call "proper functions" because they clearly affect reproduction, but that are *not* necessarily intended by designers, makers, or users. Many examples of such unintended proper functions concern social, economic, or political uses of material culture. Both nineteenth-century corsets and tiny shoes for bound feet were ostensibly intended to enhance female sexuality and attractiveness. But as Marianne Thesander (1997) points out, wearing these artifacts made it impossible for women to do even ordinary housework. Consequently these artifacts were also used to display the wealth and social status of a family by providing evidence that its wives and daughters had no need to work and could afford to be routinely incapacitated. But it is doubtful that designers, makers, or users of these artifacts explicitly recognized or consciously intended this use. This phenomenon is widely recognized in the social sciences, and the terms *manifest function* and *latent function* are often used to mark intended and unintended functions, respectively. Fortunately this difficulty may be remedied by simply striking the word *intended* from our formula. Use can be assessed from the outside without appeal to intention by observing actual patterns of behavior involving material culture, including verbal reports. This wider net will catch both intended and unintended uses, and will therefore enable us to pick out both manifest and latent proper functions.

The second difficulty concerns a common phenomenon I have called "phantom function"—cases where use and consequent reproduction look perfectly normal, and the attribution of proper function therefore seems perfectly straightforward, except that the artifacts in question are not actually able to perform their alleged function. For example, part of the proper function of communion wafers is to transubstantiate into the actual body of Christ in the course of the Christian religious ritual of the Eucharist. On the assumption that this is physically and metaphysically impossible, no communion wafer has ever performed this function or ever will. Similar examples can be found in religious and ritual contexts worldwide, and include all sorts of good luck charms, love potions, protective amulets, and so on. This phenomenon is also widespread in more mundane spheres of activity. Medicines, cosmetics, and nutritional supplements seem to be particularly prone to them. A well-known example is Linus Pauling's (1970, 1996) famous claim that vitamin C in large doses prevents and cures colds, as well as a host of other ailments including cancer. This undoubtedly had a huge effect on the reproduction of vitamin C, which was (and is) packaged in larger dosages for this use. But almost four decades later the scientific jury is still out as to whether vitamin C really does what Pauling claimed for it. So some or all of Pauling's claims may well represent phantom functions of vitamin C. Appliances are also subject to phantom functionality. Griffiths (1993: 420) gives the example of the tapered tails of early racing cars that were thought—falsely—to reduce their drag coefficient.

Phantom functions pose a much greater challenge than unintended (latent) functions because they cast doubt on the idea—essential to the notion of biological fitness—that successful performance is a necessary criterion. One option is to say that phantom functional artifacts are actually function-less. But this would require us to ignore the fact that phantom functional artifacts have perfectly normal histories of use and reproduction contingent on that use. Moreover, many artifacts that do perform successfully barely perform successfully. From this point of view, phantom functional artifacts are only the limiting case of what is in fact a continuum of more or less successful performance. So to deny that phantom functions are functions would simply be an ad hoc move to save the biological model of proper function in the face of obvious disanalogies between biology and culture.

Another option is to revise our formula again to make the successful performance condition disjunctive so that merely being believed to perform successfully is also allowed.

A current token of an artifact type has the proper function of producing an effect of a given type just in case either producing this effect *or being believed to produce it* contributed to the use of past tokens of this type of artifact, and thereby contributed to the reproduction of such artifacts.

But this will work only if all phantom functional artifacts are believed by their makers and/or users to actually do what they are supposed to do. Unfortunately this is not necessarily the case. Material culture is pervasively social, and people often have reasons or motivations for using artifacts in regular ways that are not contingent on their believing these artifacts to be performing successfully. For example, deference to authority, respect for tradition, or sheer habit motivate many people to participate in the ritual of the Eucharist even though they do not believe the doctrine of transubstantiation, or do not really understand it. And religious leaders may enjoin such rituals even when they have no belief in the efficacy of the artifacts involved in order to enhance their own status or their control of their followers' behavior. Similarly the captains of industry may—and often do—reproduce and market commodities they know perfectly well cannot do what they are implicitly or explicitly advertised to do; and many completely skeptical users may acquire and use these commodities, thus ensuring their ongoing reproduction, out of a desire to be fashionable, because it is required by some authority, out of desperation because there are no other options, and so on.

So we are stuck. Fitness in biology ties reproduction to capacities the individual organism actually exercises, and to which the successful performances of its various organs and traits actually contribute. But in material culture reproduction is only sometimes contingent on capacities the individual artifact actually exercises. It may also be contingent, in whole or in part, on what human beings *do* with that artifact, regardless of its actual capacities, and even regardless of what those human beings believe about its capacities. In other words, human agents may—and not infrequently do—use an artifact *as if* it had certain capacities, even though it does not have them, and sometimes even though they do not believe it does. We are forced to conclude that there is no good analogue of fitness for material culture. In particular, successful performance is not closely tied to reproductive success in material culture as it is in biology. So contribution to successful performance does not pick out the proper functions of artifacts or their components, although it may well pick out the proper functions of biological traits.

So what does pick out proper functions in material culture? On the basis of our investigations so far, it seems that proper functions in material culture can be identified only by looking at patterns of actual use and how they affect reproduction. So in conclusion we may revise our formula once again to reflect this direction of investigation, if not the precise details that will be available only once the investigation has actually been carried out. Provisionally, then,

A current token of an artifact type has the proper function of producing an effect of a given type just in case producing this effect contributes to the explanation of historically attested, dominant patterns of use to which past tokens of this type of artifact have been put, and which thereby contributed to the reproduction of such artifacts.

One final note: alert readers have probably noticed that this way of picking out the proper functions of artifacts does not provide any account of the functions of novel prototypes, which have no history of use and reproduction. Novel prototypes, in short, have no proper functions on this view. They may well have another sort of function—known variously as causal-role, accidental, Cummins, or (my preference) system function—that does not require a history but only a role in an embedding system of some kind. For example, a pen of entirely novel design that works as intended by its designer has the system function of writing in virtue of filling that role in the system of artifacts used in that activity (paper, pencils, erasers, etc.). The only question is whether the prototype pen works well enough to be substituted for a regular pen in writing notes or signing documents; not whether it has been used for such activities in the past and reproduced on account of this use. On the other hand, if the prototype pen does not work well enough to be substituted for a regular pen—if it is a failure as far as its intended function goes—then it has neither a system function nor a proper function.

A common intuition among function theorists is that novel prototypes must have proper functions, and a common solution is to argue that the intentions of designers establish them (Millikan 1999; Vermaas and Houkes 2003). But as I have argued (Preston 2003), this seemingly plausible and innocuous move has serious repercussions. In particular it threatens the widely accepted and important distinction between historically conditioned function (what something is supposed to do, i.e., its proper function) and current function (what something in fact does on a given occasion, i.e., its system function). I argued further that without the distinction between proper function and system function it would be impossible to appropriately describe and account for the social processes involved in the use, production, and reproduction of artifacts. I will not rehearse these arguments here, but I stand by them. So in appealing to a history of use and reproduction to pick out the proper functions of artifacts, I am not carelessly ignoring the alleged proper functions of novel prototypes. Rather I wish to assert that such artifacts have no proper functions.

However counterintuitive this conclusion may seem, there is a bright side to it from the perspective of comparing biological and cultural functions. I have argued here that cultural selection is like natural selection in crucial respects, but that precisely these similarities mean that cultural selection does not pick out the proper functions of artifacts any more than natural selection picks out the proper functions of biological traits. Furthermore, I have argued that although contributions to fitness leading to reproduction may well pick out biological proper functions, fitness cannot be used to pick out the proper functions of artifacts. Cultural fitness is at best only vaguely analogous to biological fitness; and, more importantly, cultural reproduction is often independent of fitness in ways that biological reproduction is not. Finally, I have recommended an approach that instead looks to patterns of use leading to reproduction to pick out the proper functions of artifacts. And this recommendation preserves a significant analogy between biology and culture—the centrality of processes of reproduction that ensure the continuing production of tokens of standardized types, while allowing for variation leading to new types of things.

References

Aristotle (*Physics*; *Metaphysics*) *The Basic Works of Aristotle* (McKeon, R., ed.). New York: Random House.

Buller, D. J. (1998). Etiological theories of function: A geographical survey. *Biology and Philosophy, 13:* 505–527.

Cummins, R. (2002). Neo-teleology. In: *Functions: New Essays in the Philosophy of Psychology and Biology* (Ariew, A., Cummins, R., Perlman, M., eds.), 157–172. Oxford and New York: Oxford University Press.

Endler, J. A. (1986). *Natural Selection in the Wild.* Princeton, N.J.: Princeton University Press.

Godfrey-Smith, P. (1994). A modern history theory of functions. *Noûs, 28:* 344–362.

Griffiths, P. E. (1993). Functional analysis and proper functions. *British Journal for the Philosophy of Science, 44:* 409–422.

Millikan, R. G. (1993). *White Queen Psychology and Other Essays for Alice.* London and Cambridge, Mass.: The MIT Press.

Millikan, R. G. (1999). Wings, spoons, pills and quills: A pluralist theory of functions. *The Journal of Philosophy, 96:* 191–206.

Pauling, L. (1970). *Vitamin C and the Common Cold.* San Francisco: W. H. Freeman.

Pauling, L. (1996). *How to Live Longer and Feel Better.* New York: Avon Books.

Petroski, H. (1992). *The Evolution of Useful Things.* New York: Vintage Books.

Preston, B. (2003). Of marigold beer—a reply to Vermaas and Houkes. *British Journal for the Philosophy of Science, 54:* 601–612.

Schwartz, P. H. (1999). Proper function and recent selection. *Philosophy of Science, 66 (Proceedings):* S210–S222.

Schwartz, P. H. (2002). The continuing usefulness account of proper functions. In: *Functions: New Essays in the Philosophy of Psychology and Biology* (Ariew, A., Cummins, R., Perlman, M., eds.), 244–260. Oxford and New York: Oxford University Press.

Sober, E. (1984). *The Nature of Selection: Evolutionary Theory in Philosophical Focus.* Cambridge, Mass., and London: The MIT Press.

Thesander, M. (1997). *The Feminine Ideal.* London: Reaktion Books, Ltd.

Vermaas, P. E., and Houkes, W. (2003). Ascribing functions to technical artifacts: A challenge to etiological accounts of functions. *British Journal for the Philosophy of Science, 54:* 261–289.

4 How Biological, Cultural, and Intended Functions Combine

Françoise Longy

4.1 An Attractive Classification of Functions

What is the function of coffee machines?

To make coffee, of course!

So a coffee machine that cannot make coffee properly is a malfunctioning coffee machine.

Exactly!

What makes a coffee machine a coffee machine?

The fact that it has been designed and produced for making coffee.

What is the function of the kidneys?

To filter the blood.

So a kidney that cannot filter blood properly is a malfunctioning kidney.

Exactly!

What gave kidneys their function?

Nature.

This imaginary dialogue introduces some of the obvious answers one may obtain when inquiring about the function of a typical artifact or a typical biological item. The answers concerning the artifact seem easy to justify. We plan and produce objects of various sorts in order that they do something specific. The function of a man-made object is then, it seems, merely the particular effect for which it has been made.[1] It is more demanding to justify the answers concerning the biological item. One needs to make sense of the idea that nature may pick out a particular effect and turn it into a function. As is well known, one way to do this is by employing the selectionist etiological theory of functions, SEL for short. This theory, proposed by Millikan, Neander, and others in the 1980s in order to account for biological functions, identifies the function of a trait with its selected effect or the effect that, pushed by natural selection, explains the diffusion or the conservation of the trait in the population (Millikan 1984; Neander 1991). So when we begin to investigate the nature of functions, two different sorts seem to emerge depending on whether we are considering artifacts or natural entities. On the one hand, there are those that result from human intentions, on the other hand, there are those that are due to natural mechanisms such as natural selection.

Upon closer analysis, however, the artifact case proves to be more complex. Objects may now have functions that were not the functions for which they were originally made, as is the case with old cart wheels used as decorative pieces on restaurants' walls. Moreover, some artifact functions may result from a long history without anybody having apparently ever done anything with the explicit intention of obtaining the desired effect. For example, some ergonomic forms for tools have probably been gradually selected and copied without anyone planning them explicitly with the particular aim of ergonomic correctness. These cases where parallels with natural evolution can be seen have prompted many authors to suggest that SEL could be applied to artifact functions with sociocultural selection replacing natural selection.[2] As a matter of fact, in long-standing categories of artifacts like hammers, clocks, and cars, an evolution has taken place that can be attributed to innovations gradually selected by buyers and users. Regardless of whether or not the positive effect justifying the diffusion of an innovation has been foreseen by whomever introduced it, that will be the function with which the new feature will be associated. Moreover, such a sociocultural mechanism is able to explain changes of function. In fact the effect for which people buy and use a type of entity may vary historically.

However, not every artifact function, whether that concerns the whole artifact or just one of its features, can be identified with the effect for which such an artifact has been bought in the recent past (a coffee machine for making coffee) or preferred to others (a coffee machine with a drop stop for stopping drops). As Houkes and Vermaas emphasize, there are also the functions attributed to a first generation of artifacts (2003: 264–65). Such functions cannot result from any sort of sociocultural selection. Thus the only explanation seems to be that they reflect the designer's or the producer's intentions. The traditional intentionalist conception of artifact functions is apparently an unassailable spot as far as new artifacts are concerned.

To summarize, artifact functions are not as easy to analyze as one might have first thought. They may have two sorts of origin, either they may result from the intentions of some inventor or from sociocultural selection. Nevertheless, such a difference in origins might not be of great significance, since intentions are involved in both cases. In the first case, there are the inventors' intentions, and in the second case there are the buyers' and users' intentions. So, even if the situation is more complex than it first seemed, the current assumption that all artifact functions belong to one and the same type because they all depend on intentions might still hold. It requires a deeper analysis to judge whether the attractive classification of functions into artifact functions and biological functions is indeed a good one.

4.2   Etiological Theories and the Current Classification of Functions

First of all, let us be more specific about the standpoint assumed here. Our perspective is the one adopted by etiological theories of function since the time of Larry Wright's seminal article (1973). According to this perspective, to have a function is to have a property of a quite peculiar nature, a property that can serve to ground both etiological explanations and normative claims.[3] Thus, an etiological theory of biological functions is meant to elucidate what a function is in order to explicate two things: 1) why assertions such as "the function of the heart is to pump blood" may offer an explanation for the present existence of hearts (their etiology) and 2) how normative statements such as "this heart is malfunctioning" make sense. Thereby advocates of etiological theories take a realist stance towards biological functions. In contrast both with Hempel's and with Cummins' positions, they defend that functions are genuine properties that cannot be dispensed with in scientific theories since they differ substantially from non functional properties.[4] The intentionalist theory of artifact functions can also be seen as an etiological theory of function since the effect for which an artifact has been invented or a feature has been designed—its function according to such a theory—explains the etiology of the functional item (why it exists) and fixes a norm. In fact, the artifact (or the feature) envisaged exists because it has been made in order to produce that particular effect, and tokens of this artifact type that cannot produce such an effect are judged defective. Nevertheless, it is doubtful whether the intentionalist theory entails the same realist stance as the selectionist theory. However, it is not necessary to clarify this point for the discussion that follows. So, let us leave it at that.

As we argue elsewhere, an in-depth analysis shows that the nature of a function hinges much more on whether an objective selection mechanism has had a role in establishing or maintaining it, rather than on whether there has been an intentional element involved at some stage or another.[5] As a matter of fact, functions that are supposed to be simply determined by somebody's intentions (intended functions), and functions that are supposed to depend on some objective mechanism of selection (selected functions) turn out to be quite different regardless of whether intentional elements have been involved in the selection process.[6] Roughly speaking, according to classical etiological definitions, "X has selected function F" will mean X is there because previous Xs have been selected for having done F, and "X has intended function F" will mean that X has been planned or produced because someone thought it would do F.[7] Now not only do these two definitions look dissimilar but the sorts of properties they capture are quite different. Let us consider this in more detail.

A selected effect is something real and objectively ascertainable if the mechanism of selection is itself objective in that it operates on real effects. Now, both natural selection and sociocultural selection operates on real effects. (The case of "mental selection" that Wright wrongly put on a par with natural selection (1973: 163) is here left aside.) More precisely, a selected effect is a type of real effect since it is supposed that some of these effects (some token effects) have already occurred and have subsequently acted as a cause. For example, circulating blood is the selected effect of hearts because many hearts have circulated blood, and the fact that they did so caused hearts to be preserved by natural selection. So a selected function supposes a connection between existing items, let us say the Xs, and existing effects. There is no doubt that some, if not all, of the Xs have the capacity to produce the functional effect F since some of them have already produced it.

That is not the case with intended functions. The conviction that at least some Xs should have the capacity to do F in the right circumstances is not sufficient to ensure that it is effectively so, however rational and justified such a conviction might be. Rationality does not preclude errors. So, by definition, an intended function does not necessarily refer to real effects of the type of item envisaged—it refers only to rationally predictable ones. Now the difference between an actual effect and a rationally predictable effect is by no means superficial. A criterion that puts real effects on a par with rationally predictable ones, or ontological conditions on a par with epistemological conditions, can determine only a very heterogeneous class from an ontological point of view. Such would be the case with the following disjunctive definition: X has function F if F is a selected effect of the Xs or if F is an effect one can rationally expect some if not all of the Xs to have. So the class of artifact functions demonstrates no substantial unity—but rather quite the contrary if it is made up, as is currently presumed, of both intended and socioculturally selected functions.[8]

What conclusions should we draw from the discrepancy between, on the one hand, the current image of functions in which they separate easily into two homogeneous categories, the biological and the artifactual functions, and, on the other hand, the image we obtain when we deepen the analysis and avail ourselves of the current etiological theories? Should we imagine a dividing line passing elsewhere than in between artifact functions and biological functions? For instance, between selected effect functions and purely intentional ones? Should we try to discover another criterion in order to obtain the desired distinction? Or should we simply abandon the very idea that there are different categories of teleofunctions? I argue that at least as far as material artifacts and biological entities are concerned, we should renounce the idea of putting their functions into different ontological categories. This therefore implies that we must abandon the idea that we should or could have different accounts of functions of material entities depending either on the nature of the entities (natural or artificial, inert or living), or on the origin of the functions (selection or intention).[9]

In what follows, I try to show that the customary ways of distinguishing functions (natural or biological versus artifactual or cultural, or intended versus selected) are of a superficial and pragmatic nature and that no scientific classification can be based on such distinctions. More generally, I argue that the realm of biological and artifactual functions cannot be divided into smaller domains demonstrating a higher ontological homogeneity than the whole domain. However, the results of my investigations are not only of a negative nature (indicating what we have to renounce). Some are positive. In particular, in analyzing the case of biological artifacts, a new hypothesis emerges concerning the specific content by which functional attributions contribute to scientific understanding and explanation.

To support my main claim, I analyze what is going on with functions at three decisive points:

a) when the artifactual encounters the biological
b) when new functions become culturally established
c) when new artifacts are invented.

4.3 The Artifactual and the Biological

The changes brought about in plants and animals through domestication and cultivation in prehistoric times may be seen as the first examples of humans hijacking biological mechanisms to their own ends. To what extent this resulted from the pursuit of well-defined and conscious objectives remains debatable. However, the artificial selection carried out by breeders in the nineteenth century raises no such doubts: they knew perfectly well what they were doing. More recently still, humans have extended the scope of their activities with the creation of genetically modified organisms (GMO). In all such cases the organisms have traits whose functions result both from natural mechanisms and human intentional actions. As a consequence, it seems that such functions deserve to be seen as both artifactual and biological. Is it, however, possible for them to be both?

It is because such functions depend on natural selection that they deserve to be called "biological." Artificial selection does not replace natural selection—it relies on it. Artificial selection steers natural selection in a particular direction in order to realize short-term or long-term human aims. With GMO, humans intervene in another angle of the evolution process, the mutation angle, but natural selection still gets its way later, be it relative to a natural or to a controlled environment. As long as some general features of natural life and reproduction remain, natural selection will always play a part. Artificially introduced or enhanced traits usually spread because, in relation to a context in which human activities and interests matter, they make the organisms possessing their traits more fit to compete in the Darwinian struggle for existence. This clearly emerges from the cultivated wheat example considered later in this section.

However, such traits also deserve, it seems, to be called "artifactual" because without voluntary human intervention the features with the desired effects would not have appeared, would not have been selected long ago, or would not have been recently maintained. Here we need to be a little more precise about the meaning of the expression "artifact function". Vagueness does not pose much of a problem when considering inanimate objects, since they can have only artifact functions (they may have functions only if they are used or produced by intentional agents). But, with biological items the situation is more complex since there is more than one option.

First and foremost, what does *artifact* mean? Sperber convincingly defends the claim that *artifact* is a family resemblance notion rather than one that "could be defined precisely enough to serve a genuine theoretical purpose" (2007: 124). In fact classical definitions of artifacts present a choice between two possible conditions that are not equivalent: 1) to have been "intentionally made or produced for a certain purpose," and 2) to be "the product of human actions" (Hilpinen 1999). The second condition is the less restrictive of the two, since human action does not necessarily presuppose clear purposes. However, the first condition is not very restrictive either. It just supposes some purposefulness in making or producing something. According to that condition, an entity, a feature, or a function will be artifactual if it results from purposeful human action, even if a clear anticipation of the result is missing. Moreover, as we know, a series of limited short-term aims may produce unforeseen long-term effects.

So unpredictability can go hand in hand with artifactuality. Unforeseen new features and new functions may indeed appear thanks to what Darwin considered an unconscious form of artificial selection, the form that "results from everyone trying to possess and breed from the best individual animals" (1859: 34). Besides, there is a continuum of intermediate stages from unconscious artificial selection to completely planned artificial selection, from the unconscious domestication of some species in prehistoric times to the consciously pursued aim of nineteenth-century breeders or of present GMO producers. All this supports a broad application of the term *artifact* in the biological domain, one that covers the whole range of human interventions. So in the biological realm, just as in the nonbiological realm, artifactual functions should include intended functions (effects that have been clearly anticipated and have been obtained as the result of consciously planned interventions) and nonintended functions (unanticipated effects that have been obtained at the end of a series of short-term oriented interventions).

Let us continue to clarify the issue by taking a closer look at nonintended functions. When we considered the nonintended functions of inanimate objects, we spoke of socio-culturally selected functions. It is true that many of these functions can probably be accounted for by SEL with some form of sociocultural selection replacing natural selection. For instance, in the story of manufactured artifacts, sociocultural selection through economic competition plays an important part. Is this, however, the case with all non-intended functions of inanimate objects? Sperber claims that cultural functions do not necessarily suppose something similar to natural selection, and I agree with him (2007: 128). He insists on not needing replicators such as genes. One can also dispense with a true mechanism of selection that supposes competition among variants.[10] Consequently, in his definition of a *teleofunction,* Sperber supplants the notion of selection with the broader one of propagation: "an effect of type F is a teleofunction of items of type A just in case the fact that A items have produced F effects helps explain the fact that A items propagate."[11] Natural selection is then simply one of the mechanisms that can explain propagation and evolution.

This new definition, which is meant to apply indiscriminately to biological, artifactual, and cultural teleofunctions, needs some clarifying remarks. Above all, a new notion has appeared—that of a "cultural function." From what Sperber says, one can extract the following characterization: a function is cultural if the propagation on which it relies involves at some stage some mental representations. For instance, domesticated wheat has the *cultural* function of nourishing humans because the cultivation of wheat for this end propagated thanks to mental representations. Indeed, it propagated because some farming practices were consciously imitated, because people thought and talked about how to grow wheat, because books about growing wheat were written and read and so on and so forth. First, it should be noted that the replacement of a selectionist characterization of functions by a propagationist characterization has consequences only for cultural functions. As a matter of fact, the only *biological* mechanism that may explain a propagation that is "helped by the fact that Xs do F" is natural selection. Second, artifactual and cultural functions, which overlap to a large extent, remain somewhat distinct. In fact neither sort, apparently, includes the other. Behaviors typically have cultural functions, but behaviors are not artifacts.[12] On the other hand, purely intended functions of artifacts are not cultural functions, at least not of the propagated sort. In fact it is generally admitted that an artifact has a purely intended function when it leaves the hands of its creator and has not yet been reproduced and diffused.

Now that these clarifications have been made, let us resume our investigation of the functions of a cultivated or domesticated species. A good example, analyzed by Sperber, is that of wheat and barley whose cultivation began thirteen thousand years ago by sowing part of the collected seeds in chosen locations instead of simply eating all there was. Sperber explains:

It can be quite advantageous for a plant to have a large proportion of its seeds used by humans as food, provided that the remainder of the seeds serves the goal of reproduction and dispersal in a particularly efficient way. When this became the case for various species of cereals, feeding humans became a biological teleofunction of the seeds, that is, an effect that contributed to the greater reproductive success of varieties of cereal providing better food. Both the feeding function and the reproduction function of seeds are simultaneously biological and cultural/artifactual functions of cultivated cereal. The plants take biological advantage of their cultural functions and humans exploit culturally, and more specifically economically, some of the biological functions of the plants. There has been a co-evolution of the plants and of their cultural role. Human culture has adapted to cereal biology just as cereals have adapted to human culture. (2007: 133)

He proposes calling "biological artifacts" items that, like cultivated wheat, "perform their artifactual function by performing some of their biological functions" (2007: 130). As the extracted quote shows, a biological artifact is an entity that possesses one or several functions such that the functional effect has been both biologically selected for the sake of the plant and artifactually controlled for the sake of humans.

Why does Sperber write "cultural/artifactual" at one point? What precision is achieved by stating that such functions are not only biological and artifactual but also cultural? By calling them "artifactual," one stresses the fact that the features concerned have been shaped largely by the control exerted by humans on the conditions of existence and reproduction of the plant. By adding that they are "cultural," one stresses that mental representations have been involved in the various processes that have contributed to diffuse such features. However, even if the two notions are not perfectly equivalent as we have seen above and may serve to lay different stresses, they overlap to such a great extent that it is difficult to distinguish them clearly.

Functions of biological artifacts are difficult to analyze because of various entanglements. We have just identified a *conceptual* entanglement between the notion of "artifactual function" and the notion of "cultural function." But there are also *ontological* entanglements. The biological is entangled with the cultural/artifactual because each side exploits the other. Humans exploit biological mechanisms to create artifacts better suited to their needs. Plants exploit the fact that humans are able to create good conditions for the reproduction of what they (humans) like. Plants take advantage not only of the capacity humans have to understand and control natural phenomena but also of their capacity to propagate ideas, behaviors, and entities via cultural means: imitation, transposition, theft, trade, and so forth. The ontological entanglements explain why a theoretical analysis will necessarily be complex and muddled. The complexity lies within reality itself, in the fact that there are interdependencies among factors of different sorts. However, at some level of reality, things look simple enough. There is one function, and this function indicates one single causal relation. Relative to a particular sort of entity, the assertion that feature X has function F means something quite straightforward: there is a relation between having feature X and having the capacity to do F (or to result in F) and this relation is what explains the existence or the diffusion of the bearers of the X feature. Let us examine this more concretely by giving an example. Suppose feature X is a particular ratio of carbohydrates to fibers found in actual cultivated wheat—in short, the CFR feature—and F is the property of being easily digestible for humans; then the claim that feature X has function F will mean simply that most cultivated wheat fields produce CFR corn because CFR corn is easy for humans to digest.

A functional assertion, however, indicates more than just the existence of a causal relation; it gives us information about its nature. It tells us not only that fact number 1 (the easy digestibility of CFR corn) is the cause of fact number 2 (the large presence of CFR corn in cultivated land) but also that this causal connection is neither a fortuitous one nor simply a consequence of basic physical facts, that is an effect that could be drawn from the laws of physics. It tells us also that this causal relation is a stable relation that results from a persistent structure or mechanism. Thus by being told that the function of X is F, we are told that the lasting presence (or the large diffusion) of X-bearers depends *systematically*, because of some general mechanism, on the fact that some or most X-bearers have effect F, and that without such a mechanism the situation would in all likelihood be quite different since physical laws and conditions cannot account for the lasting presence (or the large diffusion) of X-bearers. However, the issue as to whether the stability of this causal connection is due to something biological, cultural, or intentional remains out of the picture. The nature of mechanism which may explain this causal connection is not specified. So, the pieces of information that a functional assertion delivers are substantial but of a very abstract nature.

In the case of cultivated or domesticated species, the causal stability underlying bioartifactual functions is in fact the result of very complex series of phenomena and mechanisms acting at different levels: how human preferences and human knowledge transform natural selection in artificial selection, how cultural mechanisms of various types diffuse and speed up artificial selection, how plants and animals exploit the new environmental conditions created by humans, and so forth. It seems to me that the whole thing could be seen as a complex mechanism involving both biological and cultural parts. How such a complex mechanism produces its output (in the case of our example, the fact that most cultivated wheat produce CFR corn) from its input (the easy digestibility of CFR corn plus a series of cultural and biological conditions) can be understood only if one retains a certain level of generality. No satisfactory causal explanation can be obtained if the data are not made to fit into the general frame of a mechanism. Moreover, a very limited amount of information will often be sufficient to determine the general structure of such a mechanism, thus satisfying our quest for an explanation. More information will often just help to fill in the structure.

Let us take again our invented CFR corn example. A single piece of information, such as "humans have always found CFR corn more palatable" or "tribes eating CFR corn suffered less from malnutrition than neighboring tribes" is sufficient to figure in rough outline the general mechanism that may explain the diffusion or maintenance of CFR corn in some area. Depending on the theoretical background, the mechanism sketched may be somewhat different. For example, those who see in food preferences unexplained data will certainly imagine a mechanism more superficial and with a more limited scope than those who see in food preferences the result of a mechanism designed by natural selection to make us favor what is more nutritious for us. However, and this deserves to be noted, no serious scientific background is needed to come up with a plausible mechanism. Often a layman's theoretical background is enough to arrive at a sensible hypothesis. It is often possible for the layman to grasp something relatively simple that may be a part (a submechanism, let us say) or a rough sketch of the much more complex mechanism responsible for the high-level causal relationship. Not much is needed to sketch in broad outline a mechanism that may explain how a property like the human preference for CFR corn may have produced steadily a high ratio of CFR corn in cultivated land.

It must be stressed that only such a top-down explanation, an explanation relying on a mechanism or structure visible only when contemplating matters from a certain level of abstraction, can meet the task of explaining a kind of stability that does not result from physicochemical laws. Without such a top-down perspective such stability will usually be inexplicable. To continue with our example, without a mechanism explaining how the better taste or digestibility of CFR corn could have acted *steadily* in favor of CFR corn, the probable instability of genetic and climatic conditions in the period concerned would make the long-lasting presence of CFR corn a mysterious and highly improbable fact. A causal explanation remaining at a lower level will not in general be able to account for the stable causal dependency the function points to, no matter how detailed it is. If, for example, you knew whether the genetic makeup plus the actual growth conditions of the type of wheat cultivated most at present induced the presence of the CFR feature, you would be in a situation to explain why 95 percent, let us say, of the studied wheat would produce CFR corn. However, this would not explain why it was also roughly the same before (95 percent of cultivated wheat producing CFR corn) when the genetic pool of cultivated wheat and the growth conditions were somewhat different. It would also give you no reason to expect the same or another ratio for the cultivated wheat not yet studied, or for cultivated wheat three hundred years from now, if a very small change in growth conditions or in the genetic pool were to modify the carbohydrate–fiber ratio of the corn.

I conjecture that this point is an aspect that could, once properly elaborated, explain and justify the importance of functional explanations as a particular type of causal explanation. Functions provide the basis for an important sort of top-down explanations. They supply the right causal frame for explaining a type of phenomena that cannot be explained satisfactorily bottom-up. Such are the phenomena that rely on the existence of complex mechanisms that produce stable causal connections while different levels of reality (physical, biological, psychological, etc.) are involved. However, this is not the place to develop this point further. The conclusion I want to draw here concerns only the interpretation of the difficulties encountered in seeking to demarcate biological functions from artifactual and cultural ones. Such difficulties do not result from a lack of clarity at the conceptual level. They derive from the fact that some functions point to stable causal connections that depend on very complex multilevel mechanisms. The attempt to separate the biological elements from the artifactual or the cultural ones would inevitably result in making every such function disappear. Mixed functions cannot divide into two or three autonomous subfunctions (biological, artifactual, or cultural), each with its specific history. The disappearance of mixed functions would thus not result in a positive simplification; it would mean becoming blind to certain high-level phenomena.

At the end of his article Sperber questions why our prototypical artifacts are tools and machines. He suggests that our failure to acknowledge biological artifacts, despite their practical importance and their number, may come from a Stone Age bias, from the long Paleolithic period when the only artifacts were tools made out of inert material. Whatever the case, it is true that biological artifacts, that is, items possessing functions that are both biological and artifactual, are numerous, and this is sufficient reason to abandon all hopes of obtaining a scientifically valid classification of functions by separating the biological from the artifactual and the cultural.

In the mixed functions of biological artifacts, however, some selection is always involved. It may therefore be thought that a better way to separate functions would be to distinguish selected or propagated functions from purely intended ones. But, as we will see now, this distinction too is inoperative.

4.4 The History of a Typical Artifactual Function

Many artifact types have a very simple history. First, an object was invented to do F; second, objects identical to it were mass produced and advertised as tools for doing F; third, people bought these objects and used them almost exclusively to do F; and fourth, such objects have continued to be produced, sold, and used as F-doers undergoing possibly some slight modifications at some time or another. Let us suppose that peelers have such a simple history. Let us suppose that one day a certain Tom Smith had the idea of creating a peeler, and that he then designed it and succeeded in convincing someone to produce it and advertise it as a tool for peeling potatoes, carrots, and so forth. Let us suppose that this was the beginning of the story of peelers that then continued as described. The function of peelers seems to raise no problems; from start to finish it has been to peel potatoes and similar vegetables. However, the trouble begins as soon as one asks whether the function is an intended one or a culturally established one. At the beginning of the story it was, it is supposed, an intended function, but what is it now? The answer appears to be that it is a culturally established function independent of the previously intended one. In fact one does not need to know the story told here to know that the tools are peelers. It would not change anything to be told that the intended function of such objects, what they were invented for, was to extract the last coat of rubber that remains stuck to some device when collecting rubber. Only recent facts are relevant for establishing the proper function of such a device. Indeed only information about the very recent past will lead one to revise one's judgment. Mary, seeing John peeling potatoes with a device looking somewhat different from the peelers she knows, will probably change her thought from "John is using a new sort of peeler" to "John has cleverly turned a rubber collector into a peeler" if she receives the information that *at present* this sort of device is usually bought by people working in rubber plantations to extract the last coat of rubber and that it is in fact what the firm producing such devices sells them for.

If the peeler's function really has changed from being an intended function to being a cultural one, how did this happen? Does an event have the capacity to turn an intended function into a cultural one? Does competition, for instance, produce such a transformation? Let us suppose that at some point a new firm launched new peelers on the market, for example, cheaper peelers, peelers with colored handles, or peelers with a better connection between the blade and the handle. Let us furthermore suppose that the competition between the two firms resulted in the closing down of one of them. What we get then is a typical case of selection in which one of two variants wins. So SEL can apply.[13] From this time onward, then, the peeling function of the peelers should be analyzed as a culturally established function. However, the idea of an event switching the nature of the function is problematic for the same reason that we have already seen: far away history does not seem to matter. Whether or not such episodes have really occurred has no effect on the fact that, nowadays anyhow, such devices have the culturally established function of being peelers. The only thing that seems to matter is the existence now and in the very recent past of a diffused and stable association between this type of object and a typical use.

Since there seems to be no particular historical event with the role of marking the frontier between intended functions and culturally established ones, the transition from the former to the latter (supposing there is one) should be gradual. In between the end marked "intended" and the end marked "culturally established," there should be a gray zone, possibly a large gray zone, where the intended mixes with or fades into the culturally established. Independent of whether this idea of a gray zone makes sense or not, it goes against the very idea of a clear-cut distinction between these two sorts of functions, and hence against the idea of separating functions into two distinct categories along these lines.[14] Our investigation into the historical development of a function has thus led us to the following negative conclusion: if indeed artifacts had first an intended function when they leave the hands of their designers and then later on a culturally established one, it would not be possible to distinguish one from another.

4.5   The Invention Period

There is at least one situation, that of invention, where it seems possible to escape the problem of mixed functions. A function attributed to something that is still in the making cannot, it would seem, refer to something other than a mental content, that is, to the imagined effect the planned object should have in the foreseen conditions. Actually, inventing includes many different aspects; it is not just a case of drawing and calculating with pen and paper. I argue that some testing procedures in fact should be viewed as giving rise to selected functions and that consequently, in the invention phase, too, the situation is not as simple as one might expect. As explained at the beginning of the article, a selected function refers to a real property (the effect that some Xs had, which led to the selection of Xs against variants), while an intended function refers to a mental content relative to the Xs (the effect that rational humans think some Xs, at least, will have in determined circumstances). Now the testing of prototypes has to do with real effects, not with imagined effects.

Suppose that several engineers have worked on airbag triggers for cars and have come up with different models. Airbag triggers must be fast and accurate, they must not trigger too late, but they must not be too sensitive and trigger when the car is passing over a pothole or when the brakes are sharply applied. Suppose the different models were submitted to an appropriate battery of tests in a car crash test lab. Suppose that one of the models, let us say the M12, came out of the battery of tests victorious. Then the effect for which M12 devices will be put in cars as car airbag triggers is real; in fact this effect showed in tests. Better still, it showed in a situation of selection, the situation of comparative tests. Consequently a classical SEL definition of function applies with no difficulty. "M12 has the function of triggering a car airbag" can be interpreted as meaning "M12 was selected for its car-airbag-trigger effect." In fact it was selected because it demonstrated a better car-airbag-trigger effect than its competitors in certain real contexts. Before we investigate to see if this definition of the M12 function really hinges on a *car-airbag-trigger* effect, let us first clarify what the point of the whole argument is. We want to show that, contrary to widespread opinion, there is no straightforward answer to the question "What is the nature of the car airbag function of M12?" In particular, the current etiological theories do not deliver the simple single answer, an intended function. According to those theories, M12 should in fact have an *intended* car-airbag-trigger function before the first battery of tests, but a *selected* one after it. Let us now resume our investigation of the above SEL definition.

Is the selective context of the tests really a selective context relative to *car properties*? The classification of an effect depends in fact on the context. Selection is necessarily relative to car-something effects if the different models are tested in real-life situations by being installed in cars that have been sold to ordinary consumers. When the selective context is artificial, however, the categorization becomes more problematic. An intuitive grasp of the problem is made easier by considering the two ends of the prototype-testing spectrum. At one end, there are the tests made in a very sophisticated car-crash test lab of a big firm where real cars are sometimes used and where the conditions of real-life car driving are very well simulated. At the other end of the spectrum there are the tests made by amateurs with poor resources. Let us suppose that a group of experts considering some amateur tests arrive at the conclusion that these are indeed so badly conceived that they cannot provide any useful information as far as cars are concerned. They provide information about what happens in conditions very different from the ones met when driving a car in a real-life situation. If the experts are right, then a selection of airbag-trigger models made on the basis of such tests will not be relative to *car*-airbag-trigger effects and so will not be able to ground any *car*-airbag-trigger function.

The previous thought experiment could apparently yield the following conclusion: the amateur case reveals the very nature of prototype testing. The argument would be that prototype testing concerns real effects, but not *real-life* effects. For this reason, it could be argued, prototype testing can ground no function relative to real-life effects; it can only justify a rational expectation about real-life effects by considering somewhat related effects. This is neither a sound argument nor a sound conclusion. The difference between real-life situations and simulations can in fact be very small, all the more so if the accuracy of the simulation conditions is itself continually being improved, as is certainly the case with the best labs for car-crash testing. The testing of new medicines also gives a good example of the continuity between artificial and real-life conditions. The testing of medicines is usually organized in different stages starting from various chemical lab tests, going on to testing on animals, and finishing with single-blind or double-blind tests supposedly carried out in real-life conditions. Once again, the lesson learned from our investigation is that no clear-cut separation is to be found. There is a continuous line going from real-life situations to artificial test conditions. At one end there is selection in real life, a little further on there are tests done in "controlled (real-life) situations" (tests made in hospitals, supervised field trials, etc.), the next stage involves the tests made in good labs of big firms or big research teams, until finally we arrive at the other end of the spectrum with the poor tests made by amateurs or in unsatisfactory conditions. This spectrum shows that there really is no ground for separating good prototype testing from real-life selection. So if the test made is a good one, there is nothing against categorizing the effects relative to which the prototypes are tested as they would have been in the corresponding real-life situations. To conclude, there is no serious objection that could be brought against the statement that M12 has been tested in relation to car-airbag-triggering effects if the tests conducted were good ones.

Inventing often involves tinkering and this provides another argument against separating the invention period from the "normal life" period when artifacts are reproduced but often also modified. There is often no clear answer to the question "Is X a new device with a new function or is X a new specialized version of an already existing device?" For instance, a car airbag trigger may result from modifying a trigger used in planes. Suppose that an airplane trigger has been modified for trains, then for trucks, and then for motorcycles, and that these modifications have been successful every time (the modified triggers have demonstrated their ability to work in real life as expected). Suppose now that the same plane trigger is modified to be introduced in cars. How should we envisage the new device? As a new car device? Or as a long-existent trigger used in different means of transport? Let us furthermore suppose that triggers must be adjusted in accordance with parameters such as weight, possible acceleration, and deceleration. Triggers might then have to be adapted (modified) for each type of car. How should the trigger adjusted for a new type of car, let us say the new Peugeot 7007, be categorized? When the Peugeot 7007 is not yet on the market, should we see it as a new device endowed with the *intended function* of "Peugeot 7007 airbag trigger," or should we see it as the Peugeot 7007 version of a device whose *culturally well-established function* is to be a car airbag trigger?

The conclusion of this section is similar to that of the previous section. There is no clearly delineated area, be it the whole realm of invention or only some part of it, that could be said to be homogeneous with regard to function, according to the current distinctions.

4.6   Conclusion

In the vast realm of teleofunctions of material entities, we have looked at three places where, according to current etiological theories, we should have found boundaries dividing one sort of function from another. Every time we found no such boundaries. We found instead mixed functions, functions that were crossing boundaries and mixing elements of various sorts. This casts doubts either on the notion of function itself or on the distinctions that present etiological theories of function back up. What is muddled and superficial here? My answer to the question is that the problem lies essentially with the distinctions imposed by current etiological theories.

Such distinctions arise from identifying functions according to their origins. This way of identifying functions is intuitive, as our introductory dialogue shows. Moreover, it has been reinforced by SEL, which has endorsed the "one type of function, one type of origin" principle, by introducing natural selection (the mechanism supposedly at the origin of biological functions) in its definition of *biological function*. However, as we suggest in our discussion about biological artifacts, another attitude and another perspective are possible. The confused origin of many functions (a mix of intentions, sociocultural mechanisms, and natural selection) does not prove that the notion of "function" refers to nothing really deep and important, and that it therefore has no scientific value. This confused origin may, on the contrary, be an argument for aiming at a more abstract notion of "teleofunction" than those provided by current etiological theories. A more abstract notion of "teleofunction," which would ignore the question of origins, could help us to understand why identifying functions is so useful when faced with relatively simple high-level phenomena that depend on complex mechanisms operating at different levels.

Notes

1. This idea lies at the core of the traditional intentionalist theory of artifacts. For a quick historic survey of this tradition see McLaughlin (2001: 42–62).

2. See, for instance, Millikan (1984: ch.1); Bigelow and Pargetter (1987: §III); (Griffiths 1993: §8).

3. Since it is debatable whether all the functions attributed are of a single sort and whether all such attributions can be associated with etiological explanations and normative (or teleological) claims, it is better to start from the hypothesis that there may be two sorts of functions, independent of each other, only one of which is related to etiological explanations and normative claims. To avoid any ambiguity, some authors have called functions of this latter sort "teleofunctions." Since I am concerned here with teleofunctions only, I can keep using the usual term *function* without fear of creating ambiguity.

4. For Hempel, functions do not correspond to a type of scientifically admissible property. For Cummins, they do—they are physical dispositions—but they make up no single type since the difference between them and ordinary physical dispositions, nonfunctional ones, is not ontological but pragmatic. See Longy (in press) for clarification.

5. This is what I expound on in Longy (in press).

6. For the sake of discussion, we temporarily endorse the distinction between intended and selected functions. Such a distinction is supposed to apply to the proper function of an item, that is, to the one attached to it as a member of an artifact type but not to the possible occasional use functions it may get in some particular circumstances, such as when a pencil is used as a hairpin.

7. See, for instance, Neander (1991: 174) for a classical definition of SEL.

8. Of course there is no objection to putting together heterogeneous things for pragmatic reasons. "Pets" is a good example of what we may call a pragmatic category. It is not, however, a category that will have a place in biology, contrary to the categories of "dogs" or "mammals." An ontological category is a category that carves nature at some of its joints and has for this reason a place in science. We cannot deal at length with this notion here, but it is discussed in Longy (in press).

9. This conclusion, a clear negation of pluralism, is the only point on which I disagree with Perlman (this volume). I agree totally with the agenda he sets for the further development of theories of teleofunction, with his four DON'Ts, but unlike him I think one has to renounce pluralism to satisfy these DON'Ts, especially the one of not drawing a hard line between natural functions and artifact functions (see note 8).

10. Sperber is not the only one to have pointed out the differences between natural and cultural mechanisms and the difficulties encountered when trying to transpose natural selection to sociocultural phenomena. In regard to this question, the recent book by Tim Lewens (2004) is of particular interest since it focuses on functions. The well-argued conclusion of Lewens is that SEL, defined as it currently is with an explicit reference to natural selection, cannot be extended to artifact functions because, to put it briefly, sociocultural selection differs largely from natural selection (2004:140–157). I agree with him on this point, but I nonetheless dispute his final conclusion that there is a simple analogy between biological and artifactual functions. In line with what Sperber does when he replaces selection by the broader notion of "propagation," I defend instead that what we should aim at is a more general and abstract notion of "teleofunction." A step toward abstraction makes it, indeed, possible to preserve both diversity and continuity. Different sorts of selective mechanisms are present in the sociocultural realm. Some of these are quite similar to natural selection while others are not. However, with a more abstract notion of "function," we are no longer obliged to draw a line between the genuine and the watered-down mechanisms of selection.

11. Sperber (2007: 128). As he specifies, a propagation is a repeated reproduction that supposes neither a definite copying mechanism nor a strong inheritance (new items do not necessarily have to "inherit all their relevant properties from previous tokens of the type" [2007: 127]).

12. It depends of course on the limits placed upon the vagueness of the term *artifact*. Here it applies only to material entities.

13. In fact things are more complex. SEL will account for the attribution of a peeling function if the selection is somehow relative to the peeling capacity. The difference in the quality of the blade-handle connection is clearly such a case. A less well connected blade will sooner loosen or make peeling difficult by wobbling. The colored handle case is more problematic. If this variant won simply because it answered better the aesthetic taste of

housewives, its selection will not, according to a strict version of SEL, provide grounds for a peeling function but rather for an aesthetic function. In my opinion, this is one more reason for finding a SEL account of cultural functions to be too restrictive. One might try to resolve the problem by adopting a weaker version of SEL, but it is jumping out of the frying pan into the fire. It is better to directly adopt the broader notion of a "propagation function."

14. A gray zone for almost every function involves much more than simply the existence of some borderline cases; this is why it cannot be reconciled with the hypothesis of two distinct categories.

References

Bigelow, J., and Pargetter, R. (1987). Functions. *The Journal of Philosophy, 86:* 181–196.

Darwin, C. (1859). *On the Origin of Species.* London: John Murray.

Griffith, P. E. (1993). Functional analysis and proper functions. *British Journal for the Philosophy of Science, 44:* 409–422.

Hilpinen, R. (1999). Artifact. *The Stanford Encyclopedia of Philosophy* (Zalta, E. N., ed.). http://plato.stanford.edu/archives/spr1999/entries/artifact/.

Houkes, W., and Vermaas, P. E. (2003). Ascribing functions to technical artifacts: A challenge to etiological accounts of function. *British Journal for the Philosophy of Science, 54:* 261–89.

Lewens, T. 2004. *Organisms and artifacts.* Cambridge, Mass.: The MIT Press.

Longy, F. (In press). Artifacts and organisms: A case for a new etiological theory of functions. In: *Functions: Selection and Mechanisms* (Huneman, P., ed.). Boston: Synthese Library.

McLaughlin, P. (2001). *What Functions Explain.* Cambridge: Cambridge University Press.

Neander, K. (1991). Functions as selected effects: The conceptual analyst's defense. *Philosophy of Science, 58:* 168–184.

Millikan, R. (1984). *Language, Thought and Other Biological Categories.* Cambridge, Mass.: The MIT Press.

Sperber, D. (2007). Seedless grapes: Nature and culture. In: *Creations of the Mind: Theories of Artifacts and Their Representation* (Laurence, S., Margolis, E., eds.), 124–137. Oxford: Oxford University Press.

Wright, L. (1973). Functions. *The Philosophical Review, 82:* 139–168.

5 On Unification: Taking Technical Functions as Objective (and Biological Functions as Subjective)

Pieter E. Vermaas

5.1 Introduction

Biological items and technical artifacts have in common that they both allow functional descriptions. Yet these descriptions seem to differ substantially, making it difficult to capture them in one uniform theory. Biological functions are typically taken as *objective nonrelational properties* of items that do not depend on biological context or the mental states of agents, whereas technical functions are seen as *subjective relations* between artifacts and their technical context including the mental states of agents. Biological functions are, moreover, typically taken as properties that items *have*, whereas technical functions are sometimes merely seen as relations that agents *ascribe* to artifacts.

These contrasts between biological and technical functions are not supported by philosophical analyses. The question of how functional descriptions are to be understood in biology and technology is not yet settled, and answers are limiting the mentioned contrasts. In the main candidates for theories of biological functions,[1] items have functions relative to contexts, such as their evolutionary pasts, the capacities of the organisms of which the items are a part, or the selective regimes they are subjected to. Biological functions thus seem not to be *nonrelational properties* but also *relations* that items have relative to context. On particular function theories the contrasts even seems to disappear. According to John R. Searle (1995), biological functions are ascribed to items relative to goals agents impose on organisms, turning biological functions also into *subjective* relations agents *ascribe* relative to their mental states. Conversely, technical functions of components of artifacts may in Robert Cummins's (1975) theory be taken as physical capacities of the components that causally contribute to physical capacities of the artifacts, turning technical functions into *objective relations* components *have* independent of the mental states of agents.

In this contribution I argue that the alleged differences between biological and technical functions to a large extent can be avoided. This argument is not a defense of the theories of Searle or Cummins. Instead I accept the main candidates for biological function theories by assuming that biological functions are *objective relations* that items *have* relative to context, and then construct a theory by which technical functions are also *relations* that artifacts *have* relative to context. I acknowledge that these latter relations are still *subjective* in an ontological sense, but defend that they are *objective* in an epistemic sense. By thus minimizing the differences between biological and technical functions, prospects for a uniform function theory improve, which I explore at the end of this contribution.

The technical function theory that I construct is drawn from the ICE-function theory (Houkes and Vermaas 2004; Vermaas and Houkes 2006), in which technical functions are relations that agents ascribe to artifacts relative to mental states. A first assessment of this constructed theory seems, however, to immediately reveal a snag, since the theory seems incapable of accommodating the phenomenon of malfunctioning artifacts. I therefore also introduce in this contribution a new approach toward understanding malfunctioning. This approach turns the constructed technical function theory into one that can adequately accommodate malfunctioning; yet it reveals also a new difference between biological and technical functional descriptions: artifacts can be taken as malfunctioning only if they can reasonably be repaired, whereas malfunctioning biological items may be irreversibly malformed.

I introduce in section 5.2 the distinctions between the epistemic and ontological senses of objectivity and subjectivity, which I adopt from Searle. Then I present in sections 5.3 and 5.4 a strategy to construct theories by which artifacts *have* technical functions from theories by which agents *ascribe* these functions. I apply this strategy to the ICE theory in section 5.5 to arrive at my theory in which artifacts have their functions as epistemically objective and ontologically subjective relations relative to the mental states of designers. The new approach toward understanding malfunctioning is given in section 5.6. I generalize the constructed theory to a uniform "ICE-like" function theory in section 5.7, and indicate its similarities with Cummins's theory.

5.2 The Subjectivity of Technical Functions

If biological functions are, by the main theories of such functions, to be taken as objective relations items have relative to context, and if technical functions are subjective relations that agents ascribe to artifacts relative to mental states of agents, then the differences between the two consist of two elements: biological functions are *objective* whereas technical functions are *subjective*, and biological functions are relations items *have* whereas technical functions are relations that are *ascribed* by agents to artifacts. These elements are related. If technical functions are analyzed as relations ascribed by agents, then the mental states of the ascribing agents seem somehow constitutive to technical functions, giving these functions a subjective character. Yet in this contribution I consider the two elements separately, starting in this section with the first.

Taking technical function as partly subjective does not seem to be problematic. Artifacts are designed and used by agents for their functions, and this introduces an acceptable relation between technical functions and the mental states—intentions and purposes—of designers and users. In many theories of technical functions, such mental states actually play a role.[2] In intentional theories this role is made explicit: in, for instance, Karen Neander's theory, the function of an artifact "is the purpose or end for which it was designed, made, or (minimally) put in place or retained by an agent" (1991: 462). In etiological theories such as Ruth Garrett Millikan's (1984; 1993) and Beth Preston's (1998), mental states are a bit more hidden: technical functions correspond in these theories (in part) to the capacities for which artifacts have been reproduced by designers or through user-demands over a period of time, relating technical functions to the purposes held by numerous designers and/or users. These roles of mental states introduce clearly a subjective component to the understanding of technical functions. Yet accepting technical functions as merely subjective is problematic, since this seems to deny, for instance, objective limitations encountered in designing and using artifacts: engineers have to take into account scientific and technological constraints when creating artifacts with specific functions, and we cannot simply use a given artifact for any function we may have in mind. Technical functions seem to be partially objective, and ignoring this leads to all kinds of problematic consequences. If in Neander's theory an agent intentionally stores a sugar cube for generating electricity by nuclear fusion, the cube has nuclear fusion as its function for this agent. Yet sugar cubes are not reported to have been ascribed this function and engineers will readily deny that an act of storage may alter that observation.

Technical functions are thus better taken as partially subjective and partially objective, which is made possible by distinguishing an epistemic and an ontological sense of the objective-subjective distinction (Searle 1995: 7–9).

Epistemic sense (applying to judgments)

A judgment is *epistemically subjective* if the facts that make it true or false are dependent on attitudes, feelings, and points of view of the makers and the hearers of the judgment.

A judgment is *epistemically objective* if the facts that make it true or false are independent of anybody's attitudes or feelings about these facts.

Ontological sense (applying to entities)

An entity is *ontologically subjective* if its mode of existence depends on mental states of agents.

An entity is *ontologically objective* if its mode of existence is independent of any mental state.

Searle's examples of epistemic subjective and objective judgments are "Rembrandt is a better artist than Rubens" and "Rembrandt lived in Amsterdam during the year 1632," respectively; the examples of ontologically subjective and objective entities are "pains" and "mountains," respectively.

With these distinctions in place, it can be observed that understanding the subjectivity of technical functions as *ontological* is not problematic; taking the existence of technical functions as depending on mental states seems fine. Understanding the subjectivity of technical functions as *epistemic* is, however, not attractive given the earlier-mentioned limitations on designing and using; judgments about technical functions do not seem to be judgments whose truth depends on the "attitudes, feelings, and points of view" of the agents making these judgments.

So if biological functions are to be objective whereas technical functions are to be subjective, there is reason to limit this contrast to the ontological sense only. In the next section I introduce an example of a theory for technical functions that meets this requirement.

5.3   Epistemic and Ontological Function Theories

The second element identified in the differences between biological and technical functions is that biological functions are relations that items *have* whereas technical functions are relations that are *ascribed* by agents to artifacts.

Taking technical functions as relations that agents ascribe seems again not to be problematic. Technical functions are, as noted in section 5.2, related to the intentions and purposes of agents. Hence it seems perfectly acceptable to maintain that these agents ascribe technical functions to artifacts relative to these intentions and purposes. This, moreover, does not rule out that technical functions are also relations that artifacts *have* relative to context; by maintaining that technical functions are ascribed, one just emphasizes, say, that technical functions come into existence in designing and using due to the intentions and purposes of the agents involved. Taking technical functions merely as relations that agents ascribe may to some still be acceptable, but it is problematic to those who wish to arrive at a uniform function theory: if it is beyond doubt that biological functions are relations that items have, then technical functions should also be relations that artifacts have.

Let us call theories in which biological or technical functions are relations that items *have* relative to context "ontological function theories," and let us call theories in which functions are *ascribed* by agents "epistemic function theories." These labels do not fully pinpoint the purport of the distinction but capture the types of tasks involved: for ontological function theories, one has to single out functions as relations between items and contexts, relations that in principle may exist "out there"; for epistemic function theories, the task is to determine the conditions under which agents are justified to describe items functionally relative to context, independent of whether or not functions are relations that those items have. Or to make a connection with Searle's two senses of the objective-subjective distinction, in ontological theories one focuses on functions as entities, which may be ontologically objective or subjective entities; in epistemic theories one focuses on ascriptions of functions as judgments, which may in turn be epistemically objective or subjective judgments.

Cast in these terms, it is not problematic to adopt an epistemic theory of technical functions since this does not rule out that an ontological theory also exists, but adopting a theory that can merely be epistemic is better avoided if one aims at a uniform function theory. That raises the question of how to avoid such exclusively epistemic theories. I do not attempt here to analyze this latter question conclusively. I rather aim at showing that for a specific class of epistemic theories of technical functions, one can construct counterpart ontological function theories in which, moreover, judgments about technical functions are epistemically objective. I first make this plausible with a simple example; in the next section I consider this construction of ontological function theories in general.

Consider first the following technical function theory.

An epistemic design function theory

Agent a justifiably ascribes the purpose ϕ as a function to artifact x relative to its design iff agent a is justified to believe that x was designed for purpose ϕ.

In this theory agents ascribe functions to artifacts on the basis of the intentions—"artifact x is to be used for purpose ϕ"—of the artifacts' designers, but it is left open whether or not these functions are relations the artifacts have. The intentions relative to which functions are ascribed do not depend on the ascribing agent a in any epistemic or ontological sense, allowing the construction of a second function theory in which no reference is made to this agent (references to the designers remain to be present, of course) and that is an ontological function theory *counterpart*—in a sense to be determined—to the first epistemic theory.

A counterpart ontological design function theory

Artifact x has the purpose ϕ as a function relative to its design iff x was designed for purpose ϕ.

The first epistemic design function theory does not imply this second ontological theory— one can without contradiction add to the first theory the further claim that technical functions are not relations that artifacts have. Yet the ontological theory can be taken as providing support to the first epistemic design function theory by implying a third epistemic function theory that is a special case of the first; and in this sense the second ontological design function theory can be taken as a counterpart to the first epistemic design function theory. This third epistemic theory is derived from the second ontological design function theory, and an appropriate theory about justification, in the following way: if an agent a is justified to believe that an artifact x was designed for purpose ϕ, then the agent *a* can on the basis of the ontological theory justifiably ascribe ϕ as a function that *x* has relative to its design. If, conversely, *a* is justified to ascribe the purpose ϕ as a function that an artifact *x* has relative to its design, then *a* can on the basis of the ontological theory justifiably believe that *x* was designed for purpose ϕ. Hence one can arrive with the second ontological design function theory at the following associated epistemic function theory.

An epistemic function theory associated with the ontological design function theory

Agent *a* justifiably ascribes the purpose ϕ as a function that the artifact *x* has relative to its design iff agent *a* is justified to believe that *x* was designed for purpose ϕ.

This third epistemic technical functions theory is a special case of the first epistemic theory, because now the theory is explicitly one about functions as relations that artifacts have. Moreover, in the second ontological design function theory and its third epistemic associate, technical functions are ontologically subjective and epistemically objective: the mode of existence of technical functions of artifacts depends in these theories on the mental states of the designers of the artifacts, and the truth or falsity of judgments about whether artifacts have specific technical functions, that is, whether they are designed by their designers for specific purposes, does not depend on the attitudes, feelings, and points of view of the makers and the hearers—the agents *a*—of these judgments.

There thus exist ontological theories for technical functions in which these functions are epistemically objective. If such ontological theories are acceptable, the differences between biological and technical functions are to a large extent avoided: technical functions are then also epistemically objective relations that artifacts *have* relative to context. The above ontological design function theory is probably not acceptable, say within archaeology, since its application presupposes that agents typically have the means to determine for which purposes artifacts were designed originally. Neander's function theory for artifacts, briefly mentioned in section 5.2, may be taken as a modification of the ontological design function theory aimed at circumventing this presupposition since in that theory technical functions are also determined by the intentions of users, making it also applicable to artifacts of which it is not clear for what purpose they were designed, but that are nevertheless currently used for specific purposes. Note, however, that in Neander's theory, technical functions can be epistemically subjective: an artifact *x* can have a purpose ϕ as a function if an agent retains *x* for that purpose, hence for that agent the truth of the judgment that *x* has this function ϕ depends on the point of view this agent takes toward the artifact. Neander's function theory thus does not limit the differences between biological and technical functional descriptions to the extent I wish to do in this contribution.

In the next section I analyze Cummins's function theory and formulate a general strategy to construct with epistemic technical function theories ontological theories in which judgments about technical functions are epistemically objective. In section 5.5 I apply this strategy to the ICE theory.

5.4   Constructing Ontological Function Theories

In Cummins's theory, functions—biological, technical, and others—are defined as follows.

Cummins's function theory

> x functions as a ϕ in s (or: the function of x in s is to ϕ) relative to an analytical account A of s's capacity to ψ just in case x is capable of ϕ-ing in s and A appropriately and adequately accounts for s's capacity to ψ by, in part, appealing to the capacity of x to ϕ in s. (1975: 762)

Here s is a "containing system" that has the functionally described item x as its part in a broad sense: s may, for instance, be a physical object or a process, and so may x, in any combination. The analytical account A refers to an explanation of the capacity to ψ of s in terms of, in part, x's capacity to ϕ.

A first remark is that in Cummins's theory, functions refer to capacities and not to purposes, as may be the case in intentional function theories. A second remark is that it is not clear if Cummins's theory is an epistemic or ontological function theory (Houkes and Vermaas 2009). The reference to the account A suggests taking Cummins's theory epistemically as one that says that agents can ascribe capacities as functions to items if these capacities figure in explanations based on account A that s has its capacity to ψ. Yet the usual understanding is that Cummins's theory identifies functions as causal contributions: functions of items are capacities that causally contribute to s's capacity to ψ. This understanding suggests taking the theory more ontologically as one about reality independent of the account A. This ambiguity becomes manifest when one explicitly interprets Cummins's theory as an epistemic theory about agents who by account A ascribe functions, and contrasts this with an interpretation in which Cummins's theory is about functions items have relative to containing systems but independent of the account A.

An epistemic interpretation of Cummins's theory

> Agent a justifiably ascribes the capacity to ϕ as a function to x relative to the capacity to ψ of s and relative to an analytical account A of s's capacity to ψ iff x is capable of ϕ-ing in s and agent a is justified to believe on the basis of A that this capacity to ϕ of x in s causally contributes to s's capacity to ψ.

An ontological interpretation of Cummins's theory

> Item x has the capacity to ϕ as a function relative to the capacity to ψ of s iff item x is capable of ϕ-ing in s and this capacity to ϕ of x in s causally contributes to s's capacity to ψ.

Note that in the epistemic interpretation, the account A provides only justification for the agent's belief that x's capacity to ϕ causally contributes to s's capacity to ψ; by Cummins's definition it need not be true that x actually contributes in this way to s's capacity. Yet it should be true that x is capable of ϕ-ing, and this stronger requirement is captured in the epistemic interpretation by the (more ontological) condition that x *is* capable of ϕ-ing.

The ontological interpretation of Cummins's theory is a counterpart to the epistemic interpretation since the ontological theory can be taken as providing support to the epistemic interpretation, provided it is the case that in the epistemic interpretation functions are relations that items have. If Cummins's functions indeed are such relations, and presumably they are, then the ontological interpretation seems to imply the epistemic one, and one can take the epistemic interpretation as the epistemic associate of the ontological interpretation.

Functions, including the technical ones, are in the ontological interpretation of Cummins's theory epistemically objective since the truth of the judgment of whether x is capable of ϕ-ing in s, and by this capacity contributing to s's capacity to ψ, depends on physics, chemistry, and biology, and does not depend on the attitudes, feelings, and points of view of agents. In the ontological interpretation, functions are also ontologically objective since their existence does not depend on mental states.

The formulation of the epistemic and ontological interpretations of Cummins's theory can be generalized by abstracting from the particular choices Cummins made for the context c relative to which items x have functions ϕ, the evidential basis E agents use for ascribing these functions ϕ, and the requirements that must hold for functional descriptions.

An epistemic function theory T_{ep}

Agent a justifiably ascribes the capacity to ϕ as a function to x relative to context c and relative to evidence E for $R_2(x\phi c)$ iff $R_1(x\phi c)$ and agent a is justified to believe on the basis of E that $R_2(x\phi c)$.

A counterpart ontological function theory T_{ont}

Item x has the capacity to ϕ as a function relative to context c iff $R_1(x\phi c)$ and $R_2(x\phi c)$.

For Cummins's theory, the choice of c, E, and the requirements $R_1(x\phi c)$ and $R_2(x\phi c)$ are the following:

c:	the capacity to ψ of s;
E:	the analytical account A of s's capacity to ψ;
$R_1(x\phi c)$:	x is capable of ϕ-ing in s;
$R_2(x\phi c)$:	the capacity to ϕ of x in s causally contributes to s's capacity to ψ.

Yet other choices can now be considered. In effect, conformance to the generalized form T_{ep} of an epistemic function theory can be taken as a sufficient condition for the existence of a counterpart ontological function theory. If an epistemic theory can be brought in the form T_{ep} by choosing c, E, $R_1(x\phi c)$, and $R_2(x\phi c)$ appropriately, then one can arrive at an ontological function theory by substituting those choices into the generalized form T_{ont}. This ontological theory T_{ont} provides support to the epistemic function theory T_{ep}: T_{ont} implies a third epistemic function theory T'_{ep} associated to T_{ont} that is a special case of T_{ep} and that has the following form:

An epistemic function theory T'_{ep} associated with the ontological function theory T_{ont}

Agent a justifiably ascribes the capacity to ϕ as a function that x has relative to context c and relative to evidence E for $R_2(x\phi c)$ iff $R_1(x\phi c)$ and agent a is justified to believe on the basis of E that $R_2(x\phi c)$.

In the next section I use this sufficient condition to construct an ontological counterpart to the ICE-function theory. But before doing this, I add brief remarks on the relations among T_{ep}, T_{ont}, and T'_{ep}.

First, if one accepts T_{ep} for specific choices of c, E, $R_1(x\phi c)$, and $R_2(x\phi c)$, then one is not necessarily committed to accepting the ontological theory T_{ont} for those choices: one can accept T_{ep} but simply deny T_{ont} by holding that functions are not real relations that artifacts have.

Second, if one accepts T_{ont} for specific choices of c, E, $R_1(x\phi c)$, and $R_2(x\phi c)$, then one also can accept the associated epistemic theory T'_{ep}. *Proof* of the "if" part of T'_{ep}: If $R_1(x\phi c)$ is the case and an agent a is justified to believe by E that $R_2(x\phi c)$, then by T_{ont} one can conclude that a is justified to ascribe the capacity to ϕ as a function that x has relative to c and relative to E. *Proof* of the "only if" part of T'_{ep} by *ad absurdum*: Assume that a may ascribe the capacity to ϕ as a function x has relative to c and relative to evidence E for $R_2(x\phi c)$. Suppose then that it is not the case that "$R_1(x\phi c)$ and a is justified to believe on the basis of E that $R_2(x\phi c)$." This supposition implies that $R_1(x\phi c)$ is not the case or that a is not justified to believe on the basis of E that $R_2(x\phi c)$. If $R_1(x\phi c)$ is not the case, then by T_{ont} agent a cannot ascribe the capacity to ϕ as a function to x relative to c and relative to any evidence for $R_2(x\phi c)$. If a is not justified to believe by E that $R_2(x\phi c)$, then by T_{ont} agent a cannot justifiably ascribe the capacity to ϕ as a function to x relative to c and relative to that evidence E. Hence the supposition cannot be true, meaning that it is the case that $R_1(x\phi c)$ and agent a is justified to believe on the basis of E that $R_2(x\phi c)$.

Third, if one accepts T'_{ep} for specific choices of c, E, $R_1(x\phi c)$, and $R_2(x\phi c)$, then one is not necessarily committed to accepting T_{ep} for those choices: T'_{ep} is only a special case of T_{ep} in which functions are real relations artifacts have.

Fourth, if one accepts T'_{ep} for specific choices of c, E, $R_1(x\phi c)$, and $R_2(x\phi c)$, then one is not necessarily committed to accepting T_{ont} for those choices: T'_{ep} allows agents a to ascribe functions to artifacts that those artifacts do not have in T_{ont}, for instance, when agents are justified to believe $R_2(x\phi c)$ on the basis of evidence E for $R_2(x\phi c)$ that is actually incorrect, such that $R_2(x\phi c)$ is actually not the case. And even if $R_2(x\phi c)$ is the case, one is still not committed to accepting T_{ont}; there may be ontological theories different from T_{ont} that are also consistent with T'_{ep}.

Fifth, T_{ont} for specific choices of c, E, $R_1(x\phi c)$, and $R_2(x\phi c)$ provides support to T_{ep} for those choices: acceptance of T_{ont} is implying acceptance of T'_{ep}, which is a special case of T_{ep}.

Sixth, if the judgment whether the requirements $R_1(x\phi c)$ and $R_2(x\phi c)$ hold does not depend on the attitudes, feelings, and points of view of the agents making the judgment, then functions in both T'_{ep} and T_{ont} are epistemically objective. For T_{ont}, this conclusion follows directly. For T'_{ep}, the conclusion follows by noting that in this case the truth of the judgments of whether $R_1(x\phi c)$ holds and of whether it is justified to believe that $R_2(x\phi c)$ by E also does not depend on one's attitudes, feelings, and points of view.

In summary and emphasizing the results that I use in the second half of this contribution: if an epistemic function theory fits the generalized form T_{ep}, one can then construct a counterpart ontological theory by the generalized form T_{ont}; in this counterpart ontological theory, functions are epistemically objective if the judgment whether the requirements $R_1(x\phi c)$ and $R_2(x\phi c)$ hold does not depend on the attitudes, feelings, and points of view of the agents making the judgment.

5.5 An Ontological ICE-Function Theory

Using the results of the previous section, I am in the position to construct an ontological counterpart to the ICE-function theory. Wybo Houkes and I proposed the ICE theory as an analysis of specifically technical functions after arguing that the main alternatives, including intentional theories, Cummins's theory, and etiological theories, failed to meet simultaneously four desiderata for theories of technical functions (Vermaas and Houkes 2003). Yet the ICE theory is explicitly an epistemic function theory that provides conditions under which agents are justified to ascribe technical functions to artifacts, and in turn evokes the criticism that it does not determine what technical functions are ontologically. An ontological counterpart to the ICE theory would meet this criticism—that is, if it is acceptable as a theory of technical functions—and, moreover, would limit the differences between biological and technical functions to that between ontological objectivity and ontological subjectivity.

The central definition in the original epistemic ICE theory reads

The ICE-function theory

An agent a justifiably ascribes the capacity to ϕ as a function to an artifact x, relative to a use plan p for x and relative to an account A, iff:

    I. the agent a has the capacity belief that x has the capacity to ϕ, when manipulated in the execution of p, and the agent a has the contribution belief that if this execution of p leads successfully to its goals, this success is due, in part, to x's capacity to ϕ;

C. the agent a can justify these two beliefs on the basis of A; and

E. the agents d who developed p have intentionally selected x for the capacity to ϕ and have intentionally communicated p to other agents u.

A use plan p of an artifact x is a series of considered actions that includes at least one action that can be taken as a manipulation of x, and that captures the use for which that artifact is designed: using x can be described as the carrying out of a use plan p for x aimed at achieving the goal associated with the plan.

With this definition, the ICE theory has by and large the form of the generalized epistemic function theory T_{ep} as given in section 5.4. The context c relative to which technical functions are ascribed is a use plan. The evidence E agents are using to justify their beliefs is formed by an account A, which typically consists of an amalgam of technological and scientific knowledge about artifacts, hands-on experience with artifacts, and information—testimony—about their use plans. The choice of the requirements $R_1(x\phi c)$ and $R_2(x\phi c)$ is less straightforward. The I and C conditions together form an epistemic condition of the form "agent a is justified to believe on the basis of the account A that . . . " These two conditions can therefore be captured by an "$R_2(x\phi c)$ requirement." The E condition is not such an epistemic condition about beliefs of the agent a ascribing functions and seems therefore best to be captured by an $R_1(x\phi c)$ requirement. Yet in current work on the ICE theory, the E condition is also phrased in the form "agent a is justified to believe on the basis of account A that . . . " (Houkes and Vermaas 2009). Hence one can take the original ICE theory as being of the form T_{ep} with

c: the use plan p for x;

E: the account A about the use and designing of x and its use plan p;

$R_1(x\phi c)$: —

$R_2(x\phi c)$: • x has the capacity to ϕ when manipulated in the execution of p;
• if this execution of p leads successfully to its goals, this success is due in part to x's capacity to ϕ; and
• the designers d who have developed p have selected x for the capacity to ϕ in p, and have communicated p to other agents u.

These choices define the following ontological counterpart of the ICE theory:

An ontological ICE-function theory

Artifact x has the capacity to ϕ as a function relative to a use plan p for x, iff:
• x has the capacity to ϕ when manipulated in the execution of p;
• if this execution of p leads successfully to its goals, this success is due in part to x's capacity to ϕ; and
• the designers d who have developed p have selected x for the capacity to ϕ in p, and have communicated p to other agents u.

The judgment whether the identified requirement $R_2(x \phi c)$ holds does not depend on the attitudes, feelings, and points of view of the agents making the judgment, showing that the technical functions advanced in the constructed ontological ICE theory are epistemically objective. Yet the existence of technical functions does depend on the intentions and purposes of the designers of the use plans for these artifacts. Hence technical functions are ontologically subjective in the ontological ICE theory.

A full assessment of the acceptability of this ontological ICE theory must consist by my own standards of an argument that it meets the four desiderata for theories of technical functions given in Vermaas and Houkes (2003). I focus here on only one because the proof that the ontological ICE theory meets this desideratum needs additional argumentation. It is called the "malfunction desideratum" and requires that function theories should accommodate the phenomenon of malfunctioning by being able to ascribe the relevant functions to artifacts that are—temporarily—not capable of performing their functions. This desideratum may be taken as a necessary condition to a stronger requirement that a function theory should model all possible aspects of malfunctioning, such as normative statements that artifacts not capable of performing their functions are nevertheless *supposed* to be capable of performing these functions. Such additional aspects of malfunctioning are not considered here, but are discussed in this volume by Maarten Franssen and Peter McLaughlin.

The original epistemic ICE theory meets the malfunctioning desideratum partially. In the case that an artifact does not have a capacity to ϕ and that the agent a is ignorant about this state of affairs, the original epistemic ICE theory allows that the agent still ascribes this capacity as a function to the artifact; assuming that the E condition is satisfied, the agent can in this case still believe that the artifact has the capacity to ϕ (thus satisfying the I condition) and justify this belief, say, on the basis of earlier experiences with the artifact (satisfying also the C condition). (Note, however, that as soon as the agent a believes that the artifact does not have the capacity to ϕ, the agent cannot satisfy the I condition anymore and thus can no longer ascribe this capacity as a function.) For the ontological ICE-theory this case is not available: if an artifact does not have a capacity, this capacity cannot be a function of the artifact.[3] Hence it seems that in the ontological ICE theory, malfunctioning artifacts do not have their relevant functions. To mend this problem I focus in the next section on the phenomenon of malfunctioning.

5.6   Malfunctioning

A function theory can accommodate malfunctioning if it, as said, can ascribe the relevant function to an artifact even if the artifact is not capable of performing that function. For a theory in which functions refer to capacities, this means that it can ascribe the relevant capacity to ϕ as a function even if the artifact is not capable of exercising this capacity.

Cummins's theory cannot accommodate malfunctioning for this reason, since in this theory capability of exercising a capacity to ϕ—of ϕ-ing—is a necessary condition to ascribing that capacity as a function. For the ontological ICE theory, this quick conclusion does not hold. In this theory a capability of exercising a capacity to ϕ is not required for ascribing that capacity as a function to the artifact; the necessary condition is rather that the artifact has the capacity to ϕ. Hence if it can be made plausible that there is a difference between *having* a capacity to ϕ and *being capable of exercising* this capacity, then there is room for arguing that the ontological ICE theory can accommodate malfunctioning.

In the domain of technology, I believe there is such room. Take a car that is not capable of being driven. This fact need not immediately lead to the conclusion that it has lost the capacity to be driven; the reasonableness of this conclusion seems to depend on additional circumstances. If, for instance, the car was set on fire and was heavily damaged, it would indeed be taken as having lost this capacity; but when the car has simply run out of petrol or when the starting motor is broken, it is somewhat harsh to claim that the car no longer has the capacity to be driven. It still has this capacity, which is demonstrated when it is again filled up with petrol or when the starter is replaced. Hence one can envisage circumstances in which a car can be taken as both having a capacity and as not being fit for this capacity to be exercised. Assuming that in all of these circumstances the car is malfunctioning, however, does not make sense. If the starting motor is damaged, the car may be taken as malfunctioning, but when it needs petrol, saying the car is malfunctioning seems like overkill.

So one can argue that an artifact can have a capacity to ϕ corresponding to its function even in cases where the artifact is not capable of exercising this capacity. Yet these latter cases should not all be taken as ones in which the artifact is malfunctioning. Hence what is needed is to distinguish these cases, and my proposal is to do so using use plans and the concepts of reparation and maintenance.

First, an artifact may be said to have the capacity to ϕ corresponding to its function if it is in a physical state in which it is actually capable of exercising this capacity or if it can be brought to such a state by repair or maintenance; in all other circumstances the artifact does not have the capacity to ϕ. This first distinction depends on how the notion of reparation and maintenance is understood, and collapses if one allows reparation or maintenance to refer to any possible or impossible transformations of the physical states of artifacts. So to rule out repairs in which completely wrecked cars are part by part transformed into their original state, and to discard magical acts of maintenance that make any artifact tick again, I adopt a normative sense of reparation and maintenance: reparation and maintenance refer to modifications of the artifacts that can be considered as technologically and economically feasible given the relevant technological state of the art and given the available resources. Changing the starter of a car and filling it up with petrol, then, count typically as reparation or maintenance; transforming a "total-loss" wrecked car into its original form does not.

Second, in the case that an artifact is in a state in which it has the capacity to ϕ corresponding to its function but is not actually capable of exercising it, the artifact may be called malfunctioning if the reparation and maintenance needed to bring the artifact to a state in which it is capable of exercising the capacity is *not* part of the use plan for the artifact; in all other cases the artifact cannot be called malfunctioning. By this second distinction a car that has run out of petrol is not malfunctioning since it is part of the car's use plan that its user—the driver—regularly fills it up with petrol. A car with a broken starting motor is, however, malfunctioning, since changing that motor is not part of the car's use plan.

With these distinctions in place, the ontological ICE theory can be taken as accommodating the phenomenon of malfunctioning: an artifact that is—temporarily—not capable of exercising the capacity corresponding to its ontological ICE function but that can be brought back into a state in which it can exercise that capacity by feasible "non-use-plan repair or maintenance," has this capacity and thus has this capacity as its function in the ontological ICE theory. Note that the original epistemic ICE theory can with this new characterization of malfunctioning accommodate this phenomenon in a much broader way. The agents ascribing a capacity as a function to an artifact can with the new characterization, for instance, simultaneously believe that the artifact has a capacity, justify this with an account, but also acknowledge that they believe that the artifact is not capable of exercising that capacity.

A final note is that this characterization turns malfunctioning into a phenomenon that is based on, first, a normative distinction between reparation and maintenance that is technologically and economically feasible, and reparation and maintenance that is not, and, second, on a division of labor between users of artifacts and expert technologists: an artifact malfunctions if it needs repair and maintenance that is feasible for expert technologists but that is not a task of its users.

5.7   A Unified ICE-Function Theory

Having argued that the ontological ICE-function theory can accommodate malfunctioning, it can be proposed as an acceptable theory of technical functions. With this theory one can then argue that technical functions are epistemically objective and ontologically subjective, showing that the difference between biological and technical functional descriptions becomes merely that biological functions are ontologically objective whereas technical functions are ontologically subjective. This last difference seems to be one not to deny. One option for this denial is to opt for the ontological version of Cummins's theory, since in that theory technical functions are also ontologically objective. Yet this option is blocked when one requires that a function theory should accommodate malfunctioning. Another option may be to bridge this last difference by taking biological functions as ontologically subjective. Adopting Searle's theory is one way of doing that. I now end this contribution by briefly showing that the ICE theory can be generalized to a uniform function theory that applies to also biological functions, providing one is ready to accept this second option.

The original ICE theory has been, as I mention in section 5.5, proposed as an analysis of technical functions by its reference to use plans, and the same holds for the ontological version. In an exploration of how this limitation can be overcome, Wybo Houkes and I have generalized the original ICE theory to a theory that can be taken as advancing a unifying analysis of functional descriptions in technology, biology, and any other domain in which functional descriptions are used (Vermaas and Houkes 2009). The generalized central definition reads as follows:

The unified epistemic ICE-function theory

An agent a justifiably ascribes the capacity to ϕ as a function to an item x, relative to a goal-directed pattern p for x and relative to an account A, iff:

I. the agent a has the capacity belief that x has the capacity to ϕ, in the execution of p, and the agent a has the contribution belief that if this execution of p leads successfully to its goals, this success is due in part to x's capacity to ϕ;

C. the agent a can justify these two beliefs on the basis of A; and

E. the agents d who designated p have intentionally identified x for having the capacity to ϕ in p and for contributing by this capacity to the success of p, and have intentionally communicated p to other agents l.

In this definition the notion of a use plan p for an artifact x has been replaced by the notion of a pattern p that consists of a series of behaviors including behaviors of the item x, and that is directed toward a goal. This pattern is singled out by agents, called the designators d, who communicate the pattern to other agents, called laypersons l, with the aim to provide information to those laypersons about the existence of this pattern and about how item x contributes by its capacity to ϕ to the effectiveness of the pattern to lead to its goals. In technology, the designators are designers and the laypersons are users. In biology, the designators are those who identified specific biological behaviors as making up goal-directed biological patterns—for example, William Harvey, who considered the circulation of blood and saw the pumping capacity of the heart as contributing to this circulation—and the laypersons are other biologists who are informed about these patterns and the contributing roles of the partaking items. And in, say, sociology, the designators are those who identified social behavior as making up goal-directed sociological patterns, and the laypersons are those who learn about these patterns.

This generalization is meant primarily as an exploration of how the ICE theory may fare when applied within, say, biology. The unified epistemic ICE-function theory has some advantages. It can, for instance, make sense of biological functional descriptions that do not rely on evolutionary theory, such as the one made by Harvey, and are less well accounted for in current theories of biological functions—etiological theories, in particular. Its disadvantages are, first, that it is an epistemic theory about agents ascribing functions, thus violating the general intuition that biological functions are relations that biological items have relative to context, and, second, that by this theory agents ascribe these functions relative to goal-directed patterns intentionally identified by other agents, thus introducing (teleological) mental states into the analysis of biological functional descriptions. The first disadvantage can be overcome by constructing the equally explorative ontological counterpart of the generalized ICE theory.

The unified ontological ICE-function theory

Item x has the capacity to ϕ as a function relative to a goal-directed pattern p for x, iff:

- x has the capacity to ϕ in the execution of p;
- if this execution of p leads successfully to its goals, this success is due in part to x's capacity to ϕ; and
- the agents d who designated p have intentionally identified x for having the capacity to ϕ in p and for contributing by this capacity to the success of p, and have intentionally communicated p to other agents l.

In this ontological version, biological functions are again relations that biological items have relative to context. Moreover, biological functions are in this theory epistemically objective—judgments about functions depend on the three conditions that concern facts whose truth is independent of the points of view, attitudes, or feelings about these facts of the makers or hearers of the judgments. Yet the second disadvantage cannot be overcome. Functions, including the biological ones, are in the unified ontological ICE theory ontologically subjective: the three conditions concern facts that refer to the mental states of the designators since they have singled out the patterns relative to which items have functions.

This reference to the mental states of the designators can to some extent be suppressed by noting that the third condition in the unified ontological ICE theory becomes redundant when the communication between designators and laypersons is considered. If this theory is applied to cases in which designators provide information to laypersons, it seems spurious to hold that the designators first tell laypersons that the pattern can be singled out, that x has a role in the effectiveness of this pattern, and second, add explicitly that they, as designators, have made all these discoveries. The third condition in the unified ontological ICE theory can in this case be taken as redundant, which simplifies this theory as follows:

The unified ontological ICE theory in designator-layperson communication

Item x has the capacity to ϕ as a function relative to a goal-directed pattern p for x, iff:

- x has the capacity to ϕ in the execution of p; and

- if this execution of p leads successfully to its goals, this success is due in part to x's capacity to ϕ.

This application of the ontological unified ICE theory resembles the ontological interpretation of Cummins's theory as given in section 5.4, providing it with a more acceptable face. One difference is that Cummins's condition that an item should actually be capable of executing the capacity corresponding to its function is replaced by the more liberal condition that the item has this capacity (thus allowing for malfunctioning as characterized in section 5.6). A limitation of the unified ontological ICE theory is that Cummins's "containing system" s should always be taken as a goal-directed pattern. This application of the ontological unified ICE theory also resembles Searle's (1995) function theory and possibly also the goal-contribution analysis of functions of Christopher Boorse (2002). Yet it is not a theory in which biological functions can count as ontologically objective, since it refers to mental states of the designators—the singling out of the patterns p—relative to which items have their functions.

Hence as it stands, biological functions may in a unified ICE-like theory remain epistemic objective relations that items have relative to context, but they are then ontologically subjective, which may for some be reason to reject this theory in the first place.

5.8   Conclusion

In this contribution I show that for a class of epistemic theories of technical functions about the *ascription* of these functions by agents to artifacts one can construct counterpart ontological theories in which technical functions are relations that artifacts *have* relative to context. In these ontological theories, judgments about technical functions are epistemically objective in the sense of being true on the basis of facts and not on the basis of the attitudes, feelings, and points of view of the makers and the hearers of the judgments. Yet technical functions remain in these ontological theories to depend on the mental states of their designers and in this sense technical functions are ontologically subjective.

With this result, one can argue that the often-made contrast that biological functions are objective nonrelational properties that biological items have independent of the mental states of agents, and that technical functions are subjective relations that agents ascribe to artifacts relative to the mental states of agents, to a large extent can be avoided. If etiological theories are taken as acceptable function theories in biology, and the ontological ICE theory as acceptable in technology, then both biological and technical functions are epistemically objective relations that biological items and artifacts have, respectively, relative to context. The difference that remains is that biological functions are ontologically objective and technical functions are ontologically subjective relations.

In this contribution I also present a new approach toward arguing that theories of technical functions can accommodate the phenomenon of malfunctioning. On this approach an artifact is taken as malfunctioning if it is in a state in which it has the capacity to ϕ, is not capable of ϕ-ing, and can be brought into a state in which it is capable of ϕ-ing by reparation or maintenance that is technologically and economically feasible and that is not part of the use plan for the artifact.

Biological and technical functional descriptions thus still differ in my analysis. Biological functions are typically ontologically objective and technical functions are typically ontologically subjective. With my approach toward malfunctioning, one can now identify a second difference: malfunctioning biological items may, in etiological theories, be items that are irreversibly malformed, whereas in the proposed approach, an artifact may be taken as malfunctioning only if it can reasonably be brought back into a state in which it stops malfunctioning. Thus there remain enough differences that prevent taking biological and technical functional descriptions as fitting a uniform analysis.

Acknowledgments

I would like to thank audiences in Altenberg, Padova, and Paris, and to acknowledge Peter Kroes and Ulrich Krohs for valuable comments. Research for this contribution builds on previous work done in collaboration with Wybo Houkes, and is supported by the Netherlands Organization of Scientific Research (NWO).

Notes

1. See, for instance, Wouters (2005) and Preston's contribution to this volume.

2. Cummins's theory may, as said, be taken as an exception.

3. This difference between the epistemic and ontological ICE theories again illustrates that acceptance of an epistemic function theory of the form T_{ep} (or T'_{ep}) does not imply that one also has to accept the counterpart ontological function theory T_{ont}. Hence one can reject the ontological ICE theory but still subscribe to the original epistemic theory.

References

Boorse, C. (2002). A rebuttal on functions. In: *Functions: New Essays in the Philosophy of Psychology and Biology* (Ariew, A., Cummins, R., Perlman, M., eds.), 63–112. Oxford: Oxford University Press.

Cummins, R. (1975). Functional analysis. *Journal of Philosophy, 72:* 741–765.

Houkes, W., and Vermaas, P. E. (2004). Actions versus functions: a plea for an alternative metaphysics of artefacts. *The Monist, 87:* 52–71.

Houkes, W., and Vermaas, P. E. (2009). Useful material: An action-theory of artefacts and their functions. Manuscript.

Millikan, R. G. (1984). *Language, Thought, and Other Biological Categories: New Foundations for Realism.* Cambridge, Mass.: The MIT Press.

Millikan, R. G. (1993). *White Queen Psychology and Other Essays for Alice.* Cambridge, Mass.: The MIT Press.

Neander, K. (1991). The teleological notion of "function." *Australasian Journal of Philosophy, 69:* 454–468.

Preston, B. (1998). Why is a wing like a spoon? a pluralist theory of function. *Journal of Philosophy, 95:* 215–254.

Searle, J. R. (1995). *The Construction of Social Reality.* New York: Free Press.

Vermaas, P. E., and Houkes, W. (2003). Ascribing functions to technical artefacts: a challenge to etiological accounts of functions. *British Journal for the Philosophy of Science, 54:* 261–289.

Vermaas, P. E., and Houkes, W. (2006). Technical functions: a drawbridge between the intentional and structural natures of technical artefacts. *Studies in History and Philosophy of Science, 37:* 5–18.

Vermaas, P. E., and Houkes, W. (2009). Functions as epistemic highlighters: an engineering account of technical, biological and other functions. Manuscript.

Wouters, A. (2005). The function debate in philosophy. *Acta Biotheoretica, 53:* 123–151.

III FUNCTIONS AND NORMATIVITY

If an entity has or is ascribed a function, then—whatever kind of entity it belongs to—it may be said to perform its function well or poorly or not at all (malfunction). So much seems to be implied by the general notion of function. If a heart has the function to pump blood, then the performance of this function may be evaluated, whether we are dealing with a biological heart or an artificial one. With the notion of "function" comes the possibility of evaluating the actual performances of functions. Apart from this evaluative dimension of functions, there appears to be a prescriptive dimension: an entity with a particular function ought to (or is supposed to) behave in a certain way under suitable circumstances. Again, a heart, biological or artificial, ought to behave in a certain way so as to realize its function. These evaluative and prescriptive dimensions of functions, grouped under the heading of the "normativity of functions," constitute the focus in this part of the book.

Although the normativity associated with functional items does not discriminate between biological and technical objects, this normativity raises different questions in both domains. With regard to biological functional items, one of the main issues is whether this normativity implies that there are somehow values or norms in (biological) nature. If that would indeed be the case, then that would run counter to the dominant picture of nature underlying the modern natural sciences. The normativity of technical functional items does not appear at first sight to be so problematic because it may be related to the normativity of intentional human action. We still need to clarify how the normativity of human actions in which objects are used can be transposed in an intelligible way to those objects themselves. In view of the different questions involved in the normativity of functional items in the biological and technical domains, it appears highly questionable whether it will be possible to arrive at a common interpretation of this normativity that may be applied in both domains. Apart from these problems, there are other problems that must be addressed in order to clarify the normativity of functional items. For instance, a generally accepted interpretation of how to interpret the notion of "normativity" as such is still lacking. All these problems form part of the background and the subject of the following chapters.

This part starts with McLaughlin's exploration of the nature of the normativity of functions. He asks where this normativity could come from. A brief review of the three main interpretations of function ascriptions (the intentional, etiological, and causal-role theories of functions) leads him to the conclusion that each of them has its own problems when accounting for the normativity of functions: neither facts about intentions, nor selection histories, nor facts about causal roles can ground normative claims with regard to functions. Next he analyses in detail three kinds of relations that might form the origin of the normativity associated with functions, namely means-ends relations, part-whole relations, and type-token relations. Means-ends relations differ from nonnormative causal relations in that means ought to or are supposed to contribute to bringing about the ends. Part-whole relations may have a normative dimension if the whole is a hierarchically organized system with a good of its own. Finally, tokens may instantiate a type better or worse, so the type-token relation is of a normative nature. Insofar as these relations play a role in function theories, the normativity of functions may be grounded in these relations.

Franssen discusses the role that the idea that functions are inherently normative plays in the debate on adequate theories of functions. One of the chief touchstones when assessing function theories is how they deal with the alleged normativity of functions. Etiological theories are generally taken to be able to account for normativity, whereas causal-role theories deny that there is anything normative about functions. So a question arises about the list of adequacy criteria for function theories: should it include the criterion that a theory of functions has to account for the inherent normativity of functions? To answer this question Franssen analyzes in detail the nature of normative statements with regard to functions. He argues that the normativity traditionally associated with functions derives from human intentionality; it is either related to the justification of beliefs about functional items or to reasons for specific actions with regard to such items. So this normativity is not inherent to functions. This means that the advantage of etiological theories over causal-role theories with regard to accounting for the normativity of functions is illusory.

Davies argues that although most of us have the intuition that the parts of living things are supposed to fulfill certain functional tasks, we should give up this intuition and stop talking about functions. He asks why we are moved to theorize about the concept of normative functions and concludes it is because we are *conceptual conservatives* regarding the concept "purpose." Conceptual conservatives are committed to preserving or otherwise "saving" concepts that strike us as especially important, including our concept "purpose" as it applies to organisms. And yet insofar as the genealogy of our concept of "normative functions" traces back to a largely theological worldview, we now regard it as false or unpromising, and insofar as we are psychologically constituted to apply this concept with undue generosity, we ought to relinquish the orientation of the conceptual conservative with respect to normative functions on the grounds that it diminishes rather than facilitates the growth of human knowledge. This casts doubt upon the main theories of functions in the philosophical literature, except for the theory of systemic functions that eschews the alleged normative dimension of biological purposes.

Light's contribution on restoration ecology rounds off this part of the book. Restoration ecology is the science that is aimed at re-creating ecosystems that have been damaged or destroyed due to anthropogenic or nonanthropogenic causes. Philosophical critics of restoration have argued that restored environments are not natural objects but rather artifacts due to their anthropogenic origins. Underlying such claims is the assumption that while few things in the world are solely natural objects or artifacts, natural objects are those things that are "relatively free of human influence." If the distinction between natural and artificial objects is supplemented with the normative claim that natural objects have intrinsic values over and beyond the instrumental value they may have for humans, then the conclusion may be drawn, as indeed critics of ecological restoration have done, that such restorations can never duplicate the value of original nature because they are not natural things. Light proposes a definition of restoration ecology that is intended to avoid confounding the natural-artificial distinction with normative issues. He accepts that restored environments are artifacts with specific functions and argues that they may have values on account of these functions.

6 Functions and Norms

Peter McLaughlin

Function ascriptions seem to involve normative questions. If the function of the governor in a steam engine is to regulate the amount of steam fed to the cylinder, then that is what it is *supposed* to do—that is what it is for. If the function of the pineal gland is to regulate circadian rhythms, then that is what it is supposed to do. If things of a particular type have a function, then some of them may perform this function better or worse than others do. Wherever we can speak sensibly of better and worse, we are introducing not just an ordering relation among things but also an evaluation of this ordering relation. It is not just that, say, $x > y$ in some neutral sense but that x is *better* at something or for something than y is. A good pruning knife does what a pruning knife is supposed to do better than a poor pruning knife does. Things that are supposed to do something and don't do it, or do it poorly, are substandard or broken or just not properly applied. Things that have functions can also *mal*function. A malfunctioning kidney or carburetor is one that does not or cannot do what it is supposed to do. When we speak of malfunctioning machines or organs, we are appealing to normativity—to real or to metaphorical norms. These norms need not be moral or even prescriptive in any strong sense, but they must involve at least some reference to a standard or type that supports evaluative judgments. A4 is a norm for typewriter paper; a body mass index of 18 kg/m^2 is a norm for fashion models. If deviation from the norm leads to the attribution of malfunction or some other evaluation, then the norm cannot be a merely statistical norm. If the left hind leg of a water buffalo departs two standard deviations from the norm for the number of freckles, it is not on that account a better or worse water-buffalo leg. On the other hand, if a washing machine departs two standard deviations from the norm for water consumption per kilo of laundry, it is on that account a better or poorer washing machine.

How does normativity enter function ascriptions? Where do the norms come from? Assuming that the facts adduced in analyzing function ascriptions are not normative, it would seem that no number of factual propositions about functions or function bearers could ground a norm—unless we have accomplished a naturalistic reduction of norms to facts. (I ignore this angle here.) Where then does the normativity of function ascriptions come from, and how is it justified? Let us examine the three main interpretations of function ascriptions—intentional functions, etiological functions, and causal-role functions—and see how they deal with normativity.

In *intentional functions* (where the function of an item is what some agent intends that it do) it is possible that intentions could justify norms—perhaps only subjective norms, though these might be intersubjectively binding. If something has been manufactured to do something, then one can reasonably assert that it is supposed to do that something: a screwdriver is supposed to turn screws. However, intentional functions are only found in artifacts, or in nature, conceived as a divine artifact. Even in these cases we could just replace "X is supposed to do Y" with "X was intended to do Y." If anything more than intent is meant by *supposed*, then it must still be justified. The fact that someone wills or intends some state of affairs does not of itself establish that state as a norm that *ought* to be attained. One could even assert that the intentionality of intentional functions alone cannot justify norms at all; even God's intentions are normatively relevant only if he intends things that a reasonable God should intend. For instance, when Descartes grounds the conservation of force in God's will, he insists that it is "consonant with reason" to attribute this conserving action to God. A reasonable God would conserve the force and matter in the universe (see McLaughlin 1993). But a reasonable God is one who conforms to our postulated norms of rational (divine) behavior. It may be that certain normatively distinguished intentions may ground the normativity of functions, but the norm-generating capacity of these intentions must first be explained.

In the case of *etiological functions* (where the function of an item is what its predecessors have had the disposition to do), proponents tend to appeal to history, specifically to a history of selection to explain the source of normativity. The etiological view of functions is generally credited with the ability to explain plausibly why we can speak of malfunction, and why we can say that some particular individual *X*, that in fact cannot perform *Y*, nonetheless has *Y* as its function. For instance, if wings were selected for flight, we may assert that a particular broken wing is supposed to enable flight but that it malfunctions. Even opponents of the etiological view tend to give it some credit here. If an entity was manufactured to do *Y* or was evolved to do *Y*, then that is its function. But if intentions cannot ground norms, why should selection be able to ground them?

Proponents of the etiological view often hope that natural selection will provide the normativity needed for their view of functions. The key element of Larry Wright's original analysis was the postulate that natural selection *for* a function can provide the normative component of function ascriptions without presupposing intentional agents: "If an organ has been naturally differentially selected-for by virtue of something it does, we can say that the reason the organ is there is that it does that something" (Wright 1973: 159). As Neander (1991b: 173) puts it: "The function of a trait is to do whatever it was selected for." Kitcher (1993: 383) agrees: "The function of X is what X is designed to do, and what X is designed to do is that for which X was selected." The idea is that something *is supposed* to do what it *was* selected for doing. Although all hearts that pump blood also make thumping sounds, the heart nonetheless was selected only for pumping, not for thumping. It is *supposed* to pump, not to thump.

There are however three serious objections to this view: 1) Nature does not in fact select traits or organs for their functions in the same way that a watchmaker selects his gears and springs for their functions. Nature cannot build organisms out of selected traits; it selects *organisms* for their traits, and this results in the proliferation of those traits or the production of new traits. 2) If we were to say that the function of a trait is to do what caused its production or proliferation, then many traits would have the function of being linked to useful traits. 3) Even if natural selection can explain the origin and proliferation of the traits, it still might not be able to explain why the traits have functions. It is not immediately evident why the causal past of a trait should determine its normative future. Thus the etiological approach still has to legitimate the normativity of function ascriptions. Causal history seems no more normatively binding than intention.

On the other hand, the *causal-role or dispositional* view of functions (where the function of an item is its causal role in the performance of some specified activity of its containing system) is thought, even by its adherents, to have difficulty in coping with malfunction: If the function of an item in a larger system lies in its contribution to the performance of some action of that system, then if it makes no contribution, it has no function. A piece of steel that cannot regulate steam quantity is just not a governor, one could say. It does not malfunction—it simply does not have the function. Similarly, one could say that a piece of tissue that cannot regulate circadian rhythms does not malfunction—it just has no function. However, let us examine this organ that looks like a pineal gland, that is located where the pineal gland is normally found, that arises embryologically just the way the pineal gland arises, but cannot regulate circadian rhythms. We would not say that such an organ is not in fact a malfunctioning pineal gland but rather not a pineal gland at all, or that it is a pineal gland without a function. Something is wrong with this argument: we don't normally identify an entity merely by its function.

This last argument is a fairly standard objection to the causal-role approach, but I think it is unfair—even if it is sometimes embraced by proponents of the approach. I believe that the dispositional view, in spite of itself, has no more difficulty in coping with malfunction than does the etiological view or the intentional view—because the normativity involved in the ascription of malfunction is not necessarily introduced by (and is thus not explainable by) intentionality or selection history.

But why should *facts* about intentions, about selection history, or about causal roles explain, justify, or even motivate assertions about what *ought* to be the case? They can't, I presume; but this means that the normativity, if such there is, is coming from someplace else. The basic problem is how to apply the fact-norm distinction to function descriptions so as to be able to ascertain the extent to which norms and normativity are introduced in seemingly purely descriptive propositions.

If God's will, the history of selection, and the contribution to a system's performance do not ground the normativity of function ascriptions, what does? I want to consider three possible alternative sources of normativity, three aspects of function descriptions, where it is possible that apparently factual statements about functions are actually normative. I examine three possible places where normativity may have been implicitly presupposed and thus have been introduced without argument: 1) means-end relations, 2) part-whole relations, and 3) type-token relations.

Let me just assert dogmatically that propositions about all three of these relations are per se normative—and see how far I get. Somewhat more circumspectly: let's investigate how far apparently factual propositions about these three kinds of relations are actually normative—or at least involve normativity in some way, shape, or form that we did not suspect before. I start with means and ends and differentiate the problem of parts and wholes before turning to types and tokens.

6.1   Means and Ends

Any part of a material system that has some effect within the system can be viewed as a means to that effect—if the effect is in turn viewed as an end. Any link in a causal chain can be viewed as a means to the next link; any part of a complex system that contributes to some performance of that system can be viewed as a means to performing that end. When we ascribe *functions* to things, we view them as means to ends. Functions are in a sense nothing more than effects considered from a means-ends perspective. In artificial systems this is often viewed as relatively unproblematic: a can opener has the purpose or function of opening cans; a pressure valve has the function of regulating pressure. The functions of artifacts are also normally the effects intended by some agent, but the same kind of means-ends relations can also be seen in nature. We can also view natural things not merely in terms of cause and effect but also as means to ends in an attempt to understand the workings of a complex system. We analyze the causal *roles* of particular things in a system or a process. Things with functions are thus conceptualized not just as causes of certain effects but also as means to certain ends. Viewing something as a cause makes no particular presuppositions as to why the thing or event viewed as a cause is there, but viewing something as a means to an end makes very definite presuppositions about why it is there. Although causes are not necessarily there for the sake of their effects, means are in fact there for the sake of their ends; they are subordinated to the ends; that's what we mean by calling them "means." Sometimes we may consider talk about means and ends simply to be a figure of speech that does not commit us to any such intentionalistic consequences. If we say that the function of the valves in the veins is to constrain the blood to flow only toward the heart, we may mean that this is what they do, that this is their effect in the system of blood circulation. But in doing so, we are viewing them from the perspective of their contribution to some performance of the containing system, and this is a fundamentally technological perspective. The structure under consideration is one that fulfills certain tasks, without which the system would itself not perform as it should. If we say that the valves are a particular *means* to blood circulation, we imply that what they do explains why they are there in the first place: it is the end to which they are means.

When the means-end relation is viewed as a chain, it can be iterated at will: A is a means to B, which is a means to C, which is a means to D, and so on. The cause of an effect can be viewed as a means to an end, and to the extent that the effect is desired (is an end), some appropriate means to it will be desired as well. The intermediate effects, too, will become ends—but only relative ends, namely relative to the end-character of that end to which they in turn are means. But this potential regress of means to the ends that ground them always stops at some point that legitimates the series; and there is in the formulation itself an expectation that there will be some final end that anchors the whole series of means and ends.

There would seem to be three basic ways of stopping (anchoring) such a regress of functions. Assuming that A is good for (doing) B, and B is good for (doing) C, and that the regress stops at C, then

1. C is something I happen to value or to be interested in. Free agents can set goals, and the function of an item is relative to such goals. The regress of means to ends can be *arbitrarily* broken off by the decision of a free agent.

2. C is something that I am (or should be) interested in *for good reasons*. There are reasonable goals that agents should take into account. The regress of means and ends can be ended for good *reasons*. There is some argument that can be adduced such that the regress of means and ends is not just arbitrarily ended but *rightly* ended. For instance, an item may have many different effects, only one of which fits well with the kind of thing we take it to be. The escapement mechanism of a clock adds to the weight of the clock, but weighing down is not what the hierarchical organization of the clock is directed at; it is directed at keeping time. The pumping, but not the thumping, of the heart plays a relevant role in the life of living creatures.

3. C is something that lies in the nature of things (or in human nature). The regress can have a *natural* end. C is something that can benefit from B. C has a good, which may or may not involve intentionality.

The first of these alternatives ends the regress of functions by appealing to our interests—cognitive or otherwise. The second appeals to warranted or justifiable conscious interests. Neither of these seems to be very problematical if one acknowledges the existence of intentional agents. It is the third alternative that is likely to cause us difficulties.

This third kind of functional regress corresponds to a standard narrative technique in children's stories. For instance, the old farmer Pettersson uses a fishing rod to get the key out of the cistern; he needs the key to open the shed, to get the tools, to fix the bicycle, to ride to town, to buy flour, to bake a birthday cake for his cat Findus. In this example, the chain of relative ends definitely stops some place with a different kind of purposiveness or means-end relation; it comes to an end with a beneficiary. The birthday cake baked by old Pettersson was good for the cat Findus, and that stops the regress: the cat need not be good for anything. This second kind of purposiveness cannot be iterated. It denotes a relation to something whose good is not merely relative to its contribution to something else. The cat Findus may also be good for catching mice or keeping company, but he has a good of his own independent of whatever usefulness he might also have for the farmer. When we view a causal chain as a series of means and ends, we *presuppose* something that stops the regress, something that has a good. And this applies whether it is an intentional agent, an organism, or simply anything that can be said to have interests—whether or not it consciously takes interest in them. We presuppose an entity somewhere down the line which has some kind of interests that (ceteris paribus) ought to be served.

Although every effective means to an end is also the cause of an effect, causes only *have* effects. It's not that they *ought* to have their effects—they just do, or they aren't causes at all. But when we view causes of effects as means to ends we presume that the means to the ends are *supposed* to facilitate these ends.

This is our first candidate for a source of normativity in function ascriptions.

6.2   Parts and Wholes

A series of cause-effect relations and means-ends relations can be viewed not only as a process (chain) but also as a system (hierarchical structure). Just as a system can be seen to be causally dependent on its parts, so too the parts can be seen as means to the end of the whole. In hierarchical systems, we can also to a certain extent iterate the means-ends relation: part A of system B contributes to some performance of B, which contributes to some performance of larger system C (which contains B), and C in turn is part of D and makes a contribution to what D performs, and so on. But we rather quickly run out of containing systems: the valves in the veins contribute to blood circulation, which contributes to the metabolism of the organism, which may be taken to have a role in the ecology of a particular region, but after that we have trouble finding an appropriate larger containing system, the performance of which is supported by the ecosystem. Nonetheless, the primary use of functional ascriptions is actually in hierarchical systems where parts are ascribed functions for the whole. In some cases the regress is stopped arbitrarily: a performance of the system is good for some external agent. For instance, the governor of the steam engine contributes to the engine's ability to deliver regular power and this is good for the factory owner. A second kind of case is where the beneficiary that stops the functional regress is actually the function bearer's containing system itself. Here the containing system is viewed not just as the next hierarchical level but as a *whole*, as the end of the encasement, as something that displays a certain (perhaps purposive) unity or integrity. Normally when we pursue the regress beyond the level of the organism, we change our perspective and say that ecosystems or some other equilibrium systems are what we happen to be interested in; or we assume that the equilibrium of the system is good for the organisms in it.

This is our second candidate for a source of normativity in function ascriptions: we implicitly view the containing system as "more" than a mere aggregate or structure—as a whole, a hierarchically organized system. The containing system as an organic whole seems to have a good of its own like the regress stopper of the chain of means and ends.

In the relations of parts and wholes there also arises a peculiar asymmetry between considerations of organisms and artifacts. In artifacts both parts and wholes have functions in the same sense: the governor of the steam engine has a function *in relation to* the system of which it is a part, and the steam engine itself serves (or was intended to serve) a purpose or function for someone or something; the gears of a flour mill have functions, just as the mill does. In organisms, on the other hand, parts or organs or traits have functions for the organism independent of the question of whether the organism itself is thought to have a function for something or someone else. Something that has a good of its own can of course also be viewed instrumentally. Just because an entity stops one particular instrumental regress doesn't mean it cannot also be a mere link in the chain or level in the hierarchy of another regress. The fact that oats are good for horses, not just for the owners of the horses, does not prevent horses from being good for riding or pulling a plow and thus being useful to their owners. The two instrumental views may however also come into conflict with each other as can be seen in the standard example that eating lamb chops is good for the sheepdog qua dog but bad for the sheepdog's owner. That is, although we may view organisms as artifacts, and by breeding and training even make them artifacts, they nonetheless retain their ability to stop a functional regress.

Unlike the case of artifacts, which can be good or bad, worse or better, it makes no sense to ask whether Fred is a good antelope or Linda is a substandard tapir. Now, a good motor will help to make one car better than another and a good set of teeth will help to make one sheepdog a better guard than another, but a strong heart or good teeth will not make one hippopotamus a better hippopotamus than another. A hippo may be better off with well-functioning organs, but it itself has no function: it has a good. And an organ or trait can be good for a sheepdog qua dog without being good for the sheepdog qua guard dog.

Even strongly intentionalistic theories of artifact functions—which allow functions to come and go with mental events, even without any physical changes in the function bearers—balk when it comes to the functions of parts within a whole (McLaughlin 2001: ch. 3). These do not come and go so easily. Even if I turn my ax into a crowbar, the wedge that fixes the blade to the handle still has the same function as before and might be thought to be supposed to do just that independent of the new goals of the agent. A functional analysis is in fact interesting only in a complex and hierarchically ordered system where the parts are more or less tailor-made for a particular role. In such a system we may well be able to identify only one function of a part that fits the hierarchical structure—though this may be due only to lack of imagination of the functional analyst. A cuckoo clock as a system may be viewed from the perspective of its capacity to act as a counterweight in a balance, and thus each of its parts may be viewed as contributing to that capacity and having this contribution as its function. But in such a case each of the parts contributes only by its individual weight, not by its special structural properties or by its integration into the organization of the complex hierarchical system. It is indeed hard to imagine what capacity of the cuckoo clock, other than the uniform motion of the hands, the escapement mechanism might contribute to for which the complex organization is at all relevant. But with biological traits the tailor-made character is often much less clear.

These last two aspects of part-whole relations—that organic wholes, as opposed to parts, tend not to have functions at all and that the functions of parts of artifacts are much more resistant to arbitrary change than are the functions of the artifacts themselves—however, may be only obliquely relevant as sources of normativity and rather indicate a source of confusion in our thinking about functions.

6.3   Types and Tokens

Normativity is also already introduced by the type-token distinction used when characterizing function bearers. Functions are ascribed to items that instantiate a particular type that has the function.

The textbook view of the type-token distinction can be misleading. C. S. Peirce uses the terminology of types and tokens to make certain kinds of distinctions, for instance, distinctions between the letter *A* and any particular concrete way of writing it. Here are five tokens of the letter type *A*: A *A* A *A* **A**. Often the type-token terminology is introduced to distinguish between sign types and their concrete instantiations or between token sentences and their content (propositions). There is no obvious connection of this notion of type and token to normative considerations. But there is also an older philosophical use of the terminology of types—systematized by William Whewell (1840: 477) with reference to the traditional distinction between habitus and privation. This more traditional concept of type, analyzed in detail by Hempel and Oppenheim (1936), sees types as embodying norms. When we view individuals as *tokens* of a type, rather than (say) as elements of a set or members of a class, we have opened up the possibility of introducing normative considerations. Whereas individuals are either members of a class (elements of a set) or they aren't, tokens by their very nature can instantiate a type better or worse. Any element of a set is (as an element) just as good as any other—or rather it makes no sense either to rank elements or to equate them. But if I conceptualize an individual as a token of a particular type, I expect it to have the typical properties or a typically broad selection of the typical properties. And these expectations can be met in differing degrees by the various tokens of the type. Any token may instantiate its type better or worse.

Thus if a type has a function F, then so do the tokens. That is, if I have *correctly* identified an artifact as a wing, a screwdriver, or legal tender, then I have identified it as something that is supposed to enable flight, drive in screws, or pay my bills. Take an individual item. Give it a name, say, "Item 14." Now Item 14 isn't *supposed* to do anything—it just is. However, either it instantiates a cuckoo clock or it doesn't. If Item 14 is a clock, then it is supposed to go *cuckoo* every hour; if not, not. Individual entities are only supposed to be or do X or Y if they are tokens of a type that typically does that and can be said to do that better or worse.

With biological function bearers, things are somewhat clearer since they are not so often classified primarily or exclusively in functional terms. For instance, in pigeons and penguins, I know whether something is a wing or not independent of whether it typically enables flying; this is a question of morphology, anatomy, and homology—not of function. If a given penguin's or pigeon's front limbs are tokens of the type of wing, then they have whatever functions penguin or pigeon wings happen to have. If an individual pigeon has wings that do not enable it to fly, then its wings malfunction. If the wings of a penguin do not enable flight, they do not malfunction, and we still call them wings. The penguin's wings are not supposed to enable flight, but the pigeon's wings are. And even where the name of the organ is functionally determined—we call a bat's wings "wings" not "remodeled paws"—the question of whether an individual item receives the name does not depend on its actually performing that function. Whether an individual is a token of a type—that is, whether it *adequately* instantiates a kind—is already a (technically) normative question. Tokens are supposed to have the functions of the type they instantiate—if the type has functions.

This is a third candidate for a source of normativity in function ascriptions. But note that the type-token relation has no essential connection to functions or function ascriptions. The type-token distinction can introduce a minimal normativity into any context in which it is used. However, in the case of the ascription of functions to parts that are integrated into a whole and contribute to some performance of the whole that is good for the whole, the type-token distinction reinforces the presupposed normativity of part-whole and means-ends relations.

References

Hempel, C. G., and Oppenheim, P. (1936). *Der Typusbegriff im Lichte der neuen Logik.* Leiden: Sijthoff.

Kitcher, P. (1993). Function and design. *Midwest Studies in Philosophy, 18:* 379–397.

McLaughlin, P. (1993). Descartes on mind-body interaction and the conservation of motion. *Philosophical Review*, *102:* 155–182.

McLaughlin, P. (2001). What Functions Explain. *Functional Explanation and Self-Reproducing Systems.* New York: Cambridge University Press.

Neander, K. (1991a). The teleological notion of "function." *Australasian Journal of Philosophy*, *69:* 454–468.

Neander, K. (1991b). Functions as selected effects: The conceptual analyst's defense. *Philosophy of Science*, *58:* 168–184.

Whewell, W. (1840). *The Philosophy of the Inductive Sciences I, II*. London: Parker.

Wright, L. (1973). Functions. *Philosophical Review, 82:* 139–168.

7 The Inherent Normativity of Functions in Biology and Technology

Maarten Franssen

7.1 Introduction

Functions are attributed routinely in biology to organs, to traits, and to forms of behavior, and in technology to artifacts. We say, for example, that the function of a particular pump is to make the water of a central heating system circulate through the pipes and radiators, and that the function of a particular heart is to circulate the organism's blood through its arteries and veins. Apart from these two core domains of functional talk, functions are attributed in the social sciences to social traits such as marriage systems, religion, and the like, although much less generally and less enthusiastically. Such social traits are partly similar to biological traits, in that some have emerged as historical accidents, and partly similar to artifacts, in that some may have been designed, either for the function that is attributed to them or for some other function that they may or may not actually perform.

Despite the apparent centrality of the notion of "function," there is little consensus, in the philosophy of the sciences and of technology, on how to understand the concept, the sort of work it does, and the conditions governing the legitimate attribution of function. My aim in this chapter is not to settle the matter as to what is the correct theory of function, or which theory is to be preferred to which on what grounds, or whether a unified theory of function is possible at all. My aim in this chapter is to investigate the role that is played, in the debate on what is and what is not an or the adequate theory of function, by the claim that "function" is an inherently normative concept. It is widely held that functions are normative in the sense that an item can have a function but at the same time be physically incapable of performing that function. In such a case, we speak of malfunction. A pump that has broken down is not able to circulate the water through the central heating system, but most people would agree its function is still to do so. This is expressed by saying that, although the device does not pump, it *is supposed to* pump or *ought to* pump. Similarly many people would say that a heart that (momentarily) fails to circulate the blood still has the function to do so and accordingly *ought to* do so. Another aspect of the normative character of function is that we distinguish between good and poor pumps, and between good and poor (or rather bad) hearts. The terms used in these expressions—*supposed, ought, good, poor*—do not belong to the descriptive vocabulary of science.

The normative aspects of the notion of "function" have served as a touchstone in the debate between rival theories. Adherents of etiological or proper-function theories (see section 7.2) have taken the normative character of function as obvious, and judge theories by the extent that they can give an account of it. Naturally these researchers emphasize that normative statements related to functions are justified on the theories they propose. Proponents of causal-role theories (see again section 7.2), which meet with difficulties in accommodating the normative character of functions, argue, on the other hand, that it is in fact a mistake to hold that the function concept is inherently normative. They seek to show that the normative statements at issue can be reconstructed as more innocuous descriptive ones, which their theories can handle. It is therefore a matter of considerable importance for the debate among the various theories of function whether the normative character of functions can be vindicated.

The structure of this chapter is as follows. In section 7.3 I investigate how exactly the theories that claim to be able to account for the normativity of function go about doing this. Next, in section 7.4, I propose a precise interpretation of the central normative statements, since, in saying that a mere object "ought to" do something, one cannot be taken literally. In my interpretation, such statements express that one is justified in holding certain expectations concerning an item's behavior. In section 7.5 I aim to show that researchers are less prepared to apply normative judgments than they are to attribute proper functions, and that they are less prepared to attribute proper functions than their theories allow them to. This shows that normativity cannot be considered to be inherent in any current, technically defined concept of function. Section 7.6 briefly extends the arguments from sections 7.3 and 7.4 to normative statements that talk of good and poor functioning. In section 7.7 I extend the analysis of section 7.4 by sketching how functions play a role in an account of "true" normativity, to which the formation of beliefs and the choice of actions by intentional beings are central. Finally, in section 7.8, I draw my main conclusion, which is that all normativity traditionally associated with the concept of "function" derives from human intentionality, either with respect to the justification of beliefs about a functional item or with respect to reasons for certain actions with respect to a functional item. None of the various notions of function as they are applied, for explanatory purposes, in biology and the social sciences, can be treated as inherently normative; only a notion of "function" that refers directly to human beliefs and human actions can. I start, however, with a brief exposé of the currently prevailing theories of function.

7.2   Rival Theories of Function

It is common to distinguish two general approaches to characterize the notion of function. On the one hand, there is the causal-role or causal-contribution (CR) view, first proposed by Cummins (1975). According to this view, the function of an item is the causal contribution that this item makes to a capacity to show a certain behavior of a larger entity, of which the contributing item is a component, in the ordinary, mereological sense. On the other hand, we have the etiological theory, or theory class, of which Wright's (1973) theory was the first representative. A decade later, however, Millikan's (1984) much more sophisticated theory of proper functions (PF) replaced Wright's proposal as the most general articulation of this view. It restricts the attribution of functions to items that are, in a precise technical sense, reproduced.

These two approaches to the notion of "function" can be distinguished by the sort of explanatory work that they make the concept of "function" do. The attribution of a function as a causal contribution answers a "how" question: *how* does a particular entity achieve a certain behavior? An answer that refers to the contribution of components of the entity implies that this entity is complex or systemlike. For noncomplex entities, the question of how its behavior comes about seems pointless; anything to be said about it can refer only directly to the laws of nature that it is subject to. On an etiological theory, in contrast, the attribution of a function is usually said to answer a "why" question: *why* does a particular item exist? The attribution of a function to the item is a sort of summary of the relevant causal history of the item, usually but not necessarily as part of a larger entity.[1]

Both the CR and the etiological theories occur in more specialized versions.[2] A special case of the causal-role theory is Boorse's (2002) goal-contribution (GC) theory. It limits the attribution of functions to the components of goal-directed systems instead of just any system, but maintains the basic outlook of the causal-role theory that functions are the causal contributions to the system's capacities, ultimately, in the case of the goal-directed systems that are called living systems, the capacity to survive and reproduce.[3]

Recently another variant of the causal-role theory has been proposed by Krohs (2004; forthcoming). It limits the attribution of functions to the components of systems-with-a-design (SD), where the relevant notion of "design" is an extension of the ordinary concept, which refers to intentional design by human beings. A system has a design in this extended sense if, roughly, its components are picked, by whatever mechanism, to become components on the basis of the fact that they are tokens of a particular type, rather than by their individual physical properties.

A special case of the etiological theory, and in particular the PF theory, is the selected-effect (SE) theory as advocated by, for example, Neander (1991). By specifying in the definition of *function* a particular causal mechanism through which an item's current presence is historically explained, being the Darwinian theory of natural selection, the SE theory of function applies only to biological items—at least, as long as the general skepticism regarding the operation of natural selection in the development of social systems lasts. All remaining theories, however, among those introduced here claim to be general theories that apply to the entire spectrum of function attributions in (scientific) practice.

I disregard here all variants of the SE theory that trade past selection for current fitness or remote-past selection for recent-past selection, since these differences are, I think, immaterial to the arguments I develop in this chapter.

7.3   How Function Theories Account for the Normativity of Functions

The twofold division of function theories sketched in section 7.2 coincides with the sharp divide concerning the possibility of *malfunction* that is mentioned in the introductory section. Only in the etiological theories—PF, SE, and variants—can it occur that it is the function of an item x to do F while x is incapable of showing the behavior that counts as performing F. This is so because on these theories the function of an item is defined by reference not to the item's causal role but to the causal contribution of related items, its—technically defined—*ancestors*. All other theories, defining the function of an item in terms of the item's causal role, cannot grant an item that fails to show the required physical behavior a function, regardless of whether other similar items are capable of showing this behavior and are therefore attributed the corresponding function.

As far as the functions of biological items are concerned, the PF and SE theories attribute functions, and therefore also malfunctions, to items that are identified as tokens of a type. By malfunction, then, the following situation is meant: 1) x is a token of the type X, 2) x is attributed the function to do F at least partly on the basis of its being a token of the type X, and 3) x is not capable of showing the behavior by which tokens of X normally or usually perform F.[4] Any theory that refers to causal interactions in defining functions—and both the CR and the PF theories do so—must attribute functions to tokens, since only tokens are causally efficacious. Types can be assigned functions in a derived sense, in terms of the functions of their tokens.

Care must be taken to distinguish between the type that an item is presumed to be a token of, in receiving a function on account of a particular theory of function, and the functional type itself, that is, the type that is defined as consisting of all tokens that have this particular function. The type to which the theory of function refers cannot be the functional type, on pains of circularity: in order to know whether x is a token of the functional type X_F, it must be known whether x has the function F, but in order to know whether x has the function F, it must be known whether x belongs to the functional type X_F. The dominant type concept in biology is indeed the one that theories of function presuppose, identified by a combination of structural (morphological and physiological), historical (developmental), and comparative (homology) features. The predominant type concept in technology, in contrast, seems much closer to the functional type: pump, knife, and so forth. It is not commonly recognized in technology that, apart from the functional type, another type is presupposed, which is identified by the physical and historical features of existing tokens of a functional type and by the design specifications associated with the type. A defense of the importance of this "narrow" type concept for artifacts is given by Soavi (this volume).[5]

Precisely with reference to the distinction between token and type, it has recently been argued by Davies (2001) that in fact the etiological theories are just as little capable of assigning malfunctions, that is, assigning the function to do F to an item that is not capable of doing F. His argument is specifically directed to an SE-type theory, which he presents as follows (p. 194):

The selected function of type T in organism O in environment E is to do F iff:

(i)   ancestral tokens of T in O performed F in E;

(ii) T was heritable;

(iii) ancestral performances of F enabled organisms with T to perform better in E than organisms lacking T;

(iv) superior reproduction caused organisms with T to out-reproduce those lacking T;

(v)   superior reproduction caused organisms with T to persist or proliferate in the population.

This matches the way the SE theory is phrased by Neander, with one crucial difference: Neander's SE definition assigns functions to tokens, not to types.[6] Davies claims that the type T in clauses (i) to (iv) is a "success type": it is the type corresponding to tokens that actually performed F. And if in clauses (i) to (iv) type T is a success type, then it must equally be in clause (v). What is explained by the attribution of the function F to T on the basis of the etiological account of function, therefore, is the persistence and proliferation in a certain population of organisms of a type T the tokens of which do in fact perform F. Accordingly the function to do F can be attributed only to tokens of T that have the capacity to perform F, and no token lacking this capacity can have the function. Ergo selected malfunctions are impossible.

Davies' argument, however, is based on a too narrow construal of the etiological theory. Apparently he reads clause (i) in the definition extracted here as stating that *every* token of T in O performed F in E, clause (iii) to state that ancestral performances of F enabled *every* organism with a token of T to perform better in E than organisms lacking T, and clause (iv) to state that superior reproduction caused *every* organism with a token of T to out-reproduce those lacking T, and that only on this reading the conditions stated in clauses (i) to (iv) explain the presence and proliferation of T in the population.

To invalidate Davies' reading, take Neander's version of the SE definition (1991: 174):

It is the/a proper function of an item x of an organism o to do that which items of x's type did to contribute to the inclusive fitness of o's ancestors, and which caused the genotype, of which x is the phenotypic expression, to be selected by natural selection.

This definition coincides with the reduction to the direct proper function of biological items of Millikan's much more general account of function. What caused the genotype of which a token x is the phenotypic expression to be selected by natural selection is what items of x's type contribute *on the average* to the *average* inclusive fitness of o's ancestors. The causal explanation of the selection does not require the positive contribution of every historical possessor of tokens of the type corresponding to x to the selection of the genotype.

Davies might choose to save his argument by claiming that an organism with a malfunctioning token of T is not the expression of the genotype that was selected, since that genotype codes for functioning tokens of T. This would be a relevant defense if all deformation resulting in malfunctioning had its origin in genotypic defects. Indeed, in his book, Davies occasionally writes as if he believes this to be the case. However, deformation can be the effect of many different causes. Even congenital deformations are not necessarily the effect of genotypic faults but can also be the result of disturbances in the biochemical circumstances in which the fetus developed. It is therefore perfectly possible for a phenotype with a malfunctioning token to be the expression of the "faultless" genotype that was selected.

As already noted, the type of which x is a token cannot be the functional type, that is, the type that is defined in terms of x's function. For biological items, however, such type definitions are hardly, if ever, a problem: they are furnished by morphological or developmental considerations or by homology, as Davies is well aware (2001, p. 199, n. 6).[7]

The PF-type theories, therefore, are able to attribute malfunctions. This does not mean, however, that only these theories are able to describe the situation where a particular item fails to do or is incapable of doing what would be necessary for performing a particular function. The CR-type theories describe it in precisely the terms that I have been using so far: in terms of type and token behavior. If, on the PF-type or etiological theories, an item malfunctions, then on any theory, it does not show the behavior that tokens of its type normally or typically show, or it does not now show—or is currently unable to show—the behavior that it used to show and that counted as performing a particular function. Proponents of the CR-type theories consider this sort of description to be adequate enough.

So there are items of which the PF-type theories say that their function is to F, but they malfunction, whereas the CR-type theories do not attribute to them the function to do F. In deciding whether one of these positions is the "correct" one, it is important to recognize which intuitions, exactly, need to be saved here. Surely we would like to remain justified, at least within certain limits, in holding that a malfunctioning (i.e., broken) pump is still a pump, and similarly a malfunctioning (i.e., deformed, diseased) heart is still a heart.[8]

For that to be the case, *heart* cannot just mean "any organ that circulates the blood through an organism's body," and *pump* cannot just mean "any artifact that circulates a liquid through a system." As mentioned, biological types can be defined and are defined in morphological and developmental terms. It would seem that for artifact types we could refer exclusively to what a particular artifact was intended to be when designed, but structural criteria are necessary here as well to keep the inflation of such types by "wishful designing" within bounds; see Thomasson (2007). As a result, structural and historical characteristics also codetermine artifact type definitions. A knife is not just anything designed for cutting. The basic point remains that any theory of function that refers to types must be able to define these types in a noncircular way.

The CR-type theories can, therefore, account for part of what we mean by attributing malfunctions. This part, however, contains nothing normative. Malfunction is judged to be a normative notion only when we take it to include saying, for example of a particular heart, that although it does not in fact contract and thereby circulate the blood, it *is supposed to* do so or *ought to* do so. The PF-type or etiological theories do not account for this normativity, however, by showing that such statements can be derived from its theoretical framework. Indeed the theory or definition contains exclusively descriptive, naturalistic terms,[9] so no statement using terms from the intentional vocabulary—like *is supposed to*—or more narrowly, from the normative vocabulary—like *ought to* or *should*— can be implied by it. What the theory does is single out, from all the behaviors that a particular item is capable of, or from all the effects of a particular behavior, one behavior or effect that is granted a special status. This behavior or effect, then, can be considered the special thing that the item is supposed to do. For example, the proper function of the nose is to allow air to reach the lungs, even when the mouth is engaged, and to have this air pass over the smell receptors at the same time, not to support a pair of glasses. The former is what the nose should do (and a snotty nose is therefore a temporarily malfunctioning one), but the latter is not, and there is nothing wrong with a nose that does not. The proper function of a drinking glass is to hold liquid and allow that liquid to be drunk with the mouth, not to be smashed against the wall to express rage or anger or joy. The latter is not what a glass should enable one to do, but the former is, and a leaking glass is a malfunctioning glass, although it is perfectly good for being smashed against the wall.[10]

To bring out the contrast between the PF-type and CR-type theories most clearly, it should then be noted that the PF-type accounts of normative functions are based on two elements. First, they interpret "function" as a special, inherently normative notion, and accordingly narrow down the attribution of functions to items to (usually) one particular "proper" function per item. On the CR-type theories, in contrast, an item can have many functions. Second, the PF-type accounts attribute functions to items incapable of the corresponding physical behavior. A CR-type theory can try to "approximate" the marking of particular causal-contribution functions as special by being adaptations, as is suggested by Krohs in his account of the SD theory (2004: 97). If this is necessary and sufficient for making such functions special in the sense of allowing a normative reading, a CR-type theory can argue that if the special function of an item x is to do F by showing behavior B, then it not only performs F but should perform F as well. A CR-type theory cannot, however, say of another item x' of the same type that it is not capable of behavior B, that this item should still show B or should perform F, since x' does not have the function to do F in any CR-type theory. And it may be questioned whether it makes sense to say that x should be capable of B-ing if it is not at the same time true that x' should be capable of B-ing.

7.4   *Being Supposed to* and *Ought to* as a Way of Expressing Justified Expectations

The sketch in section 7.3 of the way the etiological theories handle the supposed inherent normativity of the functions matches the way Millikan introduces the connection between normativity and the notion of "proper functions" (1984: 17): "Having a proper function is a matter of having been 'designed to' or of being 'supposed to' (impersonal) perform a certain function. The task of the theory of proper functions is to define this sense of 'designed to' or 'supposed to' in naturalistic, nonnormative, and nonmysterious terms." So what the PF theory sets out to do is *not* to give an account of function from which normative statements containing "is supposed to" and its relatives can subsequently be derived; it is to give a direct account of our intuition that some entities can *be supposed to* show certain forms of behavior, something that we *alternatively* express by attributing a proper function to them. What the PF theory is meant to do is to articulate this notion of "proper function" in purely naturalistic terms. By giving an account that singles out, among the many things that a particular item x does, one specific behavior B by which it performs a function F, the theory does not justify normative statements of the form "x ought to do B," or state that *therefore* we can say that x ought to do B or to do F. The theory merely recovers a function concept that is intuitively associated with these normative statements, but this recovery does not extend as far as these normative aspects.[11]

There is, however, more than enough reason to ask for a justification of such normative statements. It may be intuitively all right to say of a particular item existing in the material world—an organ or a trait or a form of behavior or a device—that it should do something or is supposed to do something or ought to do something. But can we really make sense of such statements? Certainly not literally. Only of human beings can it be said that they *should do* or *ought to do* things.[12] One cannot prescribe anything to a mere material object; to do so would be a classic example of a category mistake. It makes no sense to say that an electron ought to move in accord with the Maxwell equations. But if such "ought to" statements are not prescriptive, then what are they? How should we understand what they say?[13] The difficulty seems least severe for "is supposed to" statements. Again, only human beings suppose things, and what an item is literally supposed to do is therefore what it is supposed to do by someone. In the case of artifacts, there are candidates available to do the supposing. It is clear that either the designer or manufacturer of an artifact or the user or both have suppositions, in the sense of expectations, concerning its behavior. It is also clear that the fact that a particular object is an artifact that has been designed for a purpose furnishes strong epistemic justification for certain expectations concerning its behavior.

Indeed, as I see it, the normative statements associated with function that are phrased in terms of "supposed to" or "ought to" or "should" are to be understood as expressing that certain expectations are in order. They are, in a sense, to be taken metaphorically—they are prescriptive in form but not in content. By saying that an item "is supposed to show behavior B," or equivalently, "ought to show behavior B," the speaker expresses that he or she is *justified in expecting* behavior B to occur. Two different ways in which such expectations may be justified can be distinguished: the speaker can be *rationally justified*, more in particular *epistemically justified*, or can be *morally justified* in holding these expectations. Epistemic justification involves the standard considerations of empirical evidence and logical implication. Being morally justified in expecting something to occur can be expressed alternatively by saying that one *has a right to it* that it occurs, for example, because this occurrence was part of the content of a promise.[14] Epistemic "ought to" statements, as I call them, apply to all items with functions that have been considered so far: biological items, social items, and technical artifacts. Moral "ought to" statements, in contrast, can apply only to artifacts, because moral rights are grounded in the intentional relations among people, not in anything "merely natural."

That the epistemic "ought to" statements apply to all items with functions that have been considered so far does not imply that such statements apply only to functional items. In fact they extend to any object concerning the behavior of which we form expectations. When we say that object *o* "ought to" show behavior *B*, we generally mean that, although we cannot be certain that behavior *B* will occur, we are justified in our expectation that it will occur. Take the following example: "When I let go of this stone (holding it under water), it ought to sink." But some unexpected stream could prevent it from sinking, or it could turn out to be pumice, which can float on water. Depending on the context, it may not even seem out of order to say: "When I let go of this stone (holding it up in the air), it ought to fall." Perhaps I vaguely suspect that some trick is going to be played on me. Or it may be that I myself intend to demonstrate to a class of students that one should never hold an empirical statement to be true with absolute confidence.

Artifacts are therefore not special in that we have special expectations concerning their behavior; only the grounds on which we form our expectations are different for artifacts as compared to natural objects. Only in the case of artifacts can we have morally justified expectations, apart from the epistemically justified expectations we generally have concerning the behavior of the things in our environment. As a consequence of the existence of two different sorts of expectations for artifact performance, it may sometimes be difficult to say on what grounds we feel justified in expecting a particular artifact to show a particular behavior. Do we expect newly purchased washing machines to wash laundry without damaging it or cars picked up from repair shops to drive smoothly, for example, because we feel that is what we are entitled to given what we paid? Or can we just not imagine how a company that would break such promises could survive? Even if our society would be so "dysfunctional," so to speak, that it would be epistemically irrational to expect a newly bought or freshly repaired apparatus to actually function as advertised or to be restored to functionality, we would still say we have a right to this being the case if we held to our part of an agreement. Indeed, a difference between epistemic and moral "ought to" statements that is important in practice is that the latter typically imply that some person or persons—usually the designer, manufacturer, and/or retailer—are responsible for the disappointed expectations, meaning that they "ought to," in the full-blooded normative sense of "ought" that will be the subject of section 7.7, indemnify the user. This only applies when an artifact is used for the function it was designed and sold for, and used according to the instructions for use and in the circumstances specified therein. If someone uses an artifact according to one's own plan, based on one's own inquiry after the artifacts capabilities, then if the plan misfires, the user can only blame oneself, if anyone, for holding expectations that may not have been sufficiently justified.

The notion of "blame" can help to understand the use of prescriptive language to express the sort of expectations at issue. We form our beliefs on the basis of our interaction with other people and with nature, and we expect the answers that we receive to be trustworthy in either case. We seem to hold nature to her part of a deal we supposedly made with her when we questioned her, to use Francis Bacon's metaphor, just as much as we hold other people to the truth of what they are telling us. Because we did our best to check whether a belief about her is true, nature should see to it that it is indeed true when it seems we received a positive answer. I suggest this is the reason why we use one expression to refer to two quite different situations. In forming our beliefs about the world and acting upon them, we always run the risk of being let down, either by our fellow men and women or by nature; that is apparently how we feel about it.

If the moral interpretation of "ought to" statements concerning the behavior of objects makes sense only if they are based on claims made by people and having the force of a credible promise, then for biological items the statement that some organ or trait x "ought to do B" can mean only that one is (or was) epistemically justified in expecting x to do B. A similar interpretation for normative statements associated with the functions of biological items has been proposed by Davies (2001: 151–156). However, Davies suggests that these expectations are warranted only with respect to the behavior of complex hierarchically organized systems, and not just regarding any system to the components of which we can, on a CR-type theory, attribute functions. It is their character of being "as if designed," by which components come to be seen to exist "for" their functions, that grounds such expectations for Davies. It seems to me, however, that Davies is overly restrictive in this. Not only is there no limitation to complex systems built into any of the theories of function, CR-type or PF-type, but my examples show that we are inclined to use the "ought to" language expressing justified expectations much more widely than we attribute functions.

7.5   The Arbitrariness of the Attribution of Normative Functions in Biology

In direct relation to function attributions, on the other hand, the corresponding normative statements occur less widely than might be expected. If we witness the birth of a kitten, we expect the kitten's heart to beat and circulate the kitten's blood just as much as we expect the kitten itself to live. We can say "The kitten ought to go on living" just as well as we can say "The kitten's heart ought to go on beating." If we watch some fishes in a pond that is frozen over, or if we put a butterfly in the refrigerator in order to be able to take photographs of it more easily afterward, we can say, for instance, to a protesting child, "It ought to survive." This is difficult to understand if such normative talk is linked to the attribution of function. Biologists are extremely reluctant to attribute functions to organisms, and even more reluctant to come up with the normative judgments that would accompany such function attributions.

In fact normative judgments are already withheld from, or at least only sparingly applied to, the other biological items apart from organs that are eligible for function attributions. Functions are also applied to traits, like the forkedness of the snake's tongue or the different maturation times of the pistil and the stamina of the flowers of monoecious monoclinous plants, and also of forms of behavior, like the mating dance of the stickleback or the mantling by which birds of prey hide their catch from the hungry eyes of other birds. Insofar as the normativity of traits or behaviors is concerned, we are far more reticent in pronouncing evaluative statements. Do we call the flowers of a white deadnettle (*Lamium album*) in which the maturation times of pistil and stamina coincide malfunctioning? It sounds odd to do so.

In the social sciences we see a distinction in the way function attributions are subject to malfunction and evaluative statements that reflects the differences between artifact functions and biological functions. Many social institutions are the result of design and subsequent implementation, monitoring, and adjustment, and can accordingly be seen as social artifacts, for example, the legal system of a country. Of such institutions it is entirely in order to say that they function poorly or malfunction. In the case of social practices that have evolved "organically," this is no longer true.[15] If the Hopi rain dance does not contribute to social cohesion, does the dance malfunction? Such issues make clear how enormously questionable such claims are, even apart from the more general problem discussed in the philosophy of science as to the explanatory value of such statements. For starters, how do we know that a particular Hopi rain dance does not contribute to social cohesion? Because there is a conflict? There can be lots of reasons why rain dances can fail to give rise to social cohesion, just as there are many reasons why an electric drill can fail to drill a hole in the wall even though the drill itself is not malfunctioning.

Normative statements seem thus to be applied much more reticently than the functions that they are supposed to be associated with are. In their turn, biological functions are attributed much more reticently than the theories that are supposed to ground them would allow. On the PF theory, for example, it is just as much a proper function of foxes to feed on rabbits as it is the proper function of the stomachs of foxes to help digest bits of rabbit. Individual foxes are the members of the higher-order reproductively established family of all present and past foxes, which are produced by the members of the first-order reproductively established family of complete fox genomes, the proper function of which is to produce fox phenotypes. And it is in terms of their habit of eating rabbits that the existence of proliferation of foxes in nature is Normally (Millikan's (1984) capitalization) explained. If we are hesitant in applying the concept in this way, it is, I suggest, because we shrink from the consequences with respect to our employment of normative language. If it is the proper function of foxes to feast on rabbits, then a fox that has a distaste for rabbits is a malfunctioning fox, which ought to eat rabbits, and one fox may be a poor fox compared to another one. One may try to oppose this by saying that it is the proper function of foxes to feed on whatever is available, rather than rabbits in particular. Still, the rabbit-eating habit of foxes satisfies the definition, or if not the rabbit-eating habit then surely the rodent-eating habit or the smaller-mammal-and-bird-eating habit, to be somewhat more accurate. What is more, there are enough examples where the feeding habits of a species are so fixed that on the PF theory the corresponding functions must be attributed to organisms, and by an obvious extension to species.

The SE theory is explicitly phrased to attribute functions to the parts of organisms, at least in Neander's articulation. There is no good reason, however, why this should be read as applying only to proper parts; that is why one cannot treat the organism itself as being a limiting case of a part of it, just as any set is a member of the set of its subsets. But even if the definition would be restricted—arbitrarily—to proper parts, then it seems possible to extend the meaning of *organism* such that an organism in the ordinary sense is indeed a proper part of an organism as technically defined. Flowers of the plant species *Yucca glauca* feed moths of the species *Tegeticula yuccasella* (or rather their larvae), and they do so because previous generations of yucca plants did, since the moths are essential to fertilizing the yucca flowers. The life of yucca and moth are so intertwined, in mutual exclusiveness, that a single yucca can be considered as a proper part of a yucca-moth pair, such that the feeding of previous moth larvae by previous yucca flowers contributed to the inclusive fitness of the ancestors (other yucca-moth pairs) of this yucca-moth pair and caused (in the sense of causally contributed to) the genome of such pairs (themselves paired items) to be selected by natural selection. It is difficult to see how blocking this route can be anything but arbitrary. In scientific practice such opportunities of extending the explanatory scope of some theoretical concept is usually welcomed. If instead it meets with rejection, there is reason to be suspicious of the motivation behind the introduction of the concept.

Note that the CR-type theories also have difficulties in withholding functions from organisms, although their proponents share in the widespread aversion to attribute functions to organisms. On Boorse's GC theory, which favors a technical definition of goal-directedness in cybernetics terms, most ecosystems, and certainly symbiotic systems, are goal-directed, and the organisms making them up thereby acquire functions through their causal contributions to the enduring state of the system. On Krohs's SD theory, organisms have functions because they are type-fixed components of ecosystems and ecosystems are designed systems. At least, it is difficult to see how Krohs's definition of a design (2004: 82) makes it possible to draw a principled boundary where design stops. It may be objected that not all components of an ecosystem are type-fixed; the soil on which plants grow, for example, is not. Neither, however, are all components of an organism type-fixed; the water it contains, for example, is not. At a sufficiently elementary level—the molecular, for instance—the components of any system are no longer type-fixed (see also Krohs, this volume).

Summarizing, it is apparent, first, that the attribution of proper functions in biology is not taken as far as the corresponding etiological theories allow, nor as far as other theories allow, for that matter, and second, that the normative judgments thought to be grounded by functions are phrased even more cautiously than are the attributions of proper functions. This shows that the proponents of PF-type theories, although they advertise their theories as uniquely capable of accounting for the inherently normative character of the notion of "function," do not in fact treat their preferred concept of "function" as being *inherently* normative.

7.6 Good and Poor Performances of a Function

Until now I have considered one type of normative statement concerning functional items only, the type that says of a malfunctioning or dysfunctional item that it nevertheless "is supposed to" or "ought to" do what it cannot do. In the case of artifacts, the notion of "malfunction" is generally understood to mean the plain failure of a device to do what it is designed to do, that is, supposed to do, on sufficiently firm grounds, by its designer.[16] In biology, however, malfunction may not be such an isolated phenomenon but may be viewed as an extreme value on a scale running from well to poor functioning. In the same vein, a well-functioning, or shortly, good, specimen of its (functional or historical) kind does exactly, or at least closely enough, what it is supposed to do, and a poorly functioning item, or a poor specimen of its kind, shows behavior that is unlike what it is supposed to do, or that is only capable of doing something that falls significantly short of what it is supposed to do.

Functioning well or poorly can be seen as (clusters of) objective positions on a scale that has total lack of performance (the worst performance) on one end. What then is at the other end of the scale? In this respect there is again an important difference between the case of technical artifacts and the case of biological items. With artifacts we can often describe an ideal token of a particular functional type while not a single actual token of that type comes even close to that ideal. This is especially so for artifacts that are based on new operational principles. Around 1900 it was fairly clear what people expected of an airplane; it was clear, for instance, what the list of functional requirements of an airplane should look like. All actual primitive airplanes were far removed from this ideal, however. That does not necessarily mean that, for example, the airplane of the Wright brothers was a poor airplane, although it certainly would not have been considered a good airplane just by being able to stay in the air. Similarly, during the late 1930s and early 1940s, the few people who were working to develop the jet engine had a fair idea what such an engine would ideally be able to do, but at the same time all prototypes that were built disintegrated or exploded within a few minutes. Evaluative judgments of particular devices are thus partly based on the performance of other representatives of the functional kind but also partly on the distance between the device's performance and the (imagined) performance of an ideal device.

In biology, an ordinal or quantitative scale for function performance is determined completely by the distribution of actual performances, and even then it is often very difficult to come to a delineation of the typical or normal performance, as is shown by the controversies concerning the definition of health and illness in medicine. It may be tempting to believe that natural selection will find an optimum for the performance of any function fairly quickly, and that once found, this performance sets a standard comparable to the ideal knife in the case of technical artifacts. This, however, is completely dependent on the timely occurrence of the right mutations and of the possibility of a mutation in the first place. The famous case of the peppered moth (*Biston betularia*) in the English Midlands, where in the 1930s the light-colored standard form was replaced by a black mutant, is a case in point. There seemed to be no reason to doubt the adaptedness of the original light-colored peppered moth in the soot-infested woodlands around Manchester until the black variety, which was evidently better camouflaged on the darkened tree stems, appeared.[17] However, if the mutation would never have occurred, would we have found it lacking? Was the light-colored variety already poorly adapted to the current environment before the black variety appeared? Would we have recognized an empty niche, waiting to become occupied? But then why is an extension of the human visual sensitivity beyond the 400 to 800 nm spectrum not similarly considered an empty niche, a level of adaptation that waits to be improved upon? Should people not be able to smell better, hear better, or run better than they do? Once one has started off on this course, one may quickly find any organism a sorry failure.

For artifacts, then, a point of reference can be set independently of all actual items, such that all items can perform worse than the norm. The idea of an ideal knife, an ideal pump, or an ideal thermometer makes sense, whereas the idea of an ideal liver, an ideal rabbit, or an ideal mating dance seems pointless.[18]

In the present context, the important question concerning a scale of relative performances of function is: what makes such a scale normative? Carrying the analysis of section 7.4 one step further, I suggest that normativity enters only when the notion of justification is added. We can objectively classify the performances of items on a scale and compare them, and define a norm either by establishing what the average performance, in the right circumstances, is among members of the population (see for a similar suggestion Krohs 2004: 100) or by introducing an ideal performance. None of this is yet normative, however, except in the very meager sense that we can base our expectations concerning the future behavior of items on the outcome of this exercise. Such a measured norm becomes normative only when we make it a norm governing our actions. Once we have ordered a collection of functional items into relatively poor ones and relatively good ones, ceteris paribus, we are justified in preferring, or ought even to prefer, a better item to a worse one in using it for its "normal" purpose. And once we have ordered existing items with respect to an ideal and we have seen that they are all far removed from the ideal, we are justified in looking for a new specimen, or even ought to look for a new specimen. (The ceteris paribus clause is there because, for more detailed purposes, the criteria can be reshuffled; there are a lot of things you may want to use a knife for that are better done with a knife far removed from the ideal knife than with a knife that comes close to it.) Again this makes no sense for biological items in general, only for those that are available for being used by us for a purpose of ours.

7.7 The Normativity of Function as the Presence of Reasons to Use

The interpretation that I give in section 7.4 of normative statements seemingly supported by functions refers to the justification of beliefs. In section 7.6 I argue that a norm or standard is normative only insofar as it governs human action. Therewith we have reached the native country of normativity, where life is all about the justification of beliefs and of courses of action. Only by traveling there can we come to understand how the notions of "functioning," "functioning poorly," and "malfunctioning" can be seen as normative notions. There, "ought" refers exclusively to what people ought to believe, ought to desire or aim for, and ought to do. The notion that a person ought to believe or to do something is currently most often explicated in terms of the *reasons* this person has for believing or desiring or doing something. By saying that someone ought to do Z, we mean that this person has a compelling or conclusive reason to do Z, or that, taking into account all reasons the person has for or against the doing of Z, the balance of reasons pleads for doing Z.[19]

It is the concept of a "reason," or rather the relation of being a reason for something, that is thought central and primitive. Other concepts spanning up the normative domain, such as "good," "bad," "right," "wrong," or "ought to," are at a minimum candidates for being defined in terms of reasons. The overall notion of the "normative" has itself been characterized in terms of reasons in the following way by Dancy (2006a): a normative fact is the second-order fact that another fact (usually, but perhaps not necessarily, a first-order fact) is a reason for someone to do one of the things that one can have reasons for, that is, to believe something, to aim for something, or to act in a certain way. This is very broad. Elsewhere, however, I have shown (Franssen 2006) that this can be applied to characterize the normative dimension—normative in the present sense related to what people can believe, aim for, or do—of technical artifacts.

On this account, the functions of technical artifacts can be shown to be related directly to reasons for a special form of acting, that is, using some object as an intermediate to achieve a particular goal. On this account, the evaluative claims such as "x is a good F-er" or "x is a poor F-er" can be interpreted as truly normative statements if they are taken to express not the first-order fact that the performance of x can be placed somewhere on an ordinal scale of performances of (possible) tokens of the (functional) type but the second-order fact that *because of* the first-order fact that x has certain physical features, which determine its performance relative to the performance of other tokens of the functional type X, if someone has a reason for F-ing, then this person has a reason to use x for F-ing.[20] Or to use a paradigm rather than abstract language, "This is a good knife" expresses the second-order fact that *because of* its physical characteristics, if someone has a reason for cutting something, this person has a reason to use this knife for that task. For malfunction, by a similar but shorter account, "This is a malfunctioning knife" expresses the second-order fact that *because* of the physical characteristics this knife has, which make it the case that it lacks the capacity to cut adequately, one has a reason not to use this knife for cutting. The reference to the reason that one must have for cutting in the first place is now redundant.

This account brings out the normativity associated with the functionality of technical artifacts only at the most general level. It goes no further than claiming that the characteristics given to artifacts by their designers present us with some reason to use the artifacts, except when these characteristics are absent, notwithstanding their design, in which case we have a reason to avoid them. Nothing is said on how good or strong such reasons are. It is definitely false that the statement "This is a good knife" can be read as expressing that anyone who has a reason for cutting something *should* use this knife to do it, since there may be a better knife available. Nor can "This is a poor knife"—even in the case of an extremely poor knife—be interpreted as expressing that one *should not* use this knife for cutting, because in the absence of alternatives, using this knife still may be the best option if one is pressed hard enough.

As a second-order fact, the fact that "This is a malfunctioning knife" can be subsumed under the second-order fact "This is useless as a knife," which expresses the fact that because of its physical characteristics, any person has a reason not to use this object for cutting. It is the first-order fact of the particular features the object has that distinguishes between an object's being a malfunctioning knife and its not being a knife at all and lacking any property that would give it knifelike capacities. Among the features that a malfunctioning knife has and an object that is not a knife at all lacks is the historic feature of having been designed as a knife.

Similarly, the second-order fact that "This is a good knife" can be subsumed under the broader second-order fact "This is useful as a knife." Something that is not a knife, in the sense of not having been designed as a knife, can still have features that make it the case that someone who has a need for cutting has a reason to use it for cutting. For most artifacts, if something that was designed for the particular task at hand is available, provided it is in working order, there will usually be a conclusive reason to use this item, rather than an arbitrary object that happens to be able to do the job though it was not designed for it. Other circumstances where we must make do with what is available are nevertheless far from rare.

The mere attribution of function, "This is a knife," cannot be subsumed under the fact that "This is useful for cutting," because of the existence of malfunction: a particular knife may be a broken or otherwise useless knife. It may be thought that "This is a knife" expresses the second-order fact that because this object has certain features (including its design and manufacture history), someone who has a reason for cutting has a reason to use a token of the narrow, design-historical type to which it belongs.[21] It is questionable, however, whether this is indeed true. If most tokens of the type are malfunctioning, I would say this is false. My suggestion is to read the statement "This is a knife" as expressing a different sort of normative fact, which adds reasons for believing something to the reasons for doing that dominate the previous cases. On this reading, "This is a knife" expresses the normative fact that because this object has certain features (including its design history), if one has a reason for cutting, one has a reason to believe that one has a reason to use this object for cutting. This is weaker than saying that one is justified in believing that one has a reason to use it for cutting, because there might be other reasons that speak against this belief, such that on the balance of reasons one should not believe that one has a reason to use it for cutting, and believing so would then not be justified.[22]

So a statement of the form "This object is of functional kind F" expresses both a first-order fact—the object was designed for the purpose of F-ing—and a second-order fact— the object's being designed for the purpose of F-ing gives one a reason to believe that it will be useful for F-ing. The fact that it was designed for the purpose of F-ing makes a difference to the question what to believe and, subsequently, how to act for people who have a reason for F-ing. First-order facts can make such a difference only to beings who have goals and who can use objects for purposes. In the absence of a context of action, such facts, including biological facts such as "x is the wing of bird y and contributes to y's capacity to fly or contributed to this capacity in y's ancestors, by which y came to exist and to have wings," express no normative facts. This is so even on the GC theory, with the assumption that biological organisms can truly be characterized as goal-directed systems. The fact that the heart's regular contraction contributes to an organism's goal to survive does not make a difference to the organism's question of how to act in order to survive, because there is no such question.

7.8   Normativity and Biological Functions

What does this analysis contribute to our understanding of the concept of a "function"? It shows how this notion can be linked directly to the concept of "normativity" that is central to philosophy. Normativity belongs to a domain where central notions are the beliefs and actions of people, and central concerns the extent to which such beliefs and actions can be justified. This view on what normative statements mean allows for no extrapolation beyond the domain of intentionality and consciousness. The account here helps to understand the way we extend some of the normativity associated with function talk beyond this domain. I concede that we have no precise ideas on where to draw the boundaries of this domain, but that is not essential for judging various other domains as being clearly elsewhere.

The examples presented in section 7.5 show that we attribute proper functions—the kind of function that supports normative judgments—to biological objects only up to a certain level, in a way that is quite arbitrary according to the only theory we have available to explain or justify the attributions of normativity-supporting proper functions. The level at which we stop is recognizable as the level of the individual organism. We attribute proper functions to organs that contribute to the continued existence of an individual organism, but we do not attribute functions to organisms that contribute to the continued existence of a species or of other organisms.[23] The reason why we do so is, I suggest, linked to the attributions of functions to artifacts. Artifact functions figure in a context where we unproblematically use a mental or intentional description: artifacts serve purposes—we use them to achieve some of the goals we have. Here there is no need to give a naturalized account of such talk of functions apart from the need one might feel to naturalize the intentional description of human thought and action in its entirety, which is an issue I do not touch upon here. Apparently we see in the individual biological organism a thing to which we can extend our description of ourselves as intentional beings: individual organisms can be thought to have the purposes of staying alive and flourishing (as we have), but that is as far as we are prepared to go. Our normative talk concerning bodily organs fits this picture, as we apparently see them as supporting the purposes of organisms, similar to the way external objects support our lives.

The justification for speaking of a "good" heart seems additionally to be inspired by the vision of being able to "improve" upon existing hearts. And in fact, thanks to medical technology, the dividing line between biological organs and artifacts is blurring, now at an increasing rate. Open-heart surgery and the implementation of pacemakers are two ways to "improve" the performance of a poorly functioning or "bad" heart. Our medical practice invites us to incorporate our organs into the artifactual realm.[24]

However, there are major obstacles in extending the analysis of artifacts in this direction, along the lines followed in this chapter, in that it would require us to make sense of the idea that we *use* our bodily organs to realize certain goals. On one hand, it may be all right to say that I use my hand to pick up a pencil from my desk, since the "handling" of my hand answers to much the same features as my handling of the pencil. On the other hand, it makes no sense to say that I use my heart to circulate my blood through the arteries and veins of my body. I cannot refrain from doing so, hesitate in doing so, do it slowly or carelessly, nor any of the other things that characterize the intentional action of using something. Linguistic practice is fairly accurate in this; the only borderline case is perhaps "use your head."

Talk of better and worse livers, hearts, and so forth, can be justified only insofar as we regard these items as falling within the engineering domain, as being quasi-artifacts, so to speak, being amenable to improvement and redesign. When we abandon this perspective, there is no normativity to be found in nature and no normativity inherent in function talk.

Consequently I conclude that the advantage that the PF and SE theories have always claimed over their rival theories—of uniquely being able to account for malfunction—is illusory. The normativity of artifacts is not what it has seemed to most people. We have functions as causal contributions to a system capacity, functions as causal contributions to surviving and reproducing behavior, functions as causal contributions to designs, functions as causal contributions to a selection process, functions as intended behavior, and functions as behavior made use of, but none of these functions uniquely supports normative "ought to" statements—applied to functional items rather than persons—to the extent that we see a point in uttering such statements. This should not be seen as an invitation to add a new function concept to this collection, functions as normativity-supporting behavior. No definition in naturalistic terms of such a concept seems to be possible. Besides, what work would this concept do, apart from covering our normative talk in certain situations? I argue that we can account for our use of this normative talk in other ways.

The etiological theories remain strong contestants, however, in the battle for the "true account" of function, as far as the explanatory role of function attribution is concerned. I do not discuss this role in this chapter, nor, therefore, whether there is such a "true account." Millikan's PF version of the etiological theory and Krohs's SD variant of the CR theory are currently the only theories that can account for the attributions of functions to biological items as well as to social items and to artifacts. The way the PF theory attributes functions to artifacts is, however, quite abstract and is in a sense a by-product of the theory's task of naturalizing intentionality, for which it was developed.[25] Boorse claims that his GC theory is a third theory capable of dealing with biological items and artifacts in one stroke, but as it seems to me, the theory faces great difficulties in accounting for the functions of technical artifacts since there are no plausible candidates for the goal-directed systems of which both artifacts and their users would be components. The SD theory currently looks like the best candidate the CR perspective has to offer in order to defend function as a special sort of causal role, which is necessary if the concept is to escape elimination. To gain wider support, I suggest that the theory's central notion of "type fixation" be given a stronger naturalistic footing.

Acknowledgments

I am grateful to the attendees of the workshop on comparative philosophy of technical artifacts and biological organisms at the Konrad Lorenz Institute and in particular to Ulrich Krohs and Tim Lewens for their comments and suggestions.

Notes

1. It may be questioned whether the distinction between answering a "how" question versus answering a "why" question is the best way to characterize the different outlooks of the two approaches. Etiological theories are perhaps better seen as also answering, by the attribution of function, a "how" question: *how* did it come about that this item now exists having this particular feature? The main difference, then, is the adoption of a synchronic (CR) versus a diachronic (PF) point of view as the one required for the purpose of explanation. This is not the place to elaborate this point, however.

2. The way in which I now apply this twofold classification to existing theories may not be as universally accepted as is the primary distinction.

3. The case of nonliving goal-directed systems is complex; societies and cultures do not live but are commonly considered to be able to survive and reproduce. McLaughlin (2001) seems to be of the opinion that the functions that biologists and social scientists assign are just these contributions to the survival and reproduction of organisms and societies. Wouters (2003), however, sees a difference between goal-contribution functions and causal-role functions in biological practice.

4. Throughout this chapter, small letters refer to tokens and capital letters to types in all cases where the distinction is relevant.

5. I am ignoring here any reason why people would want to distinguish *types* from *kinds*.

6. A second difference is that Davies thinks that a sixth clause is necessary in order to have normative statements follow logically from the definition. Further on in this section (7.3) and in section 7.4, I discuss that this is not the way that etiological theories seek to recover the normativity of functions. As the extra clause does not affect both his argument and my rejection of it, I have left it out.

7. Actually the leading PF and SE theorists seem to be confused concerning the importance of the distinction between functional and (roughly) morphological types or kinds. Millikan (1984: 17) says: "It is the "proper function" of a thing that puts it in a biological category," which, I think, is simply false for most, if not all, bio-

logical categories. Neander has since backed down from a similar view in her early work, but it is unclear how far she has moved on. See Neander (2002) for a recent account.

8. A pump that is "broken" by being completely smashed to pieces is usually not considered a pump anymore, nor is any blob at the place where ordinary people have their eyes an eye. Cf. Davies (2001: 177).

9. This at least is what its proponents claim. Millikan's PF theory, however, makes heavy use of the word *Normal*, meant as a technical term, not just in the notion "Normal explanation," which she defines, but also "Normal member," "Normal condition," "Normal property," and "Normal production" (Millikan 1984: 24–34). It is tempting to single out this notion of Normal as grounding the normativity of proper functions, but how would one substantiate this, since on the one hand, there is no way in which it allows the derivation of "should" or "ought to" statements and on the other hand, Millikan presents it as just one of the many concepts by which she builds her naturalistic, i.e., nonnormative, theory of function, and not as the linchpin. Some call this notion of Normal a normative notion (e.g., Rowlands 1997), but then what does this mean, and is it compatible with Millikan's claim to give a naturalistic account of proper function? Although this point deserves to be investigated in more detail, I do not do so here.

10. Some things are not an item's proper function but are necessary accompaniments of its proper function, such that normative statements are in order: it is not a heart's proper function to make a bumping noise, but nevertheless, taking this sort of talk for granted momentarily, a heart ought to make a bumping noise all right.

11. Indeed a theory that would extend that far could be accused of committing the naturalistic fallacy. Since the status of the naturalistic fallacy is very controversial, I do not elaborate this point here. Note, however, that if it is considered legitimate to say that x ought to do B *because* it is the proper function of x to do B, this "because" must be read constitutively, similar to the way it has to be read in "This pawn cannot go to that field because it is against the rules of chess for it to go there." But if the relation between proper function and "ought to" is that intimate, then any naturalization of proper function must run into problems.

12. In section 7.7 I give an account that has "ought to" statements as central.

13. How little thought seems to be given to what we precisely express by these statements can be seen, to take just one example, by the following quotation from McLaughlin (2001: 4): "Does the attribution of function presuppose a valuation of the end towards which it is a means—at least in the sense that the function bearer is supposed to perform its function?" It is precisely the question what a statement like "x is supposed to perform F" has to do with a normative notion like value, but McLaughlin seems little interested in posing it.

14. The two forms are not completely independent. One hardly can be morally justified in expecting something to happen if on epistemic grounds one believes it to be extremely improbable or impossible. Promises of impossible things should be rejected. This may lead one to question the equation of being morally justified in expecting something to be the case with having a right to it that this something is the case. See also note 22.

15. A more accurate discussion would introduce the distinction between manifest and latent function. Supporting social cohesion is the latent function of the Hopi rain dance, the one that social scientists are interested in. Additionally the rain dance may be considered as an institution designed and implemented for the purpose of promoting rainfall, which is its manifest function. We do not as easily say that this institution functions well or poorly as we do in the case of a country's legal system, partially because it may be doubted whether the rain dance was ever designed in the way legal systems are.

16. Although it is rarely done, I think that one can sometimes be justified in applying the notion of "malfunction" also to cases where someone uses an object incidentally for some purpose or other. This object may be an artifact, but one that was designed to be used for a different purpose than it is used for on this occasion. In such cases, the user of the object can be seen as the designer of his or her own private artifact.

17. Here it is assumed that wherever the dark variety is frequent, it is because of its superior performance in being inconspicuous to its predators. It has been challenged whether Kettlewell's famous experiments from the 1950s indeed showed this to be the case (see Rudge 1999).

18. Perhaps we find this less obvious in the case of an ideal liver than in the case of an ideal mating dance, which again testifies to the special position of bodily organs in biology as soon as normativity is an issue. Would an ideal mating dance be one that would end in copulation in virtually all cases it was performed? That would probably not be good for the animal itself, and perhaps also not good for the species.

19. Raz (1975), however, uses "ought to" in a weaker sense. He says that "p ought to do Z" even when p just has one reason to do Z.

20. An *F*-er is a device designed to have the function *F*. *F*-ing is the act of using the device for its intended function *F*. So an *F*-er is designed for the purpose of *F*-ing, or to be used for *F*-ing.

21. This is how I interpret Dancy's suggestion in Dancy (2006b: 60–61).

22. Note that when an object is a malfunctioning knife, this would ipso facto be a reason not to believe that one has a reason to use it for cutting. The question of what one is justified to believe therefore cannot be settled by the two facts that an object is known to be designed as a knife and that it is malfunctioning, since we then merely have two reasons canceling each other. Note also in section 7.4 that where a distinction is made between what one is rationally justified in believing and what one is morally justified in believing, the notion of being rationally justified in believing something can be analyzed in terms of the various reasons one has for believing it, but this may not work for the notion of being morally justified in believing something.

23. Cf. Hardcastle (2002: 149): "We prefer to think in organismic terms."

24. Cf. the following quotation from Lewens (2004: 108), who, as an adept of a CR-type of function theory, denies a diseased pair of kidneys a function and explains our insistence on its malfunctioning rather than lacking a function "because they persist in failing to provide a benefit that we desire and that we have grown to expect." Here there is a reference to expectations but also desires that we have concerning the behavior of our organs, similar to expectations of and desires for the proper functioning of our tools. The next step is the suggestion that we should consider repairing or replacing our failing kidneys. See also Lewens (2007) for a view that is similar to mine concerning several of the issues addressed in this chapter.

25. It has been argued (Vermaas and Houkes 2003) that the PF theory cannot account for all aspects of artifact functions. I do not find the argument convincing and do not regard the difficulty as one central to the debate on function.

References

Boorse, C. (2002). A rebuttal on functions. In: *Functions: New Essays in the Philosophy of Psychology and Biology* (Ariew, A., Cummins, R., Perlman, M., eds.), 63–112. Oxford and New York: Oxford University Press.

Cummins, R. (1975). Functional analysis. *Journal of Philosophy, 72:* 741–765.

Dancy, J. (2006a). Nonnaturalism. In: *The Oxford Handbook of Ethical Theory* (Copp, D., ed.), 122–145. New York: Oxford University Press.

Dancy, J. (2006b). The thing to use. *Studies in History and Philosophy of Science, 37:* 58–61.

Davies, P. S. (2001). *Norms of Nature: Naturalism and the Nature of Functions.* Cambridge, Mass.: The MIT Press.

Franssen, M. (2006). The normativity of artefacts. *Studies in History and Philosophy of Science, 37:* 42–57.

Hardcastle, V. G. (2002). On the normativity of functions. In: *Functions: New Essays in the Philosophy of Psychology and Biology* (Ariew, A., Cummins, R., Perlman, M., eds.), 144–156. Oxford: Oxford University Press.

Krohs, U. (2004). *Eine Theorie biologischer Theorien: Status und Gehalt von Funktionsaussagen und informationstheoretischen Modellen.* Berlin: Springer.

Krohs, U. (forthcoming). Functions as based on a concept of general design. *Synthese.*

Lewens, T. (2004). *Organisms and Artifacts: Design in Nature and Elsewhere.* Cambridge, Mass.: The MIT Press.

Lewens, T. (2007). Functions. In: *Philosophy of Biology: Handbook of the Philosophy of Science, vol. 3* (Matthen, M., Stephens, C., eds.), 525–547. Amsterdam: North-Holland.

McLaughlin, P. (2001). *What Functions Explain: Functional Explanation and Self-Reproducing Systems.* Cambridge: Cambridge University Press.

Millikan, R. G. (1984). *Language, Thought, and Other Biological Categories: New Foundations for Realism.* Cambridge, Mass.: The MIT Press.

Neander, K. (1991). Functions as selected effects: the conceptual analyst's defense. *Philosophy of Science, 58:* 168–184.

Neander, K. (2002). Types of traits: the importance of functional homologues. In: *Functions: New Essays in the Philosophy of Psychology and Biology* (Ariew, A., Cummins, R., Perlman, M., eds.), 390–415. Oxford: Oxford University Press.

Raz, J. (1975). *Practical Reason and Norms.* London: Hutchinson.

Rowlands, M. (1997). Teleological semantics. *Mind, 106:* 279–303.

Rudge, D. W. (1999). Taking the peppered moth with a grain of salt. *Biology and Philosophy, 14:* 9–37.

Thomasson, A. L. (2007). Artifacts and human concepts. In: *Creations of the Mind: Essays on Artifacts and their Representation* (Margolis, E., Laurence, S., eds.), 52-73. Oxford: Oxford University Press.

Vermaas, P. E., and Houkes, W. (2003). Ascribing functions to technical artefacts: a challenge to etiological accounts of function. *British Journal for the Philosophy of Science, 54:* 261–289.

Wouters, A. G. (2003). Four notions of biological function. *Studies in History and Philosophy of Biological and Biomedical Sciences, 34:* 633–668.

Wright, L. (1973). Functions. *Philosophical Review, 82:* 139–168.

8 Conceptual Conservatism: The Case of Normative Functions

Paul Sheldon Davies

What moves us to theorize about purposes in living things? Why are we concerned to theorize about normative functions that allegedly belong to parts of plants and animals? We appear to have the intuition that these things are "supposed to" fulfill certain functions. So when we try to explicate this intuition, what are we up to?

I am not asking why we theorize about functions at all. The question is why we theorize about the alleged *normativity* of functional properties. What are we doing when we try to preserve the intuition that functional traits have the property of "being supposed to" perform specific functional tasks even when, due to physical incapacitation, they cannot perform those tasks? The theory of systemic functions, first articulated by Robert Cummins and later developed by Ron Amundson and George Lauder in the context of biology, is a compelling theory of functions that eschews the imputation of normative properties.[1] The theory of systemic functions therefore is not the target of my discussion. I have in mind rather the theory of so-called proper—that is, normative—functions developed by a small army of theorists over the past thirty years or so.[2] The question, then, is what are these theorists up to when they try to preserve the admittedly powerful intuition that some functional traits are supposed to do certain things?

My answer is that the urge to theorize about purposes in nature is a product of the fact that we are conceptual conservatives regarding the concept "purpose." One source of our conservatism is the long, deep roots that this concept has in our intellectual and theological ancestry. A second source is our own psychology, the cognitive and affective capacities that cause us to apply the concept "purpose." We are, by virtue of our cultural history and our psychological constitution, prone to see or feel purposiveness where none exists, and the force with which we see and feel living things as purposive convinces us that there is something important there about which we must theorize. The evident power of this illusion illustrates the retarding effects that conceptual conservatism has on our attempts to know the world.

The specific aim of this chapter is to flesh out and defend this answer to my opening question. My larger aim, which extends well beyond this chapter, is to generalize from the retarding effects of "purpose" and suggest that, with respect to a wide range of similarly dubious concepts—especially concepts in terms of which we understand ourselves—we need to abandon our commitment to conceptual conservatism.[3] This is to call for reform in our basic orientation toward philosophical reflection. The call is justified insofar as conceptual conservatism is antithetical to progress in knowledge.

8.1 Progress in Knowledge

I assume that, as a matter of historical fact, human knowledge has grown in the past few centuries like never before. There has been progress in human knowledge even though there is nothing intrinsically progressive about it—even though the growth in human knowledge is the product of any number of happy accidents. Not that the growth of knowledge is cumulative, linear, or otherwise tidy, but rather that we as a collective know more about reality today than at any other time in the history of life on earth. Not that our knowledge is infallible or even particularly impressive when compared to our ignorance, but rather that we understand and can control the world with unprecedented success. We may not be any wiser in employing what we know—it is an interesting question whether there has been progress in politics comparable to progress in knowledge, or whether growth in one depends on growth in the other—but that there is progress in human knowledge since the rise of modern science is hard to contest.

I also assume that progress in human knowledge is unavoidably destructive of what has gone before. As old theories or old methods of inquiry give way to new ones, we are forced to put a good many things behind us. By studying the history of science—by reflecting on what contributed to the growth of human knowledge in the recent past—and also by studying the psychological capacities that underwrite human inquiry and the infirmities that thwart it, we have discovered that some methods of inquiry are more effective than others. We have discovered, in particular, that some strategies and expectations make us more effective at generating accurate predictions and informative, fruitful explanations. We make progress, then, by accepting new theories in place of old theories, but also by putting behind us methods of inquiry that have proven relatively fruitless.

What, then, of contemporary philosophical inquiry? Does it contribute to the growth of human knowledge? The answer, I think, is that to a surprising extent it does not. A good deal of contemporary philosophical inquiry is deeply conservative by its very methods and, in consequence, antithetical to progress in knowledge. To see this, consider two questions. First, how do philosophers gauge their own progress? What qualifies as progress and what qualifies as lack of progress in philosophical reflection? The answers can be gleaned by observing what philosophers say and do. And as most of us practice it today, the overarching goal of philosophical reflection is to identify concepts of apparent importance—conceptual categories that bear most intimately on how we understand ourselves or our place in the cosmos—and then try to preserve those concepts by integrating them with the burgeoning knowledge provided by the sciences. This is to aim at a "wide reflective equilibrium" of one sort or another, to produce an internal balancing of apparently important concepts with one another and with the latest findings from the natural and social sciences.

There are of course different ways of trying to integrate our traditional concepts with one another and with our evolving scientific knowledge, but that is just to say that being a conceptual conservative can be a matter of degree. A mad-dog, foaming-at-the-mouth conservative is perhaps best described as a conceptual imperialist, as committed to the view that some concepts have a kind of primacy, a kind of dominion, over all other concepts and all methods of inquiry, and that, in consequence, such concepts must be retained at any cost. A less rabid conservative is committed only to preserving certain concepts as far as possible consistent with scientific findings.

For present purposes these differences do not matter. My focus is on the core commitment of all conservatives, not the differences in degree among them. I focus on the deeply rooted assumption that the aim of philosophical reflection is the integration of traditional, humanistic concepts with the concepts and claims of our best sciences, for that assumption, it seems to me, conflicts with the pursuit of progress in knowledge. If I am right about this, then the answer to my second question is not flattering. The question is, under what conditions does progress in contemporary philosophical reflection contribute to progress in human knowledge? The answer is that apart from the dumb luck that serendipitous conditions sometimes produce, contemporary philosophical reflection, insofar as it honors the core commitment of all conservatives, has little to add to the pursuit of human knowledge. That, at any rate, is what I want to argue.

8.2   Elements of Reform

When viewed from a distance, the preservation of traditional concepts may appear a fine undertaking. It may appear noble to try to marry our humanistic concepts with our scientific knowledge. How else might we achieve an integrated or unified view of ourselves and the world? The quest for integration and unification is a grand ambition that hearkens back to the large-canvass works of the great theologians including, for example, Reinhold Niebuhr's *The Nature and Destiny of Man*. But there is one rather vexing problem. As human knowledge progresses, it is increasingly clear that the humanistic concepts bequeathed to us by our intellectual forebears cannot be sustained. They cannot play the role they used to play in the framing of our intellectual tasks. This is because the contest between our humanistic and scientific worldviews is no longer a struggle between equally powerful antagonists. The balance has tipped away from the authority of our humanistic concepts. And the cause of this change is clear: the growth of scientific knowledge.

The traditional concept "purpose" illustrates the loss of authority in our humanistic concepts. The illustration rests upon progress in (a) knowledge of our cultural history and especially (b) knowledge of our psychological constitution. To illustrate this loss of confidence, I introduce three directives for philosophical inquiry that elaborate on our knowledge in (a) and (b). Although I cannot defend these directives here—though see Davies (2009)—I can point out that they are utterly banal and that such banality ought to minimize disagreement between competing parties. I also wish to emphasize that, despite being relatively uncontroversial, these directives affect the way we frame our inquiries into a variety of issues, including the alleged purposiveness of living organisms. The larger view, then, is that the adoption of these directives is one reform that we philosophers must undertake if we wish to make our reflections relevant to the pursuit of knowledge.

8.2.1   History of Culture

Among the concepts bequeathed to us by our intellectual ancestors, some have proven themselves to be dubious by descent. A concept is dubious by descent if it descends to us from a worldview that we, in light of our growing knowledge, no longer regard as true or promising, and if it has not been vindicated by progress in some well-developed scientific theory. The traditional concept "free will" as elaborated by Chisholm (1964) and others is dubious in this sense, since it derives mainly from our theological ancestry and has not been vindicated by any contemporary scientific theory.[4] This is not to assume that all theological claims are false, only that none has shown itself relevant to the enormous progress in knowledge since the rise of modern science. The increasing irrelevance of theological concepts and claims to progress in science is as near a brute fact as we are likely to find in the study of history. It thus is rational to frame our intellectual problems and solutions in such a way that we leave out or neutralize the potentially retarding effects of dubious concepts. We need, that is, the following directive for inquiry:

(D)   For any concept dubious by descent, do not make it a condition of adequacy on our philosophical theorizing that we preserve or otherwise "save" that concept; rather, bracket the concept with the expectation that it will be explained away or vindicated as our knowledge of the world progresses.

At minimum, we should not assume that a successful theory of the relevant domain must somehow preserve or account for concepts that are dubious by descent. And as I now try to demonstrate, the genealogy of our concept of "normative functions" is sufficiently dubious to warrant the application of the directive in (D).

We know that the recent cultural roots of "purpose" and other related concepts trace back to the argument from design for the existence of the Judeo-Christian God. This argument was enormously influential in England near the turn of the nineteenth century, thanks in part to the writings of William Paley (which were formative for the young Charles Darwin). For Paley, the apparent design in nature could not be a product of blind causal mechanism but had to originate from the creative efforts of an intelligent God (Paley 1802). At more or less the same time in Germany, an idea with a comparable theological load was defended by Immanuel Kant and several of his Romantic successors. The idea, which migrated to France via the work of Georges Cuvier, concerns the existence of a prior, abstract archetype that serves as the font of all living forms.[5] As Johann Wolfgang von Goethe observed, there appear deep commonalities in the structure of all (or most) animals, and a distinct set of commonalities in all plants. These common structures are best explained, according to Goethe, by positing a very small number of antecedent types containing two essential ingredients: the basic form, a kind of template, that all descendents would embody, and an intrinsic creative drive to perpetuate that form (Richter 1985–1998). And these originating and motivating types, while allegedly efficacious in the natural order, could not have originated in nature as circumscribed by the Newtonian view, since the resources of Newton's mechanics were too paltry to determine the self-perpetuating nature of living things.

Consider, for example, the view of Johann Friedrich Blumenbach, an illustrious contemporary to Kant. Like Kant and his successors, Blumenbach was intent on understanding scientifically the apparent purposiveness of living things. There was consensus at that time that Newtonian mechanics could not explain the most striking features of living things, namely, reproduction, growth, and regeneration. The laws of motion and gravity underdetermined the highly specific purposes exemplified by the parts of plants and animals. Blumenbach proposed to fill this gap by positing a nonmechanical, form-giving power, a precursor to Goethe's archetype. He did so, moreover, on the basis of what appeared to be a compelling line of Newtonian reasoning. Newton had insisted that it was rational to accept the existence of gravity as a fundamental feature of the universe on the grounds that it explains so much of the phenomena we observe, even though he had no account of its mechanical origins or constituents. Likewise, Blumenbach asserted that it is rational to accept an archetypal, form-giving power on the grounds that without it we could not explain what is most distinctive of living things (Blumenbach 1781). Blumenbach could not explain the origins or constituents of his form-giving power in mechanistic terms— indeed he seems to have believed that this power exists prior to and somehow animates all the mechanisms of life—but he nevertheless thought we are justified in positing such a power, since otherwise we would be unable to explain the capacity of living things to reproduce, grow, and regenerate.

There is, then, an analogy between the views of Blumenbach and Paley, for both theorists, in order to explain the apparent purposiveness of living organisms, posit a centralized source and indeed an agentlike source of creative power that is difficult to square with a naturalistic worldview. That is, both theories, though claiming to explain the purposiveness of living things, offer very little in the way of explanatory power. This is clearest in Paley's view, where all the theoretical difficulties are dumped into the lap of the Judeo-Christian God. Paley offers no real explanation of the emergence and perpetuation of living forms, except to say that they must come from a deity clever enough to design and manufacture such forms. Blumenbach's view, though less transparent than Paley's, offers more or less the same "explanation." The emergence and perpetuation of living forms, according to Blumenbach, comes from a nonmechanical power potent enough to cause the perpetuation of living forms, though nothing about the actual workings of this power is ever revealed to us.

The claim that Blumenbach's theory is analogous in this way to Paley's theory conflicts with the views of some contemporary scholars. Peter McLaughlin (1990), for example, claims that Blumenbach's postulation of a nonmechanical life force was not a step outside the natural order—and thus utterly disanalogous to Paley's appeal to God—but rather a way of expanding and enriching our view of the natural order. The natural order, as conceptualized by our late-eighteenth-century predecessors, was constituted mainly from the apparently irreducible elements of Newton's mechanics. Blumenbach's view, according to McLaughlin, seeks to expand our view of the natural order by including a fundamental and irreducible formative power. This formative power is irrelevant to the physics of nonliving things, to be sure, where Newton's mechanics carry the explanatory burden, but it is essential to our ability to produce a reductive, scientific explanation of the perpetuation and purposiveness of living forms, where Newton's mechanics are not enough. And it is precisely the analogy to Newton's claim concerning gravity that Blumenbach gives in support of his formative power.[6]

McLaughlin may be right that this is how Blumenbach *intended* us to interpret his postulation of a formative power. I am not convinced, however, that this really *is* Blumenbach's view, since I am not convinced that his positing a formative power was defensible even relative to the standards of his own day.[7] There is, after all, considerable distance between Newton's argument for gravity and Blumenbach's argument for a formative power. Newton insists we should accept that gravity is real because 1) doing so helps explain the behavior of all observable objects by positing a property of attraction that is relatively simple, and 2) this relatively simple property of attraction may, with further progress in knowledge, yield to a mechanistic explanation. The content of Blumenbach's proposal is quite different, for neither (1) nor (2) are true of it.

Begin with (2). Blumenbach is explicit that his formative power, whatever else it might be, is entirely nonmechanical. This is no small difference. Newton was not giving us a fully fleshed account of what gravity is; he was admitting that, though he had not yet discovered the source of gravity, he expected that it would someday yield to a mechanistic explanation. (It is true, of course, that Newton also admitted that we would need to appeal to the influence of God if the search for a mechanistic explanation failed. But the appeal to God is a last resort.) Blumenbach, however, was not expressing his or our ignorance; he was instead *positively* asserting that his formative power was *not* mechanical. He did so, moreover, without giving us any idea what a nonmechanical formative power might *be*. We have some idea what the realm of mechanics comprises because we are told, for example, that the primary constituents of matter are indivisible particles, and *particle* is given at least some explication. By contrast, Blumenbach gives nothing comparable to help us understand the realm of the nonmechanical. We are told only what the formative power is *not*.

What, then, are we to make, relative to the standards of justification of Blumenbach's own day, of an alleged formative power that is explicitly *nonmechanical* and *unexplicated*? Bearing in mind that the natural order as conceived in the late eighteenth century was *constituted* by the mechanical view, we should conclude that Blumenbach's postulation was tantamount to the bald assertion of *a mystery*. Newton was not offering us a mystery; he was confessing his ignorance and placing his bets on further inquiry. Blumenbach, by contrast, was asking that we accept an unexplicated force from the silent realm of the nonmechanical. On his view, the nature of life is explained by positing a hermetically sealed mystery. And this makes it difficult to see how Blumenbach's view can be construed properly as an "expansion" of an emerging naturalistic worldview epitomized by Newtonian methods.

Now consider (1). Part of the power of Newton's argument is indeed the fact that gravity, whatever its source, is a relatively simple property of attraction. It is simple in the same sense that his laws of motion are simple. Newton's laws quantify over properties of matter that are universal and irreducible; they apply without exception to the simplest constituents of matter. The properties posited by these laws, then, are simple relative to the full range of behavior that they explain at the level of compound objects. Likewise, the property of gravitational attraction is a universal and irreducible property of the simplest constituents of matter, yet it helps explain the behavior of all observable objects. It is in this sense that gravity is relatively simple. The same, however, cannot be said of Blumenbach's formative power. This too is no small difference. The apparent purposiveness of living things, as we have seen, is introduced by Blumenbach to fill the explanatory gap left by Newton's mechanics. But consider how the gap is filled. Not by specifying additional mechanisms, since the formative power is nonmechanical. The gap is filled instead by positing a formative power that comprises the following powers: it *motivates* the mechanical parts of cells, tissues, organs, and so forth to metabolize, reproduce, regenerate, et cetera, and it *imposes a template* of the species' form on all the processes of growth and reproduction. This is to attribute to the formative power a broad range of remarkable powers. They are so remarkable, in fact, because they *resemble* the phenomena they are supposed to explain! Growth, for example, is explained by appeal to a nonmechanical and unexplained power for growth! Just as Paley dumps the theoretical difficulties involved in explaining the perpetuation of living forms into the lap of the divine, Blumenbach dumps them into the lap of an unexplicated, nonmechanical power—a hermetically sealed mystery. The only difference, so far as I can see, between the two thinkers is that Paley cloaks his mystery in the shroud of God while Blumenbach cloaks his in the shroud of Newton's argument for gravity.

But suppose I am mistaken. Suppose McLaughlin is right that Blumenbach's view, relative to the standards of his own day, did indeed expand their view of the natural order. It nevertheless remains that Blumenbach's formative power, relative to what is known today, belongs to a worldview we no longer regard as true. Like Paley's God, Blumenbach's formative power serves the theoretical function of a surrogate agent, that is, a *center of command and control* with respect to the perpetuation of living forms. Paley's God, after all, creates and perpetuates the forms of life by virtue of the intentional capacities of any creative agent, namely beliefs, desires, and intentions. But desires and beliefs are just beneath the surface of Blumenbach's formative drive. This drive, as I have said, is a motivating source. It provides the creative urge that drives living things to grow, reproduce, and so forth, much like the desires or the will of Paley's God. This drive is also a form-giving source. It provides the species-specific architecture that directs development, and in that regard it plays the same role that beliefs play in the mind of a creating God. And all this, of course, is something we no longer accept as true. According to contemporary theories of development, reproduction, and evolution, the forms of life that exist are not the products of any center of command and control but are rather the temporary and evolving products of a host of decentralized causal factors.[8] So even if Blumenbach's appeal to Newton's argument for gravity is more credible than I suggest, there nonetheless *is* an equally powerful analogy between Blumenbach and Paley. And that analogy, relative to today's sciences, places both views beyond the pale of a naturalistic worldview.

Kant of course famously refused to commit himself to any ontological claim concerning the source of these originating and motivating types. He insisted instead that such sources are beyond the reach of human cognition and thus that we must settle for the regulative claim that in order to investigate living things we must conceptualize them "as if" created by a superlative form of intelligence.[9] On this view, human knowledge in biology rests upon our seeing the living in terms that, for all we can know, do not truly apply—an acceptable cost in light of Kant's critique of human knowledge.[10] Some of Kant's contemporaries and successors, however, were not so modest.[11] The most likely source of Goethe's abstract, archetypal forms, for example, is presumably the creativity of the divine. Not that Goethe's God acts in the world by trumping laws of nature but rather that the deity so structured the world that living things were formed and continue to be perpetuated by a power that operates from within every living being. And that is to say that several of Kant's contemporaries and successors, like Paley and Blumenbach, were ultimately driven outside the natural order in their attempt to understand the living.

These, then, are some of the most obvious historical roots of our concept "purpose."[12] And it is telling that they descend to us from a range of theological worldviews we no longer regard as true or promising. This, as I say, is not to assume that all theological claims are false. It is to point to the undeniable fact that progress in modern biology has been marked by the growing irrelevance of theological concepts to our knowledge of living organisms. Biologists today can explain all the phenomena—all of them—that inspired Paley, Blumenbach, Kant, and Kant's successors to posit a surrogate agent, a center of command and control, that is difficult to square with a naturalistic point of view. The appeal to a form-giving agent that operates outside the mechanical realm was perhaps understandable, perhaps robustly scientific, in the eighteenth and early nineteenth centuries, prior to the publication of work by Darwin and Wallace. But it is utterly unscientific in light of progress in science during the twentieth century (more on this presently). Instead of being a concept we ought to preserve, "design" is a concept that is dubious by virtue of cultural descent.

The question then is, if the worldview from which "purpose" descends to us no longer plays a role in our best developed scientific theories, why should philosophers—especially philosophers familiar with these historical changes—nevertheless aspire to preserve this notion? This concept is dubious by virtue of descent from a worldview we no longer regard as true or promising; this is something we know or something we believe on the basis of excellent historical evidence. So why, in light of its dubious genealogy, are we so plainly conservative regarding this concept? Is it not the case that, contrary to the conservative orientation in contemporary philosophy, we now know too much about our own history to continue letting this concept serve as a parameter of our intellectual tasks?

8.2.2 History of Science

A second lesson learned from the history of science is that we make progress in understanding natural systems as we analyze inward and synthesize laterally. High-level systemic capacities—capacities, for example, such as reproduction, growth, and regeneration—are rendered explicable and predictable as we analyze into relevant low-level systems and identify components and relations among components that instantiate these capacities. This is what occurred throughout the twentieth century in biological theory. As biologists succeeded in analyzing the mechanisms of reproduction, metabolism, and the like, the underdetermination theses of Paley, Blumenbach, and Kant became increasingly implausible, as did the more general theological presuppositions of their worldview. Moreover, any proposed taxonomy of low-level mechanisms is constrained by synthesizing laterally. We confirm or disconfirm a given taxonomy, at least in part, as we look for coherence across well-confirmed theories in related areas of inquiry.

There is a wealth of case studies describing the central importance of inward analysis and lateral synthesis in the growth of scientific knowledge.[13] And one crucial lesson we learn from these case studies is that as our knowledge of natural systems progresses, the relevant conceptual categories evolve. The concepts in terms of which we understand natural systems—their high-level capacities and their low-level mechanisms and relations—are altered as our knowledge grows. Lessons from the history of science therefore make it rational to expect substantive alteration in our conceptual categories as we analyze inward and synthesize laterally. This expectation can be expressed as a directive for inquiry:

(E)   As we make progress understanding natural systems—as we analyze inward and synthesize laterally—expect that the concepts in terms of which we understand high-level systemic capacities will be altered or eliminated.

The history of human knowledge exhibits a very general pattern. For natural systems we understand poorly, the discovery of low-level mechanisms implementing high-level capacities often forces us to revise or discard the concepts in terms of which we understand the high-level capacities. The evolution of our concept "natural purpose" is illustrative. We should, in light of (E), calibrate our hunches, our intuitions, and our expectations so that the potential for conceptual alteration becomes our default orientation toward inquiry.

8.2.3   Human Psychology

Though lessons from the history of science are invaluable for understanding ourselves, I also want to focus on what we are learning about our psychological constitution, for some of these lessons bear directly on the felt importance of the concepts "design" and "purpose." Consider, for example, the "theory of mind." The psychologist Alan Leslie hypothesizes that humans develop at a very early age a set of capacities that cause us to "see" certain objects as endowed with mental states. One capacity is a selective-attention mechanism attuned to objects that exhibit characteristic features of persons or, more generally, of minded agents. Another is a set of conceptual categories akin to "belief," "desire," "intention," and more. The basic idea is that the attention mechanism causes us to pay preferential attention to things that might be agents and also triggers the application of mental concepts, causing us to "see" those objects as endowed with beliefs, desires, and more (Leslie 2000). Evidence for this hypothesis draws on work in several areas, including the study of autism.

Autistic children suffer deficits, some quite severe, in social intelligence. They appear far less capable than nonautistic children of seeing persons as mental agents. And as Leslie describes, autistic children as old as twelve regularly fail the false belief test—a test designed to detect the ability to attribute to another agent a mental state clearly distinct from one's own. In contrast, nonautistic children as young as four years old regularly pass the test, as do Down syndrome children as young as ten years old. And Leslie is careful to point out that the deficits in social intelligence associated with autism do not appear to be the effects of general cognitive deficits, as might be caused, for example, by mental retardation. The fact that 25 percent of autistic children are not mentally retarded but still suffer impairments in social skills and language is surely relevant. So too are experiments suggesting that autistic children, though unable to conceptualize persons as minded agents, nevertheless possess the cognitive sophistication required to comprehend the contents of nonmental representations, namely photographs (Leslie and Thaiss 1992).

All this suggests that we are natural-born detectors of objects that exemplify features characteristic of minded agents. We are cognitively and affectively outfitted very early in development to preferentially attend to certain sorts of stimuli and, in response to those stimuli, conceptualize the relevant objects as cognitively and affectively endowed. This is so much a part of our basic orientation, of how we cannot help but see the world, that only rarely are we in a position to notice it in ourselves. We notice it when confronted, for example, with the devastating deficits of severe autism.

The theory of mind theory may help us understand the role of "design" and "purpose" in our psychology. The suggestion is that the set of mental categories posited by the theory of mind theory might be the mechanisms that also apply the concepts "design" and "purpose." After all, to see an object as a mental agent is to conceptualize it as acting on some intention, as moving toward some goal, and that, in the usual course of things, is to see the agent as endowed with various means toward those goals, with strategies for acting that are purposive or functional. As Deborah Kelemen suggests, our natural disposition to see the parts of living things as purposive may be a by-product of the disposition to see various objects in the worlds as minded (Kelemen 2004). And just as we are fooled by our own capacities—sometimes we feel that a mindless object (a car, computer, or caterpillar) is a quasi-agent to be reckoned with—so, too, we sometimes feel certain objects are purposive even when they are not. This, at any rate, is a feature of our minds that potentially distorts the way we see or feel certain things in the world without our noticing it.

Now even if you have a healthy skepticism toward contemporary cognitive psychology, the point here is significant. It is plausible that some of our most basic cognitive and affective capacities enable us to anticipate and navigate our environments under a limited range of conditions. It is also plausible that we find ourselves, often enough, in conditions outside those limits. The resulting false positives may be, from an evolutionary point of view, a cost worth paying. The false positives may be a nuisance or even deleterious in some instances, but the presumption is that being possessed of different psychological structures would be far worse. However, from the point of view of trying to acquire knowledge, the false positives are intolerable. They are intolerable because they systematically lead us away from the truth by virtue of the effects of our own psychology, effects we tend not to notice because they operate well beneath the level of conscious awareness. If so, then progress in knowledge is limited by the retarding effects of our own psychological architecture or, less pessimistically, by the bounds of our best efforts to creatively think or feel our way around our own structural limitations.

One way false positives occur is when there exists a constitutional conflict between cognitive systems. Consider Daniel Wegner's hypothesis that the feeling of having freely willed an action is produced by a system distinct from the actual lower-level processes that cause the action. The emotion of authorship—the felt sense that one's own intention was the direct cause of one's action—is produced by a system attuned to specific conscious thoughts or perceptions. In particular, it is attuned to the co-occurrence of conscious thoughts about doing action A and subsequent conscious perceptions of oneself doing A (Wegner 2002). And yet, as Wegner's experiments suggest, the mechanisms that in fact cause us to perform action A belong to a distinct system of the mind, a system attuned to various nonconscious inputs.[14] If that is correct, we must face the possibility that the feeling of authorship sometimes causes us to conceptualize our selves as having acted freely when in fact we did not. The concept "free action," insofar as it is susceptible to such false positives, qualifies as dubious by psychological role.

The same may be true of the concepts "purpose" and "design." It is, at the very least, a plausible hypothesis that we are endowed with entrenched capacities that control the application of concepts clustering around the notions "purposive" or "end-directed," and that these capacities generate analogous constitutional conflicts. It thus is important that we become self-conscious about the conditions under which such concepts may lead us astray. We need a directive for inquiry to the following effect:

(P)   For any concept dubious by psychological role, do not make it a condition of adequacy on our philosophical theorizing that we preserve or otherwise "save" that concept; rather, require that we identify the conditions (if any) under which the concept is correctly applied and withhold antecedent authority from that concept under all other conditions.

This, more than (D), is difficult to implement. It is difficult because the capacities engage nonconsciously, because the capacities are entrenched and constitutive of the way we orient ourselves to the world, and because the concepts, even when they reach the level of conscious reflection, are central to how we portray ourselves as agents. Implementing (P) will require that we devise strategies for thinking our way around certain naturally distorting dispositions of thought and feeling.

We must be prepared, moreover, to discover that we are ill equipped to withhold certain concepts from our theoretical endeavors, our concept of "purpose" included. We must be prepared to discover that we cannot help but employ certain concepts whenever we try to understand the world. But we also must exercise due caution toward claims concerning the allegedly essential constituents of human thought. Some concepts may indeed be essential to our capacity to grasp the world, but the history of theology and philosophy (Kant's views included) should make us skeptical that we have discovered and correctly described any such constituents. We should insist, at minimum, that we employ our most reliable methods of inquiry when trying to discover the elements of cognition. And this means employing the directives in (D), (P), and especially (E).

8.3 Normative Functions

Armed with the directives for inquiry just described, I return to my opening question: Why are we concerned to theorize about the apparent purposiveness of living things? Here are two possible answers:

1. We are concerned to develop a theory of natural purposes because natural purposes in fact exist; we want to understand the origin and nature of a real feature of the natural world.

2. Natural purposes do not exist but we nevertheless are concerned to develop a theory of natural purposes because we are psychologically constituted in such a way that we cannot help but conceptualize the parts of plants and animals at least metaphorically in the same way we conceptualize artifacts.

The first answer asserts that there is nothing puzzling about our concern to theorize about normative functions since such norms are clearly part of the natural world. The second answer says that it is a mistake to attribute such norms to the natural world, but that, because of the structure of our psychology, the attribution of these norms is indispensable to the very existence of evolutionary theory. Both answers have been defended in the recent literature. Yet neither is defensible from the perspective provided by my directives for inquiry. The only way to defend these answers is to first defend something indefensible, namely, the commitment to conceptual conservatism.

Begin with the second answer. Michael Ruse has recently argued that although Darwin effectively destroyed our belief in any literal design in the living realm, we nevertheless cannot do without the metaphor of design. This is a thesis concerning human psychology. We are condemned to seeing the parts of plants and animals "as if" created by an intelligent mind for some specific purpose. And it is a good thing, according to Ruse, that we cannot help but see the living as purposive, since evolutionary biology would otherwise be impossible for us. We would lose the capacity, he claims, to ask "why" and "what for" questions. We would be helpless to initiate the search for adaptationist explanations: "Without the metaphor, the science [of evolutionary biology, at minimum] would grind to a halt, if indeed it even got started" (Ruse 2003, 285). Ruse defends this thesis in part by appealing to a handful of case studies, including the double lens of the trilobite, but mostly by appealing to the long historical roots of the concept "purpose." Indeed the bulk of his discussion traces the genealogy of this concept from Plato and Aristotle through Kant, Paley, and Darwin. And Ruse is clear that the stubborn persistence of the concept, evidenced by its long historical roots, is supposed to lead us to the conclusion that "purpose" is indeed a concept we cannot do without.

If, however, my directives for inquiry are plausible, then we cannot endorse Ruse's metaphor. We should agree with Ruse on two points, namely, that we humans are inclined to see living things as purposive and that living things are not, in fact, purposive. But also, contrary to Ruse, we ought to resist the urge to conceptualize the parts of plants and animals "as if" created by an intelligent mind. That, after all, is the lesson learned from the directives in (D) and (P). The lesson learned from (D) is that the long historical roots to which Ruse appeals in support of his metaphor should instead lead us to reject any such metaphor. We should withhold antecedent authority to our concept "purpose" precisely *because* it descends to us from a worldview (or a set of worldviews) we no longer regard as true or promising. The lesson learned from (P) is that the felt intuitive force of our concept "purpose" is no grounds for trying to preserve the concept. To the contrary, we should withhold antecedent authority precisely *because* we have discovered that the concept has undue force in our psychology. Instead of continuing, therefore, to conceptualize the parts of plants and animals in terms of normative functions, we should, in light of what we know about our intellectual history and our cognitive capacities, withhold antecedent authority to this concept. We should also try to extend our current concepts or create new ones, in an effort to think or feel differently about living things, that we might contribute to progress in knowledge.

The basic point against Ruse is in fact on the surface of his own discussion. Darwin killed literal design in the realm of the living; that, according to Ruse, is something we know to be true. We have therefore the cognitive capacity to see that living things are devoid of purposes. At the same time, we know from the study of our own psychology that we are inclined to see purposes among the parts of plants and animals even though none exists. We have therefore the cognitive inclination to see part of the world falsely with regard to "purpose." And all this is to say that one part of our psychology enables us to see that another part tends to lead us astray. Why, then, surrender to the part that we *know* is leading us away from the truth? To do so is to engage in a particularly destructive form of conceptual conservatism. It is surely not among the methods likely to contribute to the growth of human knowledge.

And there is a further point against Ruse's metaphor. It is simply not true, despite his assertions to the contrary, that evolutionary theorizing would grind to a halt were we to forego the attribution of normative functions. We have, after all, an alternative theory of functions that eschews the imputation of functional norms, namely the theory of systemic functions. As I argue elsewhere, we can conceptualize functions as nothing more than the effects of systemic components that contribute to some higher-level systemic capacity, and we can, at the same time, justify all the function attributions we wish to make in the course of theorizing about the evolutionary history of life on earth. The only thing we lose is the imputation of the alleged norms of performance, that is, the property of "being supposed to" perform a given functional task.[15]

We have, then, two basic parameters with which to frame our inquiry. We have an alternative theory of functions—the theory of systemic functions—and we have the directives for inquiry described earlier—directives based upon the historical and psychological dubiousness of the concept "purpose." It is, in light of these parameters, naïve to insist that we set ourselves the task of trying to preserve these concepts. Adopting the task of trying to preserve these concepts is tantamount to turning our back on the growth of knowledge. It is to refuse to put behind us what we know is dubious by descent and by psychological role.[16]

The first answer to my question "What are we up to when we try to preserve our concept of normative functions?" is given by the theory of "proper" functions. The answer is audaciously simple. The answer is that we are quite right to try to preserve our concept of "normative functions," on the grounds that the parts of plants and animals are literally purposive. There really is something that hearts, hands, and eyes are "supposed to" do, thanks to the mindless designing capacities of natural selection. On this view, the designing powers of natural selection have conferred literal functional norms, literal standards of performance against which actual performances are "properly" evaluated. We are properly moved to theorize about natural purposes because living things are properly purposive.

As I argue in *Norms of Nature* (2001), however, the theory of proper functions faces several problems. One is that it is redundant on the theory of systemic functions. Another is that it fails to specify the mechanisms or processes within natural selection capable of producing such standards of performance. This is to say that the theory flouts the lessons concerning analyzing inward and synthesizing laterally that motivate the expectation in (E).[17] Here, however, the point against proper functions is more general. The theory fails because it flouts the directives in (D) and (P). We know that the concept "purpose" is dubious by descent; indeed, as I have pointed out, most of Ruse's (2003) discussion testifies to its deeply dubious nature.[18] We also have excellent theoretical and experimental grounds for the claim that we are constitutionally conflicted regarding these concepts and, in particular, that we are prone to apply them even when no purposes exist. It thus is naïve to take at face value conceptual intuitions that tempt us to view the living realm as rife with normative functions. The intuition that, for example, an incapacitated heart is "supposed to" circulate blood even when it cannot is hardly an argument for thinking we ought to preserve this intuition in our philosophical theories. To the contrary, given what we know of our largely theological history, and given what we are learning about the retarding effects of certain parts of our psychology, such intuitions must be treated with caution. It is even reasonable to suspect that the stronger the intuitions we feel concerning the purposiveness of living things, the more likely we are being led astray.

8.4 Conclusion

There is nothing parochial about the concepts "design" and "function" that makes them dubious, in which case I conclude more generally that any concept dubious by descent or by psychological role ought to be divested of its former authority in the way we formulate our intellectual tasks. We philosophers ought to give up the orientation of the conceptual conservative, at least with respect to dubious concepts, and develop an orientation that is progressive, an orientation designed to contribute positively to the growth of human knowledge. The directives in this chapter are offered as a small first step.

Acknowledgments

The general theoretical framework of this chapter was developed during the 2003–2004 academic year, while I was happily on leave from teaching, thanks to the generous support provided by a National Endowment for the Humanities Fellowship. The particular occasion for this chapter arose when I was invited by Ulrich Krohs and Peter Kroes to participate in the 15th Altenberg Workshop in Theoretical Biology, hosted by the Konrad Lorenz Institute in Altenberg, Austria. The conference was intellectually stimulating as well as socially and aesthetically pleasing, and for that I am most grateful to Ulrich, Peter, and the entire KLI staff. I am further grateful to Ulrich and Peter for helpful comments on my conference presentation. I owe a particular debt to Peter McLaughlin for constructive criticisms and suggestions that resulted in the exegetical digression on Blumenbach and Paley. Peter sent me a copy of his excellent book *Kant's Critique of Teleology in Biological Explanation* and made a concerted effort to save me from any number of inaccuracies and interpretive blunders. Errors that remain are entirely the fault of my obstinacy, over which my less obstinate capacities appear to have little control.

Notes

1. See Cummins (1975; 1983) and Amundson and Lauder (1994). I explicate and defend the theory of systemic functions in Davies (2001).

2. Some early formulations include Ayala (1970), Enç (1979), and Brandon (1981; 1990). Later versions include Millikan (1984; 1989), Matthen (1988), Neander (1991), Griffiths (1993), Kitcher (1993), Papineau (1993), Godfrey-Smith (1994), Price (1995), Allen and Bekoff (1995), Walsh and Ariew (1996), Buller (1998), Preston (1998), Post (2006), etc. And several philosophers have helped themselves to normative functions in theorizing about language, knowledge, mind, and morals. See, for example, Millikan (1984), Lycan (1988), McGinn (1989), Post (1991), Papineau (1993), and Dretske (1995). A recent anthology (MacDonald and Papineau 2006) contains defenses and criticisms of teleosemantics, an approach to mental content that rests upon proper functions. It may be worth noting that Plantinga (1993) appeals to proper functions in defense of a theologically based epistemology. That should give pause to defenders of normative functions who insist that they are naturalists.

3. Defending this larger thesis is the aim of Davies (2009).

4. Chisholm (1964) defends the startling thesis that free actions are the effects of agents capable of initiating sequences of efficient causation, including sequences that lead to the full range of human behavior, where these agents are not themselves subject to any efficient causes.

5. Toby Appel's (1987) book is a marvelous history of the development of biology in the works of Cuvier and Geoffroy St. Hilaire in late-eighteenth- and early-nineteenth-century France.

6. See McLaughlin (1990), 1.

7. Look (2006) clarifies some of the difficulties in interpreting Blumenbach's theory of a formative drive.

8. On this view of development and evolution, see the essays in Oyama, Griffiths, and Gray (2001) and the splendid exposition in Jablonka and Lamb (2005).

9. Kant's strategy for reconciling the attribution of natural purposes with the efficient causality posited in Newton's mechanics is explicated in McLaughlin (1990).

10. Kant's "as if" approach to natural purposes has been resuscitated by Ruse (2003) and critically assessed in chapter 4 of Davies (2009).

11. See the marvelous discussion in Richards (2002).

12. There are of course further historical roots of the concept "purpose," including those that figure in the thought of Aristotle and Plato, and it is admittedly difficult to know how to weight the various roots that led to our present-day concept. I take it for granted, however, that the effects of our eighteenth- and nineteenth-century predecessors are at least as efficacious as our ancient predecessors.

13. Bechtel and Richardson (1993), which develops several interesting case studies, is exemplary.

14. The discussion in Wegner (2002) illustrates what appears to be a deeply entrenched constitutional conflict of enormous importance for understanding human agency.

15. See Davies (2001), especially chapter 3.

16. I develop these two parameters more fully in Davies (2009).

17. See, in particular, chapters 3 and 5 of Davies (2001).

18. And on some theories—the theory of cultural evolution defended in Richerson and Boyd (2005), for example—the perpetuation of conceptual categories across generations is, in part, a causal consequence of our psychology. The directives in (D) and (P), that is, need to be applied in tandem. Richerson and Boyd focus on our unique capacities for imitation, but the considerations discussed by Leslie, Kelemen, and Wegner are also relevant.

References

Allen, C., and Bekoff, M. (1995). Biological function, adaptation, and natural design. *Philosophy of Science, 62:* 609–622.

Amundson, R., and Lauder, G. (1994). Functions without purposes: The uses of causal role function in evolutionary biology. *Biology and Philosophy, 9:* 443–469.

Appel, T. (1987). *The Cuvier-Geoffroy Debate: French Biology in the Decades Before Darwin.* Oxford: Oxford University Press.

Ayala, F. (1970). Teleological explanations in evolutionary biology. *Philosophy of Science, 37:* 1–15.

Bechtel, W., and Richardson, R. (1993). *Discovering Complexity: Decomposition and Localization as Strategies in Scientific Research.* Princeton, N.J.: Princeton University Press.

Bergson, H. (1907). *Creative Evolution.* (Mitchell, A., trans.) New York, N.Y.: Henry Holt and Company, 1911.

Blumenbach, J. F. (1781) *Über den Bildungstrieb und das Zeugungsgeschäfte [On the Formative Power and the Activity of Procreation].* Göttingen: Johann Christian Dieterich.

Brandon, R. (1981). Biological teleology: Questions and explanations. *Studies in the History and Philosophy of Science, 12:* 91–105.

Brandon, R. (1990). *Adaptation and Environment.* Princeton, N.J.: Princeton University Press.

Buller, D. (1998). Etiological theories of function: A geographical survey. *Biology and Philosophy, 13:* 505–527.

Chisholm, R. (1964). Human Freedom and the Self, The Lindley Lecture, University of Kansas.

Cummins, R. (1975). Functional analysis. *Journal of Philosophy, 72:* 741–760.

Cummins, R. (1983). *The Nature of Psychological Explanation.* Cambridge, Mass.: The MIT Press.

Davies, P. S. (2001). *Norms of Nature: Naturalism and the Nature of Functions.* Cambridge, Mass.: The MIT Press.

Davies, P. S. (2009). *Subjects of the World: Darwin's Rhetoric and the Study of Agency in Nature*. Chicago, Ill: University of Chicago Press.

Dretske, F. (1995). *Naturalizing the Mind*. Cambridge, Mass.: The MIT Press.

Enç, B. (1979). Function attributions and functional explanations. *Philosophy of Science, 46:* 343–365.

Godfrey-Smith, P. (1994). A modern history theory of functions. *Noûs, 28:* 344–362.

Griffiths, P. (1993). Functional analysis and proper functions. *British Journal for the Philosophy of Science, 44:* 409–422.

Jablonka, E., and Lamb, M. (2005). *Evolution in Four Dimensions: Genetic, Epigenetic, Behavioral, and Symbolic Variation in the History of Life*. Cambridge, Mass.: The MIT Press.

Kant, I. (1790). *Critique of Judgment*. (Bernard, J. H., trans.) Hafner Press, 1951.

Keil, F. C. (1994). The birth and nurturance of concepts by domains: The origins of concepts of living things. In: *Mapping the Mind: Domain Specificity in Cognition and Culture* (Hirschfeld, L., and Gelman, S., eds.), 234–254. Cambridge, UK: Cambridge University Press.

Kelemen, D. (2004). Are children "intuitive theists?": Reasoning about purpose and design in nature. *Psychological Science, 15:* 295–301.

Kitcher, P. (1993). Function and design. *Midwest Studies in Philosophy, 18:* 379–397.

Leslie, A. (2000). "Theory of mind" as a mechanism of selective attention. In: *The New Cognitive Neurosciences*, 2nd edition (Gazzaniga, M., ed.), 1235–1247. Cambridge, Mass.: The MIT Press.

Leslie, A., and Thaiss, L. (1992). Domain specificity in conceptual development: Neuropsychological evidence from autism. *Cognition, 43:* 225–251.

Look, B. (2006). Blumenbach and Kant on mechanism and teleology in nature: The case of the formative drive. In: *The Problem of Animal Generation in Early Modern Philosophy* (Smith, J. E. H., ed.), 355–372. New York, N.Y.: Cambridge University Press.

Lycan, W. (1988). *Judgment and Justification*. New York, N.Y.: Cambridge University Press.

MacDonald, G., and Papineau, D. (eds.). (2006). *Teleosemantics*. Oxford: Oxford University Press.

Matthen, M. (1988). Biological functions and perceptual content. *The Journal of Philosophy, 85:* 5–27.

McGinn, C. (1989). *Mental Content*. Oxford, UK: Basil Blackwell.

McLaughlin, P. (1990). *Kant's Critique of Teleology in Biological Explanation*. Lewiston/Queenston/Lampeter: The Edwin Mellen Press.

Millikan, R. (1984). *Language, Thought, and Other Biological Categories*. Cambridge, Mass.: The MIT Press.

Millikan, R. (1989). In defense of proper functions. *Philosophy of Science, 56:* 288–302.

Neander, K. (1991). Functions as selected effects: The conceptual analyst's defense. *Philosophy of Science, 58:* 168–184.

Niebuhr, R. (1941). *The Nature and Destiny of Man: A Christian Interpretation (Volume I: Human Nature)*. New York, N.Y.: Charles Scribner's Sons.

Oyama, S., Griffiths, P., and Gray, R. (eds.). (2001). *Cycles of Contingency: Developmental Systems and Evolution*. Cambridge, Mass.: The MIT Press.

Paley, W. (1802). *Natural Theology: Or Evidences of the Existence and Attributes of the Deity Collected From the Appearances of Nature*. London: reprinted Farnborough, G., 1970.

Papineau, D. (1993). *Philosophical Naturalism*. Oxford, UK: Blackwell Publishers.

Plantinga, A. (1993). *Warrant and Proper Function*. Oxford, UK: Oxford University Press.

Post, J. (1991). *Metaphysics: A Contemporary Introduction*. New York: Paragon House.

Post, J. (2006). Naturalism, reduction, and normativity: Pressing from below. *Philosophy and Phenomenological Research, 73:* 1–27.

Preston, B. (1998). Why is a wing like a spoon? A pluralist theory of functions. *The Journal of Philosophy, 95:* 215–254.

Price, C. (1995). Functional explanations and natural norms. *Ratio (New Series), 7:* 143–160.

Richards, R. (2002). *The Romantic Conception of Life: Science and Philosophy in the Age of Goethe.* Chicago, Ill.: University of Chicago Press.

Richerson, P., and Boyd, R. (2005). *Not by Genes Alone: How Culture Transformed Human Evolution.* Chicago, Ill.: University of Chicago Press.

Ruse, M. (2003). *Darwin and Design: Does Evolution Have a Purpose?* Cambridge, Mass.: Harvard University Press.

von Goethe, J. W. (1985–1998). *Sämtliche Werke nach Epochen seines Schaffens (Münchner Ausgabe),* 21 vols. (Karl Richter et al, eds.). Munich: Carl Hanser Verlag.

Walsh, D., and Ariew, A. (1996). A taxonomy of functions. *Canadian Journal of Philosophy, 26:* 493–514.

Wegner, D. (2002). *The Illusion of Conscious Will.* Cambridge, Mass.: The MIT Press.

9 Ecological Restoration: From Functional Descriptions to Normative Prescriptions

Andrew Light

Restoration ecology is the science and social practice aimed at re-creating ecosystems that have been damaged or destroyed by anthropogenic or nonanthropogenic causes. Ecological restorationists have attempted to re-create a wide variety of ecosystems including tallgrass prairies, oak savannahs, wetlands, forests, streams, rivers, and even coral reefs. Also included in restoration is the reintroduction of species. These projects can range from small-scale urban park reclamations, such as the ongoing restorations in urban parks in cities like New York and Chicago, to huge wetland mitigation projects encompassing hundreds of thousands of acres, such as the current US$8 billion project to restore Florida's everglades ecosystem.

As a scientific practice, restoration ecology is governed primarily by academic disciplines such as field botany, conservation biology, landscape ecology, and adaptive ecosystem management. As an exercise in environmental design practice, most restoration in the field is orchestrated by landscape architects and environmental engineers. But a range of other academic disciplines has been attracted to restoration both as an object of study and as an opportunity to apply one's ideas on the ground. Environmental anthropologists and sociologists have written extensively on the social dimensions of restoration and how they help or hinder the development of human communities and the relationships among those communities and the animals and ecosystems around them (see, e.g., the essays in Gobster and Hull 2000). Environmental historians have actively shaped the ends of restoration by asking pressing questions concerning why we choose to go back to a certain temporal landmark when we restore rather than to another (see, for example, Reece [2006] on the work of T. Allen Comp).

Philosophers too have been attracted to restoration initially focusing on the issue of whether a restored ecosystem was really a part of nature or rather some kind of technological artifact. Indeed the most influential work by environmental philosophers on this topic, surely that of Robert Elliot and Eric Katz, have largely consisted in arguments that ecological restoration does not result in a restoration of nature, given their definitions of what *nature* is, and that further they may even harm naturally evolved systems considered as a subject worthy of moral consideration (Elliot 1982, 1997; Katz 1996, 1997, 2002).

These criticisms stem directly from what has been the principal concern of environmental ethicists since the inception of the field in the early 1970s, namely to describe the nonanthropocentric (non-human-centered) and noninstrumental value of nature (see Brennan 1998; Callicott 2002; Light 2002a). One of the basic presumptions of the field has been that if nature has some kind of intrinsic or inherent value, then a wide range of duties, obligations, and rights may be required in our treatment of it similar to the obligations owed to humans when they are described as entities that have intrinsic rather than only instrumental value. This is much the same way that we think about the reasons we have moral obligations to other humans according to many ethical theories. Kant's duty-based ethics argued that humans have value in and of themselves such that we should never treat them only as a means to furthering our own ends but also as ends in themselves.

Setting aside for the moment the validity of this general claim, one immediate observation we can make is that it seems to rely on a discernible line between those things in the world possessing this sort of value (natural things) and those things that do not have this value (nonnatural things, namely artifacts). Without such a distinction, then, it would appear to be the case that we have some kinds of moral obligations to all environments—natural or nonnatural—incurring a very large set of conflicting moral obligations to those environments. Here, then, is the source of one of the chief worries of environmental ethicists such as Elliot and Katz about restored environments: they would appear to be marginal cases that test our ability to draw this kind of line because they may be hybrid objects. They look like naturally evolved things, maybe even act like them, but they are made by humans and so must be artifacts. Elliot and Katz reply, however, that restorations can never duplicate the value of original nature because, by definition, they are not natural things. They are artifacts made by humans and that is the most important thing to recognize in determining their value. In Elliot's terms, restorations lack the "originary value" of naturally evolved entities and systems that are derived from having evolved separately from us as the product of autonomous biological, geological, and ecological processes. Instead the origins of restorations are human and, like other things made by humans, their value is instrumental; they have value for us. For Elliot, the value of a restored environment is more akin to a piece of counterfeit art than an original masterpiece.

But such a view is the best-case scenario for restorations on such accounts. Katz argues that when we choose to restore, we dominate nature by forcing it to conform to our preferences for what we would want it to be even if what we want is the result of benign intuitions of what is best for humans and nonhumans. Katz puts the point quite bluntly that "the practice of ecological restoration can only represent a misguided faith in the hegemony and infallibility of the human power to control the natural world" (Katz 1996: 222).

While there are many objections that one can raise to the criticisms of Elliot and Katz, here I want to make an attempt to reset the terms of the debate as they have been laid out. My underlying position is that we should try to tease apart a description of what a restoration is—let's call this the descriptive project—from our assessment of whether a restoration is a good thing in a social or moral sense—let's call this the normative project. Under the influence of Elliot and Katz these two projects have been run together by allowing our assessment of whether a restored environment is an artifact drive our intuitions about whether or not it holds positive moral or social value. Naturally evolved ecosystems are good, or at least contain some kind of value that must be respected in a moral sense, by virtue of them being biological rather than artifactual. In turn there is something morally or socially suspect about restorations in part just because they can be described as artifacts. Following the terms of this debate many critics of Elliot and Katz have tried to either re-describe the intrinsic value of nature in some way so as to make it applicable to a restored environment or else show how there are artifactual components of nonhuman environments that do not detract from the value of those systems (see, e.g., Gunn 1991; Rolston 1994; Scherer 1995; Throop 1997).

To my mind these debates do not get us very far either for the descriptive project of defining restorations or the normative project of determining their value. I find it unassailable that restorations are made things, and so, in that sense, human artifacts. But the fact that they are artifacts seems to me largely inconsequential for determining whether they are good or bad for us, other animals, or the environment. Therefore in the first part of this chapter, I look at various attempts at defining restorations, coming back to a claim that an understanding that may work best, and I hope does not run together the descriptive and morally or socially normative dimensions of restoration, is one that can be generated out of a functional description of restored environments. While there is certainly a normative dimension of our description of a restoration it need not entail a normative assessment of its social or moral dimensions. Whether or not something is good for us, or good for other animals or the environment writ large, should be determined by other means. And so in the second and third parts of this chapter I give a very different set of reasons for why artifacts can have positive or negative moral or social value regardless of the fact that they are artifacts. If this argument holds then we can give Elliot and Katz their contentious claim that the world can be divided between things that are more or less natural and things that are not, and undermine the claim that anything necessarily follows about the moral or social value of those things.

9.1   What Is Ecological Restoration?

The primary organization for restoration ecologists is the Society for Ecological Restoration International (SER). Over the years it has tried again and again to define restoration. Here are a few representative examples:

1990: "Ecological restoration is the process of intentionally altering a site to establish a defined, indigenous, historic ecosystem. The goal of this process is to emulate the structure, function, diversity, and dynamics of the specified ecosystem."

1996: "Ecological restoration is the process of assisting the recovery and management of ecological integrity. Ecological integrity includes a critical range of variability in biodiversity, ecological processes and structures, regional and historical context, and sustainable cultural practices."

2002: "Ecological restoration is the process of assisting the recovery of an ecosystem that has been degraded, damaged, or destroyed." (all three quotes from Higgs 2003: 107–110)

One notable thing about these different definitions is how thin they become over time. In 1990 the definition invoked what most ecologists and historians today would consider contentious categories, most especially, the notion of an "indigenous" ecosystem as if *indigenous* necessarily denoted a valuable end-state. By 1996 the definition had put the burden of defining whether something had achieved a good state on the substantive notion of "ecological integrity," though by this time admitting a range of variability in meeting that target without appeal to an indigenous system. By 2002 all substantive attempts to define restoration were scrapped so that the practice amounted only to the "recovery" of an ecosystem that had been damaged in some way. Whether this recovery created some form of integrity or historical fidelity to an indigenous system had dropped off the map of determining whether a practice was a restoration or not.

While this evolution of the definition of restoration by the SER was not directly driven by the philosophical criticisms offered by Elliot and Katz, having taken part in some of the discussions over the years about how to define restoration, I can attest that it is the case that these sorts of worries were in the air (one also gets a strong sense of this in Higgs [2003], to be discussed more later in this section). The choice essentially came down to this: either hold onto the distinction at the root of Elliot and Katz's concerns—again, things that are "natural" and things that are "artifacts"—and then show that restorations really are natural, or else reject the distinction between the natural and the artificial altogether and come up with a different way of referring to the realm under discussion that does not depend on such a distinction.

Those in the restoration community who opted to keep the distinction and prove that restorations really are natural, and so were not subject to the kinds of criticisms offered by Elliot and Katz, started what became known as the "authenticity debate." On this view we should redefine the desired end-state of a restoration in such a way that it was clear that what was wanted was to create the most purely authentic and natural landscapes possible. The debate was over how we would define those criteria for authenticity. We can see this in the language of the 1990 definition, which again puts a priority on re-creating an "indigenous, historic ecosystem." By and large what was meant by indigenous in this sense was to go back to a state that would be free of any trace of prior human disturbance. The logic here, while flawed, is clear: If what makes something artificial is that it is produced by humans, then the natural state of an area is what it was like prior to human contact. If this state can be re-created, then we succeed in creating something natural.

The history here, however, is a bit more interesting. As this debate was by and large occurring in New World countries, it played out in a context whereby many participants assumed that there was something to the notion that there either did exist a presettlement condition to much of the Americas that was free of human influence or else one could meaningfully distinguish between the "naturalness" of the Americas prior to and after the point of European settlement. I cannot fully discuss here the origins of this view, but I and others discuss it, its flaws, and its unfortunate influence on natural resource policy elsewhere (Cronon 1995; Light 2008). What is perhaps most important for the question of defining restoration is that this strategy for answering Elliot and Katz would result in the absurd claim that a truly authentic restoration could exist only in a place where there were discernable presettlement conditions. If these conditions were temporally defined as constrained by a particular wave of settlement in the course of history (in the case of the Americas, the pre-Columbian period), then authentic restorations could not exist at any other place. If these conditions were spatially defined as constrained at all by the arguable existence of human settlement in a place then authentic restorations could not exist in many places because of the climactic or geographic changes that had occurred over time. In short, in addition to the other criticisms that one could give of this kind of approach to defining restoration, the route of accepting Elliot and Katz's nature-artifact distinction and then trying to prove "authentic" restorations as natural would result in far too constrained a definition of restoration as to be practically useful.

In addition, such a view would of course be fallacious. What matters to Elliot and Katz is not whether some place created by humans is or is not like some prior state (sullied or unsullied by humans) but rather the place's origins. Is something made by humans or not? As all restorations are anthropogenic (as opposed to "natural" regenerations and the like) none of them can be natural on this definition and so the premise for the moral worries raised by Katz and Elliot follows.

So what about the option of responding to Katz and Elliot by rejecting their nature-artifact distinction altogether? As I mention earlier, several authors have tried to take on the terms of the debate offered by Katz and Elliot by denying the legitimacy of the nature-artifact distinction. Most of these debates don't appear to go very far as they too often reduce to philosophical linguistic analysis by intuition. A critic, for example, of Katz will point out that there are artifacts in the natural world that are not made by humans, such as beaver dams, and be off and running. Katz will simply deny that beaver dams are artifacts by stipulating that an artifact is something that must be made by a human. Both sides can point to numerous assumptions in the philosophical and nonphilosophical literature that do or do not assume the anthropogenic nature of artifacts but neither proof by stipulation is particularly compelling. To date neither Elliot nor Katz has cried uncle in the face of such examples though Elliot at least has significantly mellowed in his overall moral assessment of restoration.

Another kind of answer is pursued by Eric Higgs who has produced the most ecologically informed treatment of the philosophical dimensions of restoration so far (Higgs 2003). One fair way of understanding Higgs's view is that it is in part based on a claim that we should expect that restorations should have cultural components because their reference ecosystems have cultural components as well insofar as humans have evolved a variety of modes of interaction with different places. The mistake of many definitions of restoration (especially those espoused during the authenticity debate) is that they focused too much on technical proficiency and did not provide an "indication of the wider cultural context of restoration practice" (p. 108). A good restoration on Higgs's account is one that is characterized by attention both to historical fidelity to predisturbance conditions as well as to re-creation of the ecological integrity of a site.

I find nothing wrong with the general direction in which Higgs is moving. Good restorations should pay attention to natural and cultural elements insofar as we can meaningfully distinguish between those elements at any particular site. But I do feel compelled to stay true to the intuitions I expressed at the start that we should try to keep apart the descriptive and normative accounts of restoration so as to avoid what I take to be the unhelpful direction that most of the philosophical debate on restoration has gone in, again, to too easily derive our moral and social assessment of restorations from our description of the kind of thing we take them to be. In Katz's review of Higgs's book we can see how the account opens itself up to this traditional move in the literature. For Katz, something like the ecological integrity of a place immediately disappears as soon as human intentionality is introduced, and, on his view, nature is made into an artifact (Katz 2007: 216). An artifact cannot have an "ecological" integrity, good, bad, or otherwise. Moreover, it becomes clear by the end of Higgs's book that the baseline definition of restoration in terms of historical fidelity and ecological integrity eventually point us toward a broader notion of restoration ("ecocultural" restoration) where companion categories of "cultural fidelity" are added onto the normative description of what counts as a good restoration. Unfortunately this move even further blurs the lines between our descriptions and prescriptions of restoration. This is not to say there is anything necessarily wrong with Higgs's overall view. It deserves assessment in its own right separate from the definition of what counts as a restoration.

To get around these debates I believe that we should simply accept that restorations are artifacts and define them as such. But how can we identify them? What kind of artifacts are they? If we go back to the three definitions formulated by the SER there is one common element to ecological restorations that is sometimes explicit and sometimes implicit: all three definitions in some way appeal to the function of a restoration as part of its description. The 1990 definition says this explicitly. The 1996 and 2002 definitions imply it by appeal to the creation of a thing that actively does something, namely, assisting in the recovery of some state or process. The underlying intuition is that something is a restoration when it is an attempt to restore or re-create the function of a previously existing eco-system, a component of that ecosystem, or an ecosystem service provided by a reference ecosystem (such as habitat for endangered species, recharge of a water supply, etc.). Like all forms of human intervention with the environment, or environmental management, a restoration is designed to do something. What makes something a restoration is that it is an attempt to do something in relation to a set of prescribed boundary conditions: it must refer to some state that was there before and it must be governed by some intention to reproduce some discernable function of that prior state. Therefore I would propose the following: *Ecological restoration is a form of environmental intervention that attempts to re-create some aspect of the prior function of an ecological reference state.*

It is not my intention here to offer this suggestion by way of making any broader claims concerning the philosophical debates over biological or technological functions. But in addition to the intuitive appeal such a definition would ideally have for ordinary everyday use (which would be necessary for an organization largely composed of practitioners like the SER) the rich philosophical literature on functions should add to our understanding and assessment of restorations. Insofar as we can plausibly claim that there are biological and technological functions restorations will partially unite descriptions of these functions in practice. The reason is that the designed function of a restoration as an artifact must be related directly to what we come to understand as the functions, or functional organization, of their reference biological ecosystems. Note that this claim does not depend on one's answer to the issue of whether all ecosystems have functions, or whether all aspects of all ecosystems can be described as a function of that ecosystem, but rather whether we can *attribute* functions to ecosystems that could be replicated for some reason. As such the definition is agnostic on some of the stickier issues concerning biological functions. Restorationists are not trying to produce biological entities as such but rather to make something that reproduces functional attributes of specific kinds of systems in nature. To plan our restorations on our understanding of these attributes does not mean that all aspects of a restored site can be reduced to the re-creation of functional properties. Designs for restorations can include other elements, such as aesthetic components, even while our definition of the practice of restoration is first found in these functional properties.

Because a functional description of restorations would have to focus on the design elements of restoration, Philip Kitcher's discussion of function in general as design will be particularly helpful: "[T]he function of an entity S is what S is designed to do" (Kitcher 1993). But because my assumption is that restorations are indeed artifacts and should not be otherwise confused as anything else, a more promising route would be to adopt the ICE account of technological function as described in this volume and elsewhere by Houkes and Vermaas, which focuses on the role of use plans and design in describing the functions of artifacts (see Houkes and Vermaas 2004: 65). On this view the functions of objects are the direct result of the intentions and use plans of designers. This account gives us the capability of rationally discussing malfunction and other design properties. On this view we can safely describe restorations as an attempt to do something but remain safely agnostic on whether the thing they do is necessarily good or bad in relation to nature or ourselves. Still, by adopting the ICE account, we can begin to see another valuable aspect of restorations, namely how they can serve as large-scale ecological experiments that help us to understand how ecological systems do function, how they deliver critical services for ourselves and other species, and how different forms of disturbance can interfere with these functions. Again this is not to attribute a necessary plan to nature (or reduce the natural world to a plan) in the same way that we would understand the blueprint for a building but only to see that this form of environmental intervention by us is feasible as a product of our understanding of how nature "works" in some respects.

9.2   Relationships with Objects

If we assume that ecological restorations are artifacts that can be described in terms of their functional properties then can we circumvent the unsavory implications that Katz or Elliot would attribute to any artifact that would replace a natural object? I believe we can. What must be demonstrated is that artifacts, like natural objects, may possess obligation-generating normative properties. They may not be the same properties (though in section 9.3 I argue that, at least on my view, they are very similar) but the plausible case that there are such properties helps to show that there is still more work to be done once we have settled the descriptive question of whether restorations are "natural" or not.

One route to this kind of argument is to first recognize the implicit assumption on Katz and Elliot's views that our moral relationships (and explicit or implicit obligations) with artifacts cannot be as strong as the relationships we could have with natural systems on the assumption that natural systems have a direct moral value that should be respected. One thing that may be overlooked on such a view is that artifacts can bear meaning in a normative sense in a way that does not degenerate into some kind of occult view. At the very least objects can be the unique bearers of meaning for relationships among humans that hold strong normative content and in that sense we can interact with them in ways that can be described as better or worse in a moral sense.

There are lots of examples of how we can relate to one another in better or worse ways through objects. Some may find trite the examples that come to mind—the political meaning of flags for instance (I was terrified as a young Cub Scout to let the American flag touch the ground simply because I was told that it was wrong). But it would seem hard to deny that objects can stand for the importance of relationships between humans such as is the case with wedding bands. There may even be some argument to be made that we should respect some objects in their own right. To be more precise I would maintain that we can be lacking in a kind of virtue when we do not respect objects in some cases, especially, when such objects stand for the importance of relationships we have with others, or, as in the case of justifications for historical preservation, respect the creations of those who have come before us. Note that I use the term virtue here rather than the stronger language of obligation because I don't think we have obligations to objects themselves in the same way we have moral obligations, for example, to people.

One way to explain the value of everyday things is to consider the case of the destruction of an object that stands for a relationship in some way. The unthinking destruction of an object that bears the meaning of some relationship between individual humans reflects badly on the person who destroys that object. Consider the problem of replacing and replicating objects that are special to us. I have a pair of antique glasses of which I am very fond because they were the glasses that my maternal grandfather, Carmine Pellegrino, wore for much of his adult life. The glasses are a combination of a set of lenses that were no doubt reproduced at the time in large quantities and stems that he fabricated himself. The stems are nothing fancy, just bits of steel wire that he bent and shaped—he was a coal miner, not a jeweler—but it is important to me that he made them. If you were to come to my apartment and drop Carmine's glasses down the incinerator shoot and then replace them with a pair of antique glasses from a shop nearby then I would justifiably claim that something has been lost that cannot be replaced. Further, paraphrasing one of Elliot's famous examples about ecological restoration, if you were to make an exact replica of the glasses and fool me by passing them off as the original, then, while I might not feel the loss, I would nonetheless have suffered a loss of some sort even though I would not know that I suffered this loss. And if I were to find out that you tricked me with the replicas then I would justifiably feel regret and then anger!

The moral harm that may be done to me in this case is parasitic on the value of having been in a relationship with another person and not simply in some quality that is inherent to the object itself. Still, the object does play an irreducible role in this thought experiment—it is a unique entity that evinces my relationship with my grandfather that cannot be replaced even though the relationship in this case is not only represented in this object. Both the relationship and the object have some kind of intrinsic value. But surely not all relationships have this kind of value and so neither do all objects connected to all kinds of relationships. How then can we discern the value of different kinds of relationships?

One possible source is Samuel Scheffler's work on the value of relationships. Scheffler is concerned with the question of how people justifiably ground special duties and obligations in interpersonal relationships without this being only a function of relations of consent or promise keeping. Scheffler's account argues for a nonreductionist interpretation of the value of relationships that finds value in the fact that we often cite our relationship to people themselves—rather than any explicit interaction with them—as a source of special responsibilities. So for Scheffler:

... if I have a special, valued relationship with someone, and if the value I attach to the relationship is not purely instrumental in character—if, in other words, I do not value it solely as a means to some independently specified end—then I regard the person with whom I have the relationship as capable of making additional claims on me, beyond those people in general can make. For to attach non-instrumental value to my relationship with a particular person just is, in part, to see that person as a source of special claims in virtue of the relationship between us. (Scheffler 1997: 195–196)

On this view, relationships among persons can have value in some cases not because of any particular obligations that they incur, but because of the frame of action that they provide for interactions among persons. As Scheffler puts it, relationships can be "presumptively decisive reasons for action." While such reasons can be overridden they are sufficient conditions upon which you or I may act in many cases.

What I find most attractive about Scheffler's argument is that it conforms to our everyday moral intuitions about relationships—for example, it does not reduce them to explicitly voluntary events—and it makes sense of why we find some relationships morally compelling in a noninstrumental way. I call relationships that we find valuable in and of themselves in this way "normative relationships." Our actions and attitudes with respect to these relationships can be better or worse. The fact that we are in these kinds of relationships can provide better and worse reasons for action.

One of the interesting things about the relationships that we value intrinsically, though, is that most of them are symbolized in objects—wedding rings, mementos, gifts, and so forth. For this reason then, at a minimum, we can do harm, or more accurately, exhibit a kind of vice, in our treatment of objects connected to those particular kinds of relationships. Take for example the watch I am wearing as I write this chapter. This watch was given to me several years ago by my former partner's parents in Jerusalem as a way of welcoming me into their family. I cherish the occasion even though I am no longer in a relationship of the same kind with her or her parents. The watch, however, is a meaningful symbol of that event and that set of relationships. If someone were to try to take this watch from me and smash it I would have a presumptively decisive reason for stopping that person that was not limited to its value as mere property but would also include its value as a thing standing for a particular normative relationship. So, too, if I were to smash this watch myself with a hammer for no reason, I would be doing something wrong in some sense relative to the intrinsic value of that set of relationships as well. To tease out my intuitions on why it would be wrong to smash the watch I need not appeal to any obligation to the thing itself but only claim that I have presumptively decisive reasons to respect the watch because to do otherwise does harm to a connection of value involving my relationships with others in which the watch plays some role. Again it may help to think of this in terms of vice. I exhibit a kind of vice when I smash the watch. This is a minor vice but it is a vice nonetheless. My character is lacking if I do not seem to minimally care about this object when it is appropriate for me to do so.

Does this example mean that my character is necessarily flawed if I smash the watch? No. Under some circumstances it might even be appropriate to destroy an object from a past relationship out of some justified anger over the relationship. But where no such reason exists, and the object stands for a relationship still cherished, such an action would be questionable. Someone hearing me brag about smashing this watch for no reason might justifiably hesitate in forming a relationship with me. Does this example imply that the meaning or significance of the relationship that the watch represents is lost if I smash it? Certainly not, as any object is not the primary bearer of the meaning of any relationship. Does this mean that all objects bear meaning in this way? Again no. Just as the value of some relationships with others can be reduced merely to instrumental terms so too the value of some objects can be reduced merely to their use or exchange value.

Now imagine that I show you a second watch that I own—a plain cheap plastic digital watch. This is the watch that I use when I go running in the afternoons so I can see how long it takes me and I can find out if my time improves as I continue to run. I actually don't remember where I got this watch. If I smash this watch very little is implied about my character as this watch does not bear any meaning that has normative content that can reflect on my relations to others.

Finally on this point, if there is something to these intuitions, then the meaning of objects in this normative sense can fade over time. But, importantly, this is not a unique property of objects since the meaning of our relationships with other persons can also fade over time. Still, recognizing that the normative content of objects can fade deserves some attention. If I find an object in an antique store, say a watch made in 1850 with an inscription from a wife to a husband in it, would it be worse of me to smash it than it was to smash the plastic runner's watch? If I find reason to assume that this watch stood for someone else's normative relationship, even though that person is not me nor anyone that I know, is there something better or worse about my character depending on how I treat that object? I probably do not want to think about the meaning of my treatment of the antique store watch in the same way that I would the treatment of an object that has meaning in a relationship I am in now but I think there is something there that should give us pause. Whatever the meaning of the 1850 watch is we can imagine our assessed value of it as providing something akin to the reasons we might have to avoid smashing up old buildings or other historical artifacts. Still, it also might be that we have independent reasons to try to respect such objects as well, similar to the arguments I have offered so far in this section. Such issues deserve more treatment than I have space to address here though I do not believe that reflection on those issues would change the conclusions I come to in section 9.3.

Where does this discussion get us with respect to our topic at hand, ecological restorations? At least it gives us reasons we can build on to find value in restorations even if, as Elliot and Katz have it, they are only artifacts. On this account, however, their value as artifacts also depends on how they help to mediate the sort of human relationships that are presumptive reasons for action.

9.3   Restoration as a Source of Normative Ecological Relationships

There are no doubt many ways to describe the value of nature. We are natural beings ourselves and so nature has value as an extension of the value that we recognize in ourselves. The resources we extract from nature are valuable at least insofar as we value the things that we construct out of those resources as well as their role in sustaining our lives. And certainly there is something to the intuition that other natural entities and whole systems are valuable in some kind of noninstrumental sense even if we can be skeptical that this sort of value offers sufficient resources to justify moral obligations for their protection. Is there anything else?

Consider again Scheffler's argument about the value of relationships. When applied to considerations of the environment this approach resonates somewhat with the focus in environmental ethics on finding noninstrumental grounds for the value of nature. But rather than locating these grounds in the natural objects themselves an extension of Scheffler's views would find this value in relationships we have with the natural environment either (1) in terms of how places special to us have a particular kind of value for us, or (2) in the ways that particular places can stand for normative relationships between persons. On reason (1) certainly Scheffler would have trouble justifying the value of such relationships between humans and nonhumans, let alone humans and ecosystems, using his criteria, but I think there is no a priori hurdle in doing this especially if we can separate Scheffler's claim about the noninstrumental value of such relationships from the possible obligations that follow from them. Focusing just on the value of these relationships we can imagine having such substantive normative relationships with other animals whereby the value we attach to such relationships is not purely instrumental. We do this all the time with our relationships with pets. And why not further with nature, more broadly conceived, or more specifically with a particular piece of land? Because the value of such relationships is not purely instrumental reciprocity is not a condition of the normative status of such relationships, but rather only a sense that one has noninstrumental reasons for holding a particular place as important for oneself.

For some like Katz, the moral force behind such a suggestion would best be found in a claim that nature is a moral subject in the same or a very similar way that we think of humans as moral subjects. So just as we can conceive of being in relationships with other humans as being morally important, we can conceive of being in relationships with any other nonhuman subjects as important in the same way. Again, though, this claim rests on a form of nonanthropocentrism that Scheffler, and probably most other people, would find objectionable. And it would miss an important part of what I'm trying to argue for here: it is not only the potential subjectivity of nature that makes it the possible participant in a substantive normative relationship but it is the sense that nature, or particular parts of nature, can be "presumptively decisive reasons for action." Being attentive to such a relationship can be assessed as good or bad. If I have a special attachment to a place, say, the neighborhood community garden that my family has helped to tend for three generations, then whether I regularly visit it to put in an afternoon's work can be assessed as good or bad because of the history that I have with that place regardless of whether it is an artifact or not. My relationship with that place, as created by that history, creates presumptively decisive reasons for action for me in relation to that place.

The same would be true if I were in a substantive normative relationship with another person. There would be something lost or amiss if I didn't contact them for a year out of sheer indifference (for an example, see Light 2000). In such a case my indifference could be interpreted as reason to doubt that the relationship was important to me at all. So, too, something would be lost if I didn't visit the community garden for a year out of indifference. But what would be lost need not rely on attributing subjectivity to the garden. My relationship with the garden is a kind of placeholder for a range of values none of which is reducible as the sole reason for the importance of this relationship. To distinguish this kind of relationship from others, I want to call it a "normative ecological relationship," both to identify it as a relationship involving nature under some description in some way and just in case some wish to set aside for later consideration the issue of how this sort of relationship might be substantively different from other normative relationships. Critically though, because this argument does not depend on attributing something like intrinsic value to nature itself, let alone subjectivity, the metaphysical status of the object in such a relationship is not important to the justification for forming a relationship with or through it.

I should also note here that if I am in a normative ecological relationship with something this does not mean that my reasons for action derived from that relationship could never be overridden, either in the face of competing claims to moral obligations I might have to other persons or other places, or because of some other circumstances that caused me to separate myself from that place. It means only that my normative relationship to the place can stand as a good reason for me to invest in the welfare of that place. Also important is that the moral status of my relationship to such a place does not exist in an ethical or historical vacuum. If my relationship to a place has been generated out of my experience of having acted wrongly toward others at some site (let us say it is an inhumane prisoner-of-war camp where I worked contentedly as a prison guard) then my character can be justly maligned for so narrowly understanding the meaning of a place that has been a source of ills for others. Outside of such extreme cases, though, my relationship to places can exhibit the qualities that we would use to describe our relationships with others such as fidelity and commitment.

Can ecological restorations be a source of such normative ecological relationships? It seems entirely plausible if not unassailable that they can. There is sociological evidence to document this effect for those who volunteer in restorations (Miles, Sullivan, and Kuo 2000). Over the past few years I have elaborated on this evidence and used it to argue that we can maximize this potential value of ecological restorations when we open them to public participation (Light 2002b). I do not repeat those arguments here but say only that restorations can serve as opportunities for the public to become more actively involved in the environment around them and hence in the potential for work on restoration projects to encourage environmental responsibility and stewardship. In this way people can form important relationships with the restorations that they participate in producing.

No doubt some will still demur that the things produced in a restoration are nothing but artifacts but in this sense at least it doesn't matter. Assuming that a particular restoration can be justified for other ecological reasons—that it reproduces an important function of a previously existing ecosystem, such as protecting native biodiversity in an area or even simply cleaning up a site so that it is a better habitat for persons and other creatures, the issue of whether a restoration is really natural is practically moot on this account. Just as in the case of the special watches from section 9.2, the objects produced by a restoration can be valuable in and of themselves as special things to us and as place holders of important sources of meaning in our lives. This claim does not prohibit us from criticizing those restorations that are intentionally produced either to justify harm to nature or to try to fool people that they are the real thing. But such restorations, which I have termed "malevolent restorations" (Light 2000), can be discounted for the same reasons that we would discount the attachment that people have to persons or places that are morally tainted in other ways.

For all the reasons offered in this chapter the moral potential of restoration ecology, even if the objects produced by this practice are artifacts, is that restorations can foment relationships between persons and nature as well as simply among persons. What can be restored in a restoration is the function of prior ecosystems and our connection to places and to one another.

References

Brennan, A. (1998). Poverty, puritanism and environmental conflict. *Environmental Ethics, 7*: 305–331.

Callicott, J. B. (2002). The pragmatic power and promise of theoretical environmental ethics. *Environmental Values, 11*: 3–26.

Cronon, W. (1995). The trouble with wilderness. In: *Uncommon Ground: Toward Reinventing Nature* (Cronon, W., ed.), 69–90. New York: W. W. Norton and Company.

Elliot, R. (1982). Faking nature. *Inquiry, 25*: 81–93.

Elliot, R. (1997). *Faking Nature*. London: Routledge.

Gunn, A. (1991). The restoration of species and natural environments. *Environmental Ethics, 13*: 291–309.

Gobster, P., and Hull, B. (eds.). (2000). *Restoring Nature*. Washington, D.C.: Island Press.

Higgs, E. (2003). *Nature by Design*. Cambridge, Mass.: The MIT Press.

Houkes, W., and Vermaas, P. E. (2004). Actions versus functions. *The Monist, 87*: 52–71.

Katz, E. (1996). The problem of ecological restoration. *Environmental Ethics, 18*: 222–224.

Katz, E. (1997). *Nature as Subject: Human Obligation and Natural Community*. Lanham, Md.: Rowman & Littlefield Publishers.

Katz, E. (2002). Understanding moral limits in the duality of artifacts and nature: A reply to critics. *Ethics and the Environment*, *7*: 138–146.

Katz, E. (2007). Review of "Nature by Design." *Environmental Ethics*, *29*: 213–216.

Kitcher, P. (1993). Function and design. *Midwest Studies in Philosophy*, *18*: 379–397.

Light, A. (2000). Ecological restoration and the culture of nature: A pragmatic perspective. In: *Restoring Nature*: *Perspectives from the Social Sciences and Humanities* (Gobster, P., and Hull, B., eds.), 49–70. Washington, D.C.: Island Press.

Light, A. (2002a). Contemporary environmental ethics: From metaethics to public philosophy. *Metaphilosophy*, *33*: 426–449.

Light, A. (2002b). Restoring ecological citizenship. In: *Democracy and the Claims of Nature* (Minteer, B., and Taylor, B. P., eds.), 153–172. Lanham, Md.: Rowman & Littlefield.

Light, A. (2008). The moral journey of environmentalism: From wilderness to place. In: *Pragmatic Sustainability*: *Theoretical and Practical Tools* (Moore, S., ed.). Cambridge, Mass.: The MIT Press.

Miles, I., Sullivan, W. C., and Kuo, F. E. (2000) Psychological benefits of volunteering for restoration projects. *Ecological Restoration*, *18*: 218–227.

Reece, E. (2006). Art that works: T. Allen Comp and the reclamation of a toxic legacy. *Democratic Vistas Profiles 5*. Chicago: Center for Arts Policy at Columbia College.

Rolston, H., III. (1994). *Conserving Natural Value*. New York: Columbia University Press.

Scheffler, S. (1997). Relationships and responsibilities. *Philosophy and Public Affairs*, *26*: 189–209.

Scherer, D. (1995). Evolution, human living, and the practice of ecological restoration. *Environmental Ethics*, *17*: 359–379.

Throop, W. (1997). The rationale for environmental restoration. In: *The Ecological Community* (Gottlieb, R., ed.), 39–55. London: Routledge.

IV FUNCTIONS AND CLASSIFICATION

Functions play an important part in the way we carve up our world, especially that part of the world that is populated by technical artifacts. Usually these technical artifacts are classified on the basis of their functions, that is, their proper functions. If someone uses a coin to fasten a screw, that coin does not become a member of the class of screwdrivers. Technical artifacts are not classified on the basis of their accidental functions but on the basis of their proper functions. That immediately raises the question of how to distinguish between proper and accidental functions, a topic that is addressed extensively in part II. Another issue is whether the classification of technical artifacts in functional classes corresponds to the existence of real functional kinds in the world or whether such a functional classification is just a practical method of finding our way in this world. This concerns the ontology of the artifactual world. This is a topic that raises, for instance, the question of how the ontology of the artifactual world is related to the ontology of the natural world. The artifactual world, as compared with the natural world, has a ring of being "artificial," that is, of "lacking in natural quality" or even of "being feigned or faked." Indeed the artifactual world is often taken to be ontologically inferior to the natural world. For instance, the desk on which this introduction is written may be claimed not to exist in the same sense that the atoms and molecules of which it is made exist. If the artifact kind "desk" is taken to exist in the world at all, its ontological status is usually taken to be subordinated to or dependent upon the ontological status of natural kinds (for a criticism of this position see, e.g., Thomasson 2007). Such an ontological position presupposes that it is possible to make a clear distinction between natural and artificial entities. However, conceptually as well as ontologically the classification of objects as natural and artificial is problematic. So the natural-artificial distinction as well as the classification of artificial objects themselves into subclasses raises conceptual and ontological questions in which the notion of "function" often plays a central role. This part is devoted to an analysis of some problems related to the role of functions in classifying entities in our world, especially in the artificial world.

Romano discusses the user's capacity to conceptualize the function that a designer intended for an artifact and the role that that function plays in distinguishing between artifactual and natural objects. Within the field of cognitive psychology this capacity is often explained by reference to the Design Stance (Dennett 1987), which presupposes a metaintentional capacity on the part of users of technical artifacts, namely the capacity to form intentional attitudes about the intentional attitudes of the designer or maker of the artifact. He criticizes this approach and as an alternative he proposes the hypothesis that the human cognitive apparatus comprises a Functional Stance, which enables people to deal with functional knowledge, that is, with knowledge about what objects are for. He argues that the concept of "for-ness" goes beyond the concept of "causality" and that attributing for-ness to an object does not presuppose a metaintentional capacity. The Functional Stance on its own does not lead to a distinction between natural and artifactual objects, since empirical research indicates that in the early stages of human development, functions (for-ness) are attributed indiscriminately to natural and artificial objects. According to Romano, the distinction between natural and artifactual objects is to be based on inferential reasoning on the part of the user about the intentional origins of an object that is for something. Thus the Functional Stance would be a constitutive component of a more general human attitude of categorizing objects into artifacts and natural entities.

Let us assume that the distinction between artifactual and natural objects can be given a firm basis. The objects in the classes of natural as well as artificial objects may be further divided into many subclasses. What is the ontological status of these subclasses, in particular of artifactual classes? Do specific classes of artifactual objects correspond to real kinds, that is, are they part of the structure of the world? That is the problem addressed by Soavi. She defends a realist approach to artifact kinds provided that artifact kinds are functional kinds. This is rather surprising because generally one of the main reasons adopted for rejecting real artifact kinds is that artifact kinds are functional kinds. A real kind of physical objects is a kind whose items must share a set of common physical features used in explaining their behavior. A functional kind does not grant the existence of such a set because of the multiple realizability of functions. Hence if artifact kinds are functional kinds, they may bring together objects with completely different physical structures. So artifact kinds can refer only to nominal kinds. According to Soavi, the reality of artifact kinds can be defended if the notion of "function" is defined in an appropriately narrow way. From the main theories on functions she extracts three types of criteria for the classification of artifacts into functional kinds: the selectionist criterion, the intentional-use criterion, and the intentional-production criterion. For each of these three types, she sketches the ontological consequences of their adoption for artifact classification into kinds. Thereafter she suggests a strategy for a defense of real kinds for artifacts by individuating narrow functional kinds on the basis of a characterization of a function in terms of the triple (1) input-output–relations, (2) system of interaction and (3) structure. She points out that everyday functional classification terms do not correspond to such narrow functions and therefore do not individuate real kinds.

Just as natural scientists may be regarded as experts in the classifying of natural entities, so engineers may be considered to be experts in the classifying of technical artifacts. In the latter classification the notion of function plays a crucial role. The final contribution to part IV therefore turns to the notion of function from an engineering point of view. Kitamura and Mizoguchi start with the observation that in spite of the importance of the notion of function for engineering practice there is no common interpretation of it. Apart from a clear definition of the notion, engineers are very much interested in a formalization of the notion of function that would allow them to represent functions of technical artifacts in computer models. Kitamura and Mizoguchi propose a device-oriented definition of function that is related to device behaviors. They define "function" as a role played by a behavior in its use context. The types of context of use are discussed and a comparison is made with the definitions of biological functions. They also examine function definitions other than the device-centered definition. It is interesting to note that the proposed device-centered definition of function makes reference to goals in contexts of use and therefore to perspectives or viewpoints of agents. This means that their definition of function does not correspond to the narrow definition of function proposed by Soavi. Because of their reference to human intentions, functional kinds individuated on the basis of Kitamura and Mizoguchi's notion of "function" cannot be real kinds.

References

Dennett, D. C. (1987). *The Intentional Stance.* Cambridge, Mass.: The MIT Press.

Thomasson, A. L. (2007). Artifacts and human concepts. In: *Creations of the Mind: Theories of Artifacts and Their Representation* (Margolis, E., and Laurence, S., eds.), 52–73. Oxford: Oxford University Press.

10 Being For: A Philosophical Hypothesis About the Structure of Functional Knowledge

Giacomo Romano

Much philosophical reflection about artifacts, natural entities, and the comparison between these items, focuses on ontological questions. Most theorists who have been and are still interested in these topics aim at characterizing identity criteria to distinguish artifacts from natural entities.

Ontologists have considered the dependency of artifacts on the mind and their *functionality* as basic features distinguishing them from natural entities. These features have not, however, been clearly explained or adequately delineated. To shed some new light on the notion of "functionality" as well as on the notion of "mind dependency," I draw on some studies from cognitive psychology. During the past decade these studies have in fact focused on the categorization and conceptualization of artifacts. From a different perspective to the ontological one, they have thoroughly analyzed the cognitive mechanisms that make humans understand artifacts. The results of cognitive investigations might provide novel conceptual grounds for ontological questions about the artifactual-natural dichotomy, which would also be empirically justified.

However, psychological accounts do not completely explain the distinctive features characterizing artifacts either: an appeal to the vague concept of "Design Stance" is meant to explain the human attitude toward artifacts. According to the explanation proposed in the Design Stance, humans recognize artifacts thanks to a *metaintentional capacity*; the capacity to engage in intentional attitudes of second order, that is, intentional attitudes about other intentional attitudes.[1] By means of this capacity, human beings would allegedly be able to conceptualize the supposed function that an author intended to be performed by way of the artifacts that they have designed. I criticize this view because it presupposes that the recognition of a function is dependent upon it having been conceived by a designer. I also argue in favor of a different, perhaps complementary view that requires a basic scheme employed by the human cognitive apparatus to deal with functional knowledge, that is, the knowledge that a certain object is *for* something. I maintain that such knowledge is independent of intentional attributions and can be applied to artifacts as well as to natural entities, even though it is an important cognitive element for the recognition of artifacts.

My chapter is organized as follows: Section 10.1 introduces the properties that ontologists consider distinctive of artifacts versus natural entities, and scrutinizes the psychological accounts of these features in terms of the Design Stance. Section 10.2 constitutes a critique of these psychological accounts. In section 10.3 I propose my explanatory hypothesis about those properties by appealing to "functional knowledge" as distinct from metaintentional capacities. Finally, I summarize my ideas and refer to their relation to ontological questions.

10.1 Artifacts and the Stance of Design

10.1.1

Generally speaking, philosophical ideas about artifacts, natural entities, and the difference between the two share the aim and the scope of the following definition of *artifact* by Peter Simons (1995: 33):

artefact Any object produced to design by skilled action. Artefacts are continuants, that is, objects persisting in time. . . . Artefacts are not exclusively human. . . . Artefacts contrast with natural objects. Aristotle considered artefacts, defined by function rather than an autonomous principle of unity and persistence, not to be substances.

Simons's definition sets out to grasp the real nature of artifacts by listing the properties that amount to artifactually essential features.[2] Among these, several theorists (e.g., Rea 1995; Wiggins 2001; Baker 2004) have stressed the feature that marks the opposition between artifacts and natural objects as a distinctive artifactual property, not a marginal one. Such a feature in Simons's definition is implied by the notion of being "produced to design by skilled action" as well as, though less explicitly, by the notion of being "defined by function." These notions can be explained with the appeal to the dependency of artifacts on the mind. Baker (2004), for example, recognizes mind-dependency as one of the core properties of artifacts; one that does not belittle their ontological dignity, contrary to what other philosophers hold (e.g., Hoffman and Rosenkrantz 1997; Wiggins 2001). Most probably, philosophers will endorse artifacts as being mind-dependent. Generally speaking, the relation that determines mind-dependency, and thus also the identity of an artifact, is the ratio between the intentional perspective of the creator of the artifact and the conformity of the process of realization of the artifact to such a perspective. The same relation is also supposed to account for the *functional* characterization of artifacts because the function of artifacts is usually taken to be the effect intended by their creators (cf. Baker 2004).

Mind-dependency is therefore identified as the relational property (or set of properties) that determines the dependency of an artifact on the mind of its creator; but this characterization, if left unspecified, is trivial, because it does not clarify what the relation of dependency is. Ontologists have also undoubtedly worked on the definition establishing the conditions that an entity must fulfill in order to be mind-dependent. However, in order to gain a richer understanding of the relation of mind-dependency, a complementary analysis is required. The analysis most worth pursuing is a cognitive investigation of how we perceive and conceive artifacts and their difference from natural entities. This may enhance the explanatory power of previous ontological investigations into these fields.

10.1.2

I scrutinize the question of how we perceptually and conceptually approach artifacts and natural entities by referring to studies of the cognitive abilities of human beings. Most cognitive psychologists consider the identification of artifacts to be based on the capacity to recognize their mind-dependency; the same holds for ontologists. Psychologists have worked to attain a suitable and detailed explanation of the relational property of mind-dependency that we recognize in artifacts. According to them, we ascribe mind-dependency to artifacts by means of certain mental mechanisms. These mechanisms are usually considered to implement second-order intentional attitudes, that is, intentional attitudes about intentional attitudes. According to this idea, in order to identify an artifact a cognitive subject has to detect the intentional relation occurring between an author and his or her creation. This is also taken to determine the function of the artifact at stake. The capacity to recognize artifacts therefore is taken to be metaintentional because it is an intentional pattern (the one of the interpreter) about another intentional pattern (the one engaged by an author with the artifact that he or she has created). There are two different psychological accounts of this.

Some theorists (e.g., German and Defeyter 2000; German and Johnson 2002; Kelemen and Carey 2007) have labeled the metaintentional capacity at stake "Design Stance," inspired by the work of the American philosopher Daniel Dennett, who termed "Stance of Design"[3] a particular predictive strategy. To avoid misunderstandings, a sketchy characterization of Dennett's Design Stance is needed; this characterization in fact has a meaning that is different from the one adopted by cognitive psychologists.

Dennett (Dennett 1983, 1996) has formulated his philosophical framework on the basis of a methodological strategy that can be defined as the theory of stances. According to Dennett, we can predict what a certain item will be and do by taking either the *Physical Stance*, the *Intentional Stance*, or the *Design Stance*. The Physical Stance predicts how a certain object works in accordance with physical and mechanical laws. The Intentional Stance predicts the behavior of that object in terms of its rationality. The Design Stance predicts the workings or the role of the entity at stake inasmuch as it has been designed in a certain way.[4]

The approach taken by Dennett with the theory of the stances is methodological, and does not need (or seek) any direct specific correspondence with actual structures of the human mind. The theory of the stances is based on the interpretation of certain patterns that make the phenomena to which these patterns apply more intelligible. These patterns themselves are considered to be only *partially* real regardless of their ontological status, which is not a crucial matter: they are useful from a predictive point of view, and therefore their existence is justified.[5] There is no need to ascertain the counterpart of the Stance of Design in some aspects of the world because it is a practical heuristic strategy and as such it makes sense. Thus the Design Stance in principle, being only methodological heuristics, does not need any real and empirical counterpart (cf. Dennett 1999).

10.1.3

Despite the strongly instrumental significance that Dennett assigns to the Stance of Design, cognitive psychologists (cf. German and Defeyter 2000; German and Johnson 2002; Defeyter and German 2003; Kelemen and Carey 2007) have interpreted it in a more realistic sense.[6] These scholars, contrary to Dennett, do not take the Design Stance as a mere predictive device. They assume it is an actual cognitive device present in the human mind, explaining and justifying it as part of the human cognitive system on the basis of several empirical studies. According to them, the Design Stance is an effective form of reasoning that is employed by people, both adults and children after a certain age (though there are significant differences), to understand artifacts. These authors have also provided some hypotheses about the psychological genesis, the cognitive structure, and the workings of the Stance of Design. For example, German and Johnson (2002: 279–280) maintain that the Design Stance is probably a mental attitude or frame ". . . in which an entity's properties, behavior, and existence is explained in terms of its having been designed to serve a particular purpose." According to Kelemen and Carey's version of the Stance of Design (2007: 214), "an artifact is intentionally created by a designer to fulfill some function."

Generally speaking, these psychologists agree about the main points in the interpretation of the Design Stance. Each hypothesis about the Stance of Design endorses the fact that this is a cognitive framework that is applied by any human subject in order to recognize (a) that artifacts are produced intentionally by human beings (they are usually taken to be human-made), (b) they are defined according to their functional features, (c) their categorization endorses categorization extension to superordinate items (e.g., the capacity to categorize an object as a "goblet" enables one to categorize it also as a "glass"); and (d) the creators of these items also have baptism rights in relation to their creations (in psychological jargon that means that creators are the ones who coined the name, and determined the nature and identity of what they created, e.g., whether what they made was a paper boat or a paper hat).

The Design Stance, so characterized, seems to be a kind of cognitive ability that is suitable for dealing with the knowledge of specific kinds of objects, namely artifacts. Furthermore, since the concepts of "production" ("creation") and "intentionality" are involved with the cognitive competence that is provided by the Stance of Design, the hypotheses about its psychological structure require from the subjects who are supposed to apply it 1) the capacity to understand causal relations, and 2) the capacity to ascribe intentional attitudes. Both these components seem necessary to grasp the notion of something being "intentionally made." German and Johnson also proposed a controversial hypothesis about the complex pattern of reasoning (the psychological structure) that underlies the understanding of artifacts and is based on these two capacities. Such reasoning

stems from the idea that the notion of "intentionally made for purpose *x*" involves coordinating two mental states: firstly that of the maker and secondly that of a subsequent user. One way of capturing the notion of design, therefore, is as a recursive mental state, as in "the maker intends that 'the user intends that *x*.'" (German and Johnson 2002: 297)

A debated question is relative to the development of the Stance of Design from more primitive psychological components. There is in fact some disagreement about when children can be said to have fully acquired the Design Stance and the solution to this problem can also define the modalities of its acquisition as well as its basic working. German and Defeyter (2000) hold that the Stance of Design reaches maturity only after each gear of the mental apparatus (mechanisms deputed to physical/causal knowledge, to the recognition of others' minds, to naïve biological categorization) has been oiled with some practice. Children would master the Stance through its repeated application and that would provide reasoning with some progressive constraints only by the age of seven. To German and Defeyter's view, therefore, the Stance of Design is a complex and abstract scheme built on core cognitive structures (probably innate) with which every human being is endowed from birth. Thus the Design Stance is an acquired reasoning skill consisting of the application of a *useful* mental scheme that is acquired after the development of prior basic competencies. Such a hypothesis implies that the Stance of Design is neither an innate psychological faculty, nor a cognitive *ability* that develops having evolved from the specific articulation and combination of certain other human faculties.

Kelemen and Carey also propose a developmental pattern in the understanding of design:

. . . children move from understanding an artifact as a means to an intentional end (thus "for" a user's current goal), to viewing it as the embodiment of a goal (thus "for" a privileged, intrinsic, enduring, function) to finally understanding it in terms of a full-blown Design Stance—an explanatory structure that is anchored by an understanding of intended function and supports rich inferences about the artifact's *raison d'étre*, kind, properties, and future activity. (2007: 224)

Like German and Johnson, Kelemen and Carey consider the Design Stance to be the result of attitudes that are more basic, such as the systems of core knowledge that provide part of the material from which it is constructed, and . . . general theory building processes that guide the child toward essentializing and theorizing about artifacts in terms of their origins. (2007: 228)

Their idea does not differ much from the previous hypothesis. However, Kelemen and Carey provide their proposal with an additional idea that makes their picture more convincing. They assert that children have an original, primitive bias for teleological accounts of all of the phenomena, labeled "promiscuous teleology":[7] "the tendency to treat . . . objects of all kinds as occurring for a purpose" (2007: 229). Humans start with such a bias to explain not only artifactual entities but also biological entities (both whole organisms and body parts) in terms of "what they are for." *Promiscuous teleology* is the natural bias that frames the development of the conceptual tools of children who will then acquire the Stance of Design. To further clarify how these ideas differ from the ideas of German and Defeyter, for Kelemen and Carey this primitive tendency to conceive of everything as being *for* something drives children to learn (arguably at a younger age than seven, perhaps at five or even at four) the Design Stance: promiscuous teleology is an inborn drive of human beings who apply that drive as a complementary trigger for the categorization and conceptualization, in the beginning, of all kinds of phenomena, but later mainly of artifacts. For Kelemen and Carey, promiscuous teleology is the characterizing feature that leads to the acquisition of the Stance of Design. The Stance of Design therefore appears as a special cognitive ability, even though it is based on other prior cognitive faculties. It is derived from the combination of core cognitive competencies, which are bound by promiscuous teleology. For Kelemen and Carey, the Stance of Design is the result of an inborn, natural predisposition that is peculiar to human beings; it is not just an acquired skill in reasoning.

10.2 Doubts About the Design Stance

The two hypotheses, one by German and Defeyter and the other by Kelemen and Carey, explain the Stance of Design as based on metaintentional capacities. They are intriguing, even though they both require a much better formulation as well as stronger empirical confirmation. Insofar as they have been advanced, they are more speculations than actual hypotheses, and present some considerable flaws.

The first hypothesis, by German and Defeyter, is questionable, not only because of the developmental data. In fact there is now evidence of design stance understanding at least by the age of five years and perhaps even earlier, when children are not normally thought to have much competence in second-order mental reasoning (cf. Diesendruck, Markson, and Bloom 2003). It is also questionable because it is not clear that recursive reasoning is needed in order to think about design (e.g., "the maker intends that X does Y" or "the maker intends that the user does X with Y" might suffice).[8] Furthermore it does not account for the features that mark the Stance of Design as anything more than a simple reasoning skill. These are the relative quickness and precocity with which humans apply it (cf. Bloom 1998). Finally German and Defeyter do not explain how and why the Design Stance develops from primitive competencies.

The second hypothesis, by Kelemen and Carey, that hinges on the idea of promiscuous teleology seems to be more coherent; yet, unfortunately, it lacks an adequate clarification of "promiscuous teleology." Such a notion is intuitively perspicuous, but is described as an innate *bias toward purpose*: little explanation is provided for it, both in terms of rigorous characterization and in terms of its justification. Kelemen assumes this bias on the basis of some interesting experimental evidence, but she does not clearly argue her assumption, making it thus appear rather ad hoc. Moreover, the appeal to a *bias toward purpose* runs the risk of being a *petitio principii*, if no further conceptual analysis of it is provided. It seems to explain the ascription of purpose or function[9] to things with the bias or tendency of young humans to ascribe purpose or function to things. Therefore the explanatory power of promiscuous teleology is limited to a little empirical observation that young human beings ascribe purpose or function to things, but the notions of "purpose" and "function" remain unexplained. The justification of the Design Stance on the basis of promiscuous teleology might be acceptable in common terms but is relatively insignificant from a scientific and/or philosophical point of view.

There are other general remarks to make about the two proposals that I have taken into consideration. Each characterization of the Design Stance includes an appeal both to causal cognition[10] and to the capacity to identify intentionality and/or agency. However, neither is the appeal to causal cognition sufficient, nor is the appeal to the capacity of identification of intentionality necessary or sufficient, to capture the entities to which we apply the Stance of Design. The characteristic feature at stake is the functionality of those entities, a feature that I will label "for-ness" from now on (cf. Meijers and Kroes 2005: x) because it underscores those entities in terms of "what they are for."

Let us now address the problem of characterizing "for-ness" in causal terms. In order to realize that a certain item X *is for* Y we need to know more than the simple fact that X *causes* Y. Arguably the knowledge that X *is for* Y also requires the knowledge that X *causes* Y. Indeed if one knows that X *is for* Y, maybe one also knows that *in some way* or *in some sense* X *causes* Y; but equalizing the two forms of knowledge is inappropriate: the knowledge of simple *causality* does not account for the amount of information that is provided by knowing *for-ness*. Indeed functional knowledge could be thoroughly scrutinized in terms of causal reasoning and the perception of causal relations. However, none of the cognitive approaches[11] to causal cognition would be able to properly account for the perception of functionality. Causal cognition consists essentially of the psychological processes that bring forth knowledge of at least a binary relation between the cause C and the effect E. For example, one knows that fire (C) causes smoke (E),[12] but stating that fire *has the function* or *is for* producing smoke would sound wrong as well as weird, even though sometimes (cf. Kelemen 1999) we may have been willing to claim this. Likewise, we usually take for granted the fact that the function of a glass is to contain a liquid: the glass *is for* containing liquids, though we are not eager to claim that the glass *causes* a liquid containment. In fact we may need to know the *causal* properties of the glass in order to understand its function. We may, for instance, need to know that a glass has to be made of waterproof material and that it has to have a solid, hollow, and compact structure. These properties can be considered necessary to cause the containment of liquids and other incoherent substances. That is, we know that if an object, for example, a toy sponge glass, does not possess these properties, it cannot be a glass. Possessing the concept of "causality" is perhaps necessary to know for-ness; it is definitely not sufficient.

The other condition presupposed by cognitive psychologists to account for for-ness, that is, the mastery of a metaintentional capacity, is unnecessary *and* insufficient. Apparently according to some intuitions as well as to a more theoretical bias, there is agency underlying design, and "design" is the key concept that makes sense of functions and, more generally, of for-ness. Whenever a certain item is conceived as *being for* something, the *hidden hand* of some agent is presupposed. An agent is supposed to have designed that item for the purpose of *being for* something. The presupposition of an agency behind for-ness requires that those who recognize that a certain entity is for something are able to ascribe intentionality to the putative author who has designed it. For this reason, recognizing for-ness involves a metaintentional capacity. Such an idea is the starting point for the hypotheses of both German and Defeyter and Kelemen and Carey, and in a less immediate way, for the original Dennettian idea of the Stance of Design.

Indeed Dennett has stretched the concept of "design" based on the concept of "designer" far enough to identify Mother Nature as being a designer, that is, an intentional agent. He does describe the design of Mother Nature in a somewhat metaphorical way, so that Mother Nature is to be perceived *as if* she were a designer, even though she is not; but he does not clearly define the terms of his metaphor. Thus there is no clearness about what he means by "designer" and "design" when he refers to nature and its work with intentional concepts.

In fact the observation based on our experience of human facts can provide some ground for the inference that each object produced by humans is there for a reason and is therefore also made by someone. Yet there are no justified steps that endorse the inference that since human-made things are made for a reason, any object, also in nature, is made by someone because it is made for something. Undoubtedly the lexicon used to describe natural phenomena in terms of "design" and "designer" is charged with a heavy load of intentionality, which is introduced by the kinship of these notions with the conception of agency, and usually involves conscious deliberation. But nature does not consciously deliberate; to describe natural facts with intentional jargon is deceptive because it improperly depicts nature as an intentional subject. Nature is more properly described in raw causal terms.

The characterization of natural facts in intentional terms is also responsible for a crucial misunderstanding. This is the bias with which we usually consider for-ness to be dependent on agency and intentionality. For-ness instead can be considered as a feature that is both logically and psychologically independent of any intentional characterization. From a logical point of view, that a certain object *is for* something is a property that does not need to rely on the relational dependence of an agent (the supposed *designer*) and his or her perspective. Indeed the recognition that a certain object *is for* something does depend on an intentional perspective, but this is one and the same as a beholder who identifies any other property. Thus the fact that a certain object *is for* something is related to a point of view, but not necessarily that of a designer. For-ness is also identifiable from a psychological point of view without any need to appeal to an agent, the one who is supposed to have designed the object at stake as *being for* something. We can easily recognize that a shell as well as an ashtray can *be for* containing ash or other powder. The requirement of the capacity to grasp agency and intentionality is not therefore necessary to detect for-ness, and it only *seems* to be needed because of the intentional bias that makes us match the feature that a certain entity *is for* something upon *being made*—arguably because we mostly refer to things that are made by humans and that are always made *for* something. We live in artifact-saturated environments.

Let me summarize the remarks of this section. First, I displayed the weak points of each of the two hypotheses about the Design Stance that I reviewed. Then I criticized the two cognitive conditions that are presupposed by both hypotheses in order to account for functional cognition. These are, on the one hand, the prerequisites for causal cognition and, on the other hand, the idea that in order to make sense of the for-ness of a certain item we have to presuppose an agent that made it *for* something. I argued that the mastery of causal cognition is not sufficient to provide an understanding of for-ness. I asserted that the requirement to master a metaintentional ability is unnecessary. Given that I maintain that intentionality recognition is unnecessary, I also considered the hypothesis that matches the two cognitive processes discussed in this section 10.2 to be unnecessary (causal cognition plus agency cognition) in a more complex and systematic cognitive structure. Such a hypothesized complex cognitive structure (actually, like the ones proposed by the two pairs of psychologists) would be explanatorily ineffective and excessively concocted.

The aforementioned are just some reasons for believing that the accounts of the capacity to categorize and conceptualize artifacts by means of the Stance of Design are not fully satisfactory. In fact the Design Stance appears to be a rather confused explanatory principle that creates more problems than it solves. The relation between the for-ness of an artifact and the intentions of its designer, which are taken to be constitutive of the Stance of Design, are particularly dubious, because they are unspecified. Furthermore I think that these accounts are unsatisfactory because they do not explain either why we are inclined to distinguish between artifacts and natural entities or why we often misapply this categorical distinction, although we naturally categorize artifacts differently from natural entities. (Often people are convinced that certain natural items are artifacts and vice versa.)

Yet I have not argued for a different account. To propose a better alternative account than the ones reviewed, I should avoid the flaws discovered. I sketch my hypothesis of for-ness and functional knowledge in the next section where I also clarify some ideas that could be useful when giving a more satisfactory characterization of the Design Stance.

10.3   A Functional Stance for the Design Stance

The hypothesis that I put forward is mainly based on intuition, and is abstract and speculative at best. It does not aim at replacing the Stance of Design; it aims rather to make it fully coherent and complete. According to the Design Stance approach, we recognize artifacts by means of an inferential process that draws on (unspecified) assumptions about the relationship between a certain object and the intentions that its putative designer has for it.

I maintain that we do apply a cognitive scheme when we recognize that certain entities are *for* something, but this scheme is different from the one assumed by the theorists whose work I have scrutinized. Such a scheme makes us perceive the elements of two states of affairs as bound by a *catalyst*. The catalyst realizes the connective link that relates the two states of affairs and in effect is the item that implements the transformation of one state of affairs to another. This *functional* scheme also determines a basic temporal sequence according to which the elements that are involved in the states of affairs at stake change. The sequence consists of three logically ordered stages: first the antecedent state of affairs, then the process of transformation (that is realized by the catalyst or vehicle), and finally the consequent state. This sequence is different from a linear causal chain and cannot be reduced to it. In a causal chain the basic units are discrete causal links occurring between two events; the causal relata of an individual causal link are two.[13] Instead, in order to perceive the for-ness sequence, we need to perceive all of the three stages as inseparable. Grasping for-ness requires the identification of three-phasic, irreducible basic units; there are three relata of the pattern identified by the functional scheme. Furthermore, while we perceive of causation as occurring between events (or objects), we recognize for-ness as a relational feature occurring between states of affairs and a specific discrete entity (the catalyst).

For example, the claim "yeast is *for* leavening the dough" entails a ternary relation of an antecedent state (when the dough has not yet risen), a consequent state (when it has risen), and the process of transformation (through the yeast) from the antecedent to the consequent. Clearly there can be an alternative description that sounds more plausible such as "yeast *causes* the rising of the dough," but this is a description of a direct connection between two events. Instead, to claim that "yeast is *for* leavening" (although it is a strange sentence) appeals to the process of transformation for which yeast is the catalyst. The recognition of a functional feature therefore possibly involves the capacity to understand a *logic of change*.[14] Grasping for-ness is more complex than grasping causality because for-ness requires a more elaborate discrimination of time stages, states of affairs, and discrete objects; it has to take into account a relation with more terms.

Probably the relation of for-ness is recognized *on the grounds* of causal cognition because each mechanical phase that is part of the sequence at stake can be causally understood. However, causal cognition is not enough to make sense of the whole sequence. The reason for this is trivial but effective: the reduction of a functional pattern that is also rudimentary (such as the screwing dynamics of a simple screw) into a succession of causal relations is overwhelmingly complicated. Yet some might object that this reduction would be feasible *in principle*. Indeed in principle the structure of for-ness could be described in terms of causal relations and it could be logically fragmented into simpler relations. Undoubtedly such a description would appear more factual from a scientific point of view. However, this description would not account for the phenomenal picture that we realize with the perception of for-ness, which is implemented at a different and plausibly higher cognitive level. When we perceive the functional feature of a certain item, when we perceive that it *is for* something, we do not see a number of joint causal relations. At a glance we see them unified into a pattern that amounts to for-ness, even though we can *logically* (but not psychologically) distinguish the three stages. For instance, recognizing that a corkscrew *is for* drawing corks from bottles (or, that its function is to draw corks from bottles) provides a subject with a different knowledge than the knowledge that a corkscrew *causes* the removal of a cork from a bottle. The pattern that we recognize presupposes a procedural step—the transformation process—that can be grasped by the human mind only at a level that is different to the one in which individual causal links are cognitively processed. Therefore the difference between recognizing the causal chains that can be constitutive of a functional device and recognizing the functional feature of this device is not only a matter of complexity. It is also a matter of different cognitive levels: the functional one is higher than the causal one, even though the ratio between the functional image that we grasp and the perception of the individual causal links and chains still has to be made explicit.

I think of the functional scheme as a device that operates in a similar way to the one that makes us perceive the immediate succession of the elements of the frames of a cartoon as if they were moving. In fact the elements represented in the cartoon do not actually move, but we perceive them as moving. We can describe each of the elements of the pictures in a cartoon as still if we analyze them one by one. However, if the pictures are projected immediately one after the other, the human eye is not able to register the interruption between two photograms and thus the eye sees a unified flow. If we perceive and describe each of the pictures individually, we can of course redescribe them, but we lose the perception of the motion of the elements that are contained in them. Therefore I hypothesize that we could perceive a functional sequence by means of a functional scheme that probably belongs to our perceptual and conceptual apparatus and that makes us conceive of the elements of the for-ness sequence as being unified in an individual procedural path.[15]

The ability to jointly perceive an antecedent state of affairs, a transformation process, and a consequent state of affairs in the for-ness sequence makes possible or enhances the *phenomenal* understanding, so to speak, of certain operational performances. In fact detecting for-ness provides an understanding of certain paths for the realizability of some phenomena, those phenomena that are engendered by a transformation processes. This ability amounts to what can be defined as functional knowledge and I take it to be implemented with a Functional Stance, the stance realized by the application of the functional scheme. The knowledge acquired by means of simple causal cognition does not facilitate the understanding of the procedural path that is provided by functional knowledge.

That which I have defined as *functional knowledge* is what enables humans to detect that certain entities are *for* something without appealing to the agency of the author who is supposed to have designed those entities. Therefore there is no need to involve metaintentional abilities as did German and Defeyter as well as Kelemen and Carey (though less convincingly). According to these authors, such cognitive performance would make use of the capacity to detect someone's intentional attitudes about a certain entity, the entity that is intended *for something*. I argue instead that in order to grasp that an object is *for* something, no special metaintentional operation is required that assumes that someone, an agent, has intentionally made or conceived that object for something.

According to my hypothesis, we recognize for-ness in objects just because we apply our functional scheme to them. This makes them appear as catalysts of the transformation process in the for-ness relational sequence. No other cognitive device or presupposition is required to realize that some entities are *for* something. In my hypothesis, therefore, for-ness is a fairly simple, perhaps primitive, elementary feature that is revealed with the application of the functional scheme. This scheme is a tool of the human cognitive apparatus; it does not need a particular load of cognitive resources required in a metaintentional capacity, nor does it require special cognitive training. I take it to be effective already by the age of four or possibly younger—as soon as a child recognizes that an object *is for* something. One might take the functional stance by default, automatically or without active consideration, due to the lack of viable cognitive alternatives, such as the recognition of causation in case this was not cogent. The application of the functional scheme by default would thus explain why we can indistinctly recognize any kind of object, either natural or artifactual, as being *for* something; even though some of these objects are or might be *for* nothing. This functional scheme leads to a transformational procedure that very likely consists of a number of other processes. These are too many and too complex to be grasped by the human mind and may be in principle cognitively inaccessible; they merge into our broader phenomenal understanding of the world. For this reason we need a psychological scheme that helps us to understand them at a glance as a unified procedure. Such a procedure has to be rapidly and automatically recognized, possibly through the application of an unaware mechanism—hence the functional scheme.

Some phenomena that are described in cognitive literature can be considered as weak evidence in favor of the Functional Stance that we take when we apply the functional scheme. These phenomena are the capacity to detect the for-ness that appears early in childhood and is extended to any possible kind (both natural objects and artifacts), as partially reported by Kelemen and Carey 2007, as well as the universality of this capacity, observed in different cultures (Walker 1999; German and Barrett 2005). These variables seem to endorse the fact that the recognition of for-ness results from the work of a specialized cognitive mechanism that is deputed to implement functional knowledge. They seem to prove also that functional categorization is fast, unconscious, and performed by young children. However, functional categorization does not necessarily correspond to artifact categorization, as some authors seem to hold (Bloom 1998: 91). Functionality, rather than artifactuality, is recognized as a characterizing feature, but functionality is different from artifactuality. Functionality is a feature that is in fact not identified as being distinctive only of artifacts, even though very often they can be associated with artifactuality. Such an association would indeed be established by a further cognitive step that is performed by the Design Stance, a more complex cognitive device that might depend on the Functional Stance. This further step could be useful in determining the intentional *origins* of artifacts, not their functional properties, which, as I have pointed out, can be independently detected.

Functional knowledge, so characterized, may appear to be a cognitive trick because it covers our ignorance about the real workings of the phenomena that we recognize as being *for* something. It might be objected that we can reach a real understanding of them only if we are able to grasp their hidden causal mechanisms. Thus for-ness recognition would provide us with a grasp of superficial mechanisms that arguably does not reveal the real nature of the phenomena with which we are in touch. The functional stance at least endorses an understanding of phenomena from a *practical* point of view that is essential for the conducting of our daily lives, a fortiori in a technologically advanced culture. Functional knowledge is, of course, to be considered a pragmatic heuristic of common sense; it would be a part of the broad cognitive area covered by *folk competences* (such as naïve psychology, naïve physics, naïve biology, naïve mathematics, etc.). Such knowledge, although not scientifically reliable, is effective for the practical goings on of everyday life.

Indeed engineers, architects, and designers in general have to deal with a lay dimension for the realization of several products. These products have to be conceived of in terms of the intuitions, which underlie common sense for two basic reasons. First, the terms, concepts, and methodology of strict sciences do not fit the demands of the practical management of life. Strict science often deals with idealized situations and does not take into account the contingencies of everyday life. Moreover, a matter of grain size is at stake: the sciences normally approach selected, fine-grained phenomena that are described with specialized jargon. Practical life, on the contrary, deals with coarse-grained phenomena that are usually characterized according to the jargon of common sense. The second reason why designers have to take into consideration the dynamics of common sense is that, even though they may plan their projects making use of a vast amount of technical knowledge, they often have to think of the products that realize those projects as being intended for laypeople. Indeed they have to design their products in such a way as to make them understandable to laypeople. Thus they also have to take into account functional knowledge that is reasonably to be considered constitutive of common sense. Donald Norman (both an engineer and a cognitive scientist), for example, applied the principles of ecological psychology to design methodology in order to deal with the practical dimension of common sense.[16] He also systematically took into consideration several psychological studies of folk competences. On the analogy of Norman's research, functional knowledge could be seriously scrutinized as well.

The assumption of a functional scheme also seems rationally justified: in the account of the human capacity to manage practical life, such a cognitive tool is an easy explanatory principle. Indeed functional knowledge would account for the human cognitive skill to understand complex mechanisms, a skill that is pivotal in the development of technology. In fact technology cannot be properly explained only in terms of causal cognition simply by claiming that it "was originally the result largely of imaginative trial and error" (Wolpert 2003). The ability to arrange human experience and cognitive skills into a structured form of knowledge that can be accessed easily, such as with technology, requires fairly advanced and effective competencies, one of which could be that deputed to functional knowledge.

The hypothesis that our cognitive system, by means of the application of the functional scheme, makes us perceive objects as *being for* something appears reasonable. Thus it appears reasonable that we are endowed with a spontaneous classificatory bias that is engendered by our cognitive mechanism for the recognition of for-ness. We are also erroneously prompted by our habits, which are used in artifact-saturated environments to identify the things that we perceive as being *for something* and as being *made by someone*. Such an incorrect inference has induced us to think of the entities that we recognize as being *for something* as being entangled with agency and intentionality, while such matching is not necessary. In fact interpreting any functional feature as a feature that is intentionally loaded is unnecessary and misleading, so we are convinced that we need to extend the characterizing trait of artifacts, their *being made for something*, to all the things that we perceive only as being *for something*. Among these we include several natural items that in fact do not reveal any clue of an agency relative to their origins.

According to this hypothesis an important practical heuristic *devised* and put to use by our cognitive system, in combination with our experience of an artifact-saturated world, tends to confound our dichotomic categorization of artifacts and natural entities. Simple categorical disarray, however, seems to be a reasonable price for quick and frugal heuristics, such as with the functional stance. This facilitates and quickens our understanding of the workings of several natural and artifactual things, and it contributes to our application of the Stance of Design in an effort to understand the intentional origins of artifacts, and so improves our general technical capacities.

10.4 Conclusion

To clarify notions such as "mind-dependency" and "functionality," which ontologists take to be distinctive of artifacts, I refer to some cognitive studies on artifact categorization and conceptualization. Cognitive investigations appeal, however, to an indeterminate idea, that of "Design Stance," which needs to be more thoroughly explained. In fact the Stance of Design is characterized as the human approach toward artifacts in terms of inferential processes that draw on assumptions about the relationship between the function of an artifact and the intention of its designers. I argue that we need to know more in detail about the constitutive elements and how they work. I hypothesize that in order to understand the functionality, or *for-ness*, of artifacts, a more basic *stance* has to be assumed. This is the stance that we take by applying a cognitive scheme that makes us perceive functional things as *vehicles, catalysts* of transformational processes. Humans apply this scheme by default, automatically and unconsciously, mostly by means of the activation of a specialized cognitive mechanism. This explains why the human capacity to grasp for-ness is prompt and selective; and it also explains why it is not exclusively applied to artifacts but also to several natural items. This basic capacity does not therefore discriminate between artifacts and natural entities. Our distinction of artifacts from natural entities is based rather on the inferential reasoning (usually performed with the Design Stance) that makes us recognize their intentional origins. It is such reasoning that presupposes the functional stance, rather than being presupposed by it. According to my proposal, therefore, functional knowledge could be a specialized and basic constitutive component of the more general attitude of the human mind to categorize the objects of this world into artifacts and natural entities.

More thorough investigation is needed, investigation that develops further discussion on the issues considered here and arguably on other issues such as the cognitive nature of the for-ness detection mechanism (whether it is an innate modular mechanism or an acquired general capacity, how it could develop in the human mind, etc.) or the adequate logical formalization of the for-ness relation. Other empirical confirmation would be required as well. I hope that after the development of this investigation there will be suitable conditions to decide whether my proposal can be formulated into a *sound* hypothesis. This could help to explain better how the human mind identifies the features that are taken to be distinctive of artifacts with respect to natural items, and it could also make these features more certain from an ontological point of view.

Acknowledgments

I wish to thank Melissa ("Missy") Ciaravino, philosophy student at the State University of New York at Buffalo, for editing the English of a previous version of my chapter.

Notes

1. An intentional attitude of second order could be, e.g., *my belief* **that** I remember that I met Elvis, and *Jill's belief* **that** Mary thinks that Elvis is still alive. Perhaps "metaintentionality" is a special case of "metacognition" as defined by Moses and Baird, that is, "any knowledge or cognitive process that refers, motors, or controls any aspect of cognition" (1999: 533–535).

2. Simons theorizes about artifacts in other important texts also, such as in Simons (1989) and Simons and Dement (1996), but in these texts he focuses on artifacts from a strictly ontological, and more precisely *mereological* point of view, without providing a general explicit definition of *artifact* such as the one that I have quoted. However, I report Simons's definition because it is sufficiently broad to comprehend all of the other ontological definitions and characterizations of *artifact*.

3. This, in Dennett's philosophy, is interchangeably used with "Design Stance."

4. Recently there has been quite a debate about the relations between the Dennettian *Stances* and the philosophical relevance of the Stance of Design (cf. Baker 1987; Millikan 2000; Ratcliffe 2001).

5. In fact Dennett's ontological commitment relative to the *mental* is to be considered as a form of *instrumentalism* or *mild realism*.

6. In terms of *real* intentional states and attitudes.

7. Deborah Kelemen has explicitly argued in favor of this idea (Kelemen 1999a, b, c, d; 2003; 2004).

8. This remark is derived from personal communication with Professor Deborah Kelemen.

9. Kelemen, together with most of the other psychologists who have theorized about the Stance of Design, uses *purpose* without distinguishing it much from *function*.

10. That is, the reasoning that enables humans to recognize sequences of events as being causally related.

11. For an overview of the cognitive studies of causal cognition, see Cheng (1999).

12. Here I am referring to *causality* and *causation* as loosely understood terms in the common sense; I am not appealing to any philosophical and/or scientific theory. However, I take the original Humean account of "causality" (Hume 1987 [1739]) in terms of temporal priority, contiguity, and the constant conjunction of cause C to its effect E, as being rather close to the conception of "common sense"; even though the Humean account was further developed in modern regularity approaches that seem intuitively less clear to the causal reasoning of common sense.

13. Here again I am referring to "causality" and "causation" as loosely understood in common terms, not in a scientific or philosophical theory.

14. The path toward a logic of change has been broken by von Wright (1969); a relationship between the proper formulation of a logic of change and its employment in studying the cognition of functional features could be a first step toward a thorough investigation of functional knowledge.

15. One could think of a somewhat Gestalt-like principle of unification of the individual causal links to the functional picture; this is, however, pure speculation.

16. Norman specifically appealed to the work of Gibson. Of course I do not make a secret of myself being inspired by the work of Gibson (1979) in this hypothesis. However, there are some considerable differences (only sketched here in the interests of brevity) between the concept of "for-ness" and the key concept of

"affordance" employed by Gibson. According to Gibson, an "affordance" is an interaction between an animal and an environment; it is a resource that the environment offers to an animal, but the animal has to be able to perceive and use it. Moreover, affordance also exists if not perceived. For-ness is instead, in my view, a relation that is always dependent on the cognitive apparatus of an animal (the human being), therefore it is not present on its own by itself, and it is a relation that is projected between things in the world, separate from their usability.

References

Baker, L. (1987). *Saving Belief: A Critique of Physicalism*. New Jersey: Princeton University Press.

Baker, L. (2004). The ontology of artifacts. *Philosophical Explorations*, 7: 1–14.

Bloom, P. (1998). Theories of artifact categorization. *Cognition*, 66: 87–93.

Cheng, P. (1999). Causal reasoning. In: *The MIT Encyclopedia of Cognitive Sciences* (Wilson, R. A., and Keil, F. C., eds.), 106–108. Cambridge, Mass.: The MIT Press.

Defeyter, M., and German, T. (2003). Acquiring an understanding of design: evidence from children's insight problem-solving. *Cognition*, 89: 133–155.

Dennett, D. C. (1983). *The Intentional Stance*. Cambridge/London: The MIT Press.

Dennett, D. C. (1996). *Darwin's Dangerous Idea*: Evolution and the Meanings of Life. New York: Touchstone.

Dennett, D. C. (1999). The intentional stance. In: *The MIT Encyclopedia of Cognitive Sciences* (Wilson, R. A., and Keil, F. C., eds.), 412–413. Cambridge, Mass.: The MIT Press.

Diesendruck, G., Markson, L. M., and Bloom, P. (2003). Children's reliance on creator's intent in extending names for artifacts. *Psychological Science*, 14: 164–168.

German, T., and Barrett, C. (2005). Functional fixedness in a technologically sparse culture. *Psychological Science*, 16: 1–5.

German, T., and Defeyter, M. (2000). Immunity to functional fixedness in young children. *Psychonomic Bulletin and Review*, 7: 707–712.

German, T., and Johnson, S. (2002). Function and the origins of the design stance. *Journal of Cognition and Development*, 3: 279–300.

Gibson, J. J. (1979). *The Ecological Approach to Vision Perception*. Boston: Houghton Mifflin.

Hoffman, J., and Rosenkrantz, G. (1997). *Substance*: Its Nature and Existence. London: Routledge.

Hume, D. (1987 [1739]). *A Treatise of Human Nature*. Oxford: Clarendon Press.

Kelemen, D. (1999a). Beliefs about purpose: On the origins of teleological thought. In: *The Descent of Mind*: Psychological Perspectives on Hominid Evolution (Corballis, M., and Lea, S., eds.), 278–294. Oxford: Oxford University Press.

Kelemen, D. (1999b). Why are rocks pointy? Children's preference for teleological explanations of the natural world. *Developmental Psychology*, 35: 1440–1453.

Kelemen, D. (1999c). The scope of teleological thinking in preschool children. *Cognition*, 70: 241–272.

Kelemen, D. (1999d). Functions, goals and intentions: Children's teleological reasoning about objects. *Trends in Cognitive Sciences*, 12: 461–468.

Kelemen, D. (2003). British and American children's preferences for teleo-functional explanations of the natural world. *Cognition*, 88: 201–221.

Kelemen, D. (2004). Are children "intuitive theists"? Reasoning about purpose and design in nature. *Psychological Science*, 15: 295–301.

Kelemen, D., and Carey, S. (2007). The essence of artifacts: Developing the design stance. In: *Creations of the Mind*: Theories of Artifacts and Their Representation (Margolis, E., and Laurence, S., eds.), 212–230. Oxford: Oxford University Press.

Meijers, A., and Kroes, P. (2005). Introduction. In: *Philosophy of Technical Artefacts*. Joint Delft-Eindhoven Research Programme. Eindhoven: Technische Universiteit Eindhoven.

Millikan, R. G. (2000). Reading mother nature's mind. In: *Dennett's Philosophy: A Comprehensive Assessment* (Ross, D., Brook, A., and Thompson, D., eds.), 55–75. Cambridge, Mass.: The MIT Press.

Moses, L. A., and Baird, J. A. (1999). Metacognition. In: *The MIT Encyclopedia of Cognitive Sciences* (Wilson, R. A., and Keil, F. C., eds.), 533–535. Cambridge, Mass.: The MIT Press.

Norman, D. A. (1987). *The Psychology of Everyday Things.* New York: Basic Books.

Ratcliffe, M. (2001). A Kantian stance on the intentional stance. *Biology and Philosophy, 16*: 29–52.

Rea, M. (1995). The problem of material constitution. *The Philosophical Review, 104*: 525–535.

Simons, P. (1989). *Parts: A Study in Ontology.* Oxford: Oxford University Press.

Simons, P. (1995). Artefact. In: *A Companion to Metaphysics* (Kim, J., and Sosa, E., eds.), 33. Oxford: Blackwell.

Simons, P., and Dement, C. (1996). Aspects of the mereology of artefacts. In: *Formal Ontology* (Poli, R., and Simons, P., eds.), 255–276. Dordrecht: Kluwer.

von Wright, G. H. (1969). *Time, Change, and Contradiction.* Cambridge: Cambridge University Press.

Walker, J. (1999). Culture, domain specificity and conceptual change: Natural kind and artifact concepts. *British Journal of Developmental Psychology, 17*: 203–219.

Wiggins, D. (2001). *Sameness and Substance Renewed.* Cambridge: Cambridge University Press.

Wolpert, L. (2003). Causal belief and the origins of technology. *Philosophical Transactions of the Royal Society of London, A 361*: 1709–1719.

11 Realism and Artifact Kinds

Marzia Soavi

11.1 Introduction

Strong realism is the thesis according to which a structured world exists independent of human thought and knowledge—a world composed of distinct entities of different natures.[1] One of the main problems for strong realists, then, is that of establishing which of the entities that we commonly individuate are real components of the independent world and which are mere projections of our thoughts. In this chapter I analyze an argument aimed at proving that artifacts, unlike natural objects, are not constituents of the real world as conceived by strong realism. I try to show that it is possible to conceive artifact kinds as both real and functional.

Within the debate over the existence of kinds, it is instructive to distinguish between epistemological and metaphysical issues. To bring to light the tension that exists between the two approaches, I adopt the term *natural kinds* for the kinds conceived from the epistemological perspective and *real kinds* for the kinds conceived from the metaphysical angle.

Natural kinds are considered to be kinds whose instances are objects that share one or more properties that are fundamental from a certain theoretical point of view. For example, samples of the same chemical kinds share the same molecular composition or have the same atomic number. Typically, natural kinds are characterized as kinds that strongly support induction, that is, they allow for the discovery of properties that are projectable over their instances. Normally, natural kinds are contrasted with kinds whose instances do not share any theoretically relevant property—typical examples are the kind of bachelor, the kind of widow, or the kind of vixen. These are sometimes called "artificial kinds" or alternatively, "nominal kinds."[2] The idea is that items of natural kinds necessarily share properties that can explain their superficial similarities and that grant that items of the same kinds behave and react to the environment in the same way, while that is not the case for artificial or nominal kinds.

Real kinds—metaphysically characterized—are those kinds that constitute the real units of the world. Objects belonging to real kinds have the same nature. Real kinds are often contrasted with nominal kinds. These are kinds that collect objects that do not necessarily share any common nature, again bachelor, vixen, and widow are traditional examples of nominal kinds. For any metaphysics that accepts the existence of real kinds, if an object *o* is individuated as an object of a kind K then *o* is a real object only if K is a real kind. For example, we can individuate the same portion of matter as a certain amount of clay or as a statue, but if we do not admit that the kind of *statue* is among real kinds then the statue is not a real object. If we acknowledge the kind of *amount of clay*, then the amount of clay is a real object. If we acknowledge that both the kind of *statue* and the kind of *amount of clay* exist and that criteria of identity for statues and amounts of clay do not coincide, then we are bound to admit that there are possibly two objects occupying the same space at the same time.

If we accept the idea that when natural kinds strongly support induction they cannot do it by accident but by selecting precisely those objects that share the same nature, then it is easy to see why the distinction between natural and artificial kinds is assimilated in the distinction between real and nominal kinds. Natural kinds strongly support induction because they individuate real kinds, that is, they collect objects with the same nature and, ideally, the nearer they get to the individuation of some real kind the stronger their support of induction will be.

Artificial kinds, by contrast, are nominal kinds; they do not necessarily collect objects that share the same nature and they trace distinctions simply according to our needs, beliefs, or linguistic practices.

This relation has sometimes caused conflict between the epistemological and the ontological distinction, leading to the use of *natural kinds* as a synonym for *real kinds*. In such cases the qualification "natural" has nothing to do with the notion of "natural" that is normally contrasted with that of "artifact." The distinction between artifacts and natural objects is controversial and highly problematic. Generally speaking, artifacts can be considered man-made objects, mostly those produced to perform a certain function, while natural objects are not man-made. Indeed there are objects that are deemed to be natural objects that are intentionally produced, but I am not interested in the defense of this distinction here, nor am I interested in a refined version of it. What is important is to keep the two notions of "nature"—one that is compared to the notion of "artifact" and one that is compared to the notion of "artificial kinds"—clearly distinct, otherwise the expression "natural kinds" seems to lead to the trivial conclusion that artifact kinds are not real kinds while, conversely, there is nothing in the idea that an object is intentionally produced that can allow us to infer that there is no real or natural kind to which such objects belong. That is, there is nothing in the assertion that *o* is an artifact that can allow us to infer the assertion that *o* is not a real object.

Devitt claims that artifact kinds are not entitled to be real kinds because they are metaphysically unnecessary—artifacts already fall under common physical kinds (unfortunately he does not provide any example of such *common physical kinds*). According to

Devitt, those objects that are said to be tables do not share a common nature: being a table is not a real property. We just have the word *table* to name those objects that are used in a certain manner or play a certain role or are built according to certain intentions. This can be labeled, as Losonsky labels it, the "Aristotelian position on artefacts" (Losonsky 1990: 81–82). Many other authors adopting strong realism defend the idea that artifact kinds are not real kinds with different arguments. But artifacts are objects that we do individuate and classify as easily as natural beings; different artifacts seem to have a different nature just like different animals or any other natural being of an acknowledged kind. Thus once we admit the existence of kinds for natural objects, we need very good reasons to deny the existence of kinds for artifacts. Moreover, since there are well-known arguments according to which the individuation of an artifact never coincides with the individuation of physical kinds, such as the amount of matter composing it, it is not even easy to accept a reductionistic[3] solution as proposed by Devitt and others.

Among the arguments that have been put forward to deny the existence of real kinds for artifacts, there are some that are based on the idea that artifact kinds cannot be real kinds because they are functional kinds. In the present study I examine the relation between artifact functions and artifact kinds in an attempt to establish reconciliation between the functional characterization of artifact kinds and realism.

11.2 The Argument from Multiple Realizability

Why can functional kinds not be real kinds? Real kinds are supposed to collect objects that have the same nature. According to some famous examples presented by Kripke (1980) and Putnam (1970), the nature can be identified with the molecular or atomic structure for chemical substances, and with the genetic content for biological entities. That is, objects of the same real kind have the same inner structure that causes them to react with the environment and behave in similar ways, which is why real kinds strongly support induction. For functional kinds, the following principle holds:

(FK) o is an item of a functional kind K iff o has the function F.

Generally speaking, "o has the function F" means roughly that o is used for or is produced for F. This interpretation plus the widely accepted principle that functions are multirealizable leads to the consequence that objects of the same functional kind may have very different structures and be composed of different materials. Identity of function does not therefore guarantee any identity of nature, according to the Kripke-Putnam notion. Artifact kinds, such that *watch, chair*, and *pen* are kinds of this type that collect objects with no common inner structure, for this reason cannot be considered real kinds.

First, we need a more critical account of what "o has the function F" means. I do not want to analyze the general epistemological problem of what the right criteria for function attribution to artifacts are, as pluralism seems to be a perfectly feasible solution. Instead my concern is to try to understand which criteria for function attribution are involved in our classification of artifacts as kinds.

Traditionally since the work of Wright, two distinct types of criteria for function attribution have been proposed: they are the selective criterion for biological entities and the intentional criterion for artifacts—either related to use or to production. More recently this distinction has been challenged by those claiming that it is possible to adopt selective criteria for function attribution to artifacts.[4] That makes it possible to individuate three main criteria for attribution of function to artifacts: a selective criterion analogous to the criterion defined for biological entities, and the two intentional criteria of use and production criteria.

Then accepting the common functional classification of artifacts mentioned in the first section,

(FA) o is an artifact of a kind K iff o has the function F^5,

it is possible to analyze "has the function F" according to three different interpretations and to formulate the following three options:

1. o is an object of an artifact kind K iff o has been selected for F.
2. o is an object of an artifact kind K iff o is used for F.
3. o is an object of an artifact kind K iff o has been produced for F.

In (FA) and (1), (2), and (3), K and F must be suitably chosen; that is, we must grant that there is some connection between F and K. According to a pluralistic approach, all three solutions can provide a good criterion for some attribution of functions to artifacts. My interest is not in eliminating any one of them as improper in an account of the attribution of functions to artifacts but simply in exploring which notion of function can provide the basis for a definition of functional kinds that will meet the challenges of antirealists for artifacts.

11.3   First Option

The first statement corresponds to the attempt to apply the function attribution worked out for biological entities to artifacts.

1. does not require that o is directly selected for the function F. What it requires is that there is an appropriate history of selection for o, according to which it is true that o has been selected for F.

Undoubtedly natural selection necessarily involves inheritance, that is, some process granting the transmission of characters. Any general theory of selection that is thus aimed at covering both a theory of selection for biological entities and artifacts must define a general relation of inheritance that is viable for the application to both.

In line with Millikan, we can introduce a general notion of "copy relation" that is intended to cover biological inheritance and the corresponding phenomenon for artifacts. To increase our conceptual tools we can distinguish between a strong and a weak notion of "copy."

- According to the strong notion the relation of copy implies a counterfactual dependence of the features of the copy on the features of the original.
- The weak notion of copy does not require any strict counterfactual dependence of this type; it simply requires that there is a noncasual similarity between copies.

According to Millikan, any copy relation involved in a selective process has to be of the strong type. It must grant that for every determinable property such as skin color or iris color there are local laws governing the transmission or copying of the determinate characters in such a way that counterfactuals can be warranted.[6] To defend a strong selectionist approach to the attribution of function to artifacts, we need to individuate a mechanism that can grant the holding of a strong copy relation for artifacts of the same kind. Indeed artifacts do not reproduce themselves as do biological entities, but the notion of "copy" allows us to discard the material aspect of the process. Alternatively if we do not succeed with the strong notion of "copy," we can still try to create a selectionist criterion for artifact function that adopts a weak relation of copy rather than a strong one. It seems reasonable in fact to allow some discrepancy between the processes of selection for natural beings, and for artifacts. We may agree, for example, on the identity of type between the copy process for biological entities and for artifacts while accepting that there is a difference in the degree of predictability—the copying process for artifacts is simply less predictable where the transmission of characters is concerned and leaves more space for incidence of variation.

There is an important role that the relation of copy plays in the theory of selection; to illustrate it I consider the general analysis of the theory of selection provided by Lindley Darden and Joseph Cain (1989). They develop an analysis in five steps: preconditions, interactions, effects, longer-range effects, and even-longer-range effects. Preconditions are a population of entities coexisting in the same environment E and differentiated by the fact that some of them have a certain property P while others lack P. Interaction is due to the fact that E has some critical factors C and to the fact that members of the population that have P interact with C in a different way with respect to the other members. Effects are due to the fact that, for example, interaction with C brings benefits only (or more) to those members that have P. There are longer-range effects when the interaction and first effects are *followed* by an increased reproduction of members bearing P. Even-longer-range effects are, for example, the accumulation over generations of beneficial properties like P and the subsequent production of lineages of individuals.

What the general scheme of Darden and Cain does not sufficiently stress is that the fact that members with P outnumber members without P is not something that simply follows from the interaction among members of the relevant population and the environment, but that it is something that is actually caused by that interaction. That is the reason why many philosophers of biology believe that the function of P can be used to explain the actual presence of the character P, or the actual proportion of members with P in the population. And this is the crucial point.

The copying process in any selection theory is what grants the causal link between the interaction of P with E and the resultant spreading of the character P within the population. What about artifacts, though?

11.3.1   Strong and Weak Copy Relation

Most of the artifacts produced, even those produced on the basis of a new original design, have something in common with previous exemplars of artifacts with the same function belonging to the same kind. This might suggest that there is a process of copying for artifacts that is akin to biological inheritance.

Millikan and Elder defend the thesis that in the cases of both biological entities and artifacts there are local laws governing a copying process and permitting a strong copy relation. As far as biological entities are concerned, these local laws are quite well known. But what about artifacts?

Let us consider a very common kind of artifact, the kind of object normally used for drinking called *glass*. Suppose we have the following three types of glasses: glasses of type I are made of crystal and are square, transparent, and have a long stem; glasses of type II are made of plastic and are square, transparent, heavy, and have no stem; glasses of type III are made of glass and are opaque, round, heavy, and have no stem. According to the theories pertaining to some copying process, glasses of a new type, type IV, must be copies of glasses of types I, II, or III or a mixture of all three. What laws govern this copying process? The idea is that the designer of a new type of glass is influenced by previous glass design experiences. The knowledge of glass designs is passed down through generations of designers. If we know, for instance, that the designer of type IV is a Western designer who is requested to design an elegant wineglass, and that no constraints are placed on the price of the final product, then we can conclude easily that it is likely that glasses of type IV will be made of crystal and will be circular, very light, and have a long stem. Nonetheless, do we really need to appeal to the occurrence of a copying process here? Would it not be sufficient to take into account the rational, physical, and economical, as well as the social and cultural, constraints placed on the designer?

Moreover, the process behind the strong copy relation holding among artifacts must fulfill the causal role already described. For biological entities, the fact that there is a process for copying characters grants that P—the character that gives some advantages to the individual—is passed to the following generation. The increasing proportion of individuals with P in relation to individuals without P is then the consequence of the selective action of E, the environment, and the inheritance process. For artifacts, there is no automatic process of inheritance; it seems that the only copying process that can play the same role of inheritance is the intentional copying of designers. The fact that some features are copied more frequently than others is partly because such features are more effective in the performing of a certain function F and partly because designers intentionally copy those features that are more effective with respect to a certain function. Artifacts with roughly the same function F, such as glasses, are produced with those features that designers regard as apt for the performing of F. These features may or may not be copied from previous types of glasses.

Of course the introduction of new features providing new solutions to a functional problem is always possible, either intentionally or by accident, and they can play the same role that the occurrence of variations plays in the natural process of selection. If we adopt this perspective, we may have to admit that the reproductive process of artifacts is more prone to variations than the biological process, but this seems a reasonable price to pay. Nonetheless, it seems clear that according to a selective account of this kind, the function F for which a certain object (or features of a certain object) has been selected is simply the function for which it has been intentionally designed. It is the efficiency in performing F of an object with P, plus the intention of the designers in copying those objects that perform F better, that can explain why P is copied over and over again. But if that is the case, why not simply appeal to an intentional criterion for function attribution, which would allow us to attribute function not only to copied features like P but also to those that have been newly introduced?

Indeed it seems plausible to accept that the selection process for artifacts is different from the selection process for biological entities and that it is possible to subsume both the artifact and biological copy relations under the same type allowing for the existence of a weak copy relation. That is, we do not need any counterfactual dependence, and we can allow for a weaker causal role. Designs are almost never the result of a completely new discovery; designers always take inspiration from previous designs. It seems reasonable, then, to grant that there is a noncasual similarity of artifacts produced according to the same design.

Nonetheless, even if we accept a weak copy relation and we assume that some kind of suitable copying process exists, finding a similarity between the biological *copy* relations and the artifactual one turns out to be very hard. It is then very difficult to give sense to the hypothesis that in both biological and artifact cases we attribute function on the basis of a unique and common selection theory. The main differences are the following:

- For artifacts, there is no regularity whatsoever in the type of properties that can be copied nor in the degree of similarity between original and copied features. That is, it is not possible to distinguish between inheritable and noninheritable characters.
- It is not always the case that items of the same artifact kind are copies of one another. The holding of a copy relation is just a contingent historical fact that cannot be taken to be a necessary condition for artifacts to be artifacts of the same kind.
- While for biological entities the arity of the relation is widely dependent on the kind of entities being considered (kinds with sexual or asexual reproduction), for artifacts it is not even possible to decide how many arguments the relation has. It is possible to design a new type of glass drawing inspiration from one, two, three, four, and so on, different types of previous glasses.

I am not denying that sometimes a sort of weak copy relation can hold between artifacts of the same kind due to the fact that designers draw their inspiration from previous types of artifacts, but it is doubtful that this relation can be compared to the biological copy relation and it is doubtful that it can play the same causal role.

Hence my conclusion that, on the one hand, if we take artifact kinds to be general kinds like those of the glasses in the example, then the inference to the existence of a copy relation holding between artifacts is not straightforward, because either we can account for their similarity and historical development without appealing to any copying process or we must appeal to the designer's intentions.

On the other hand, if we admit the existence of a copy relation that is sufficiently weak to cover both the case of biological entities and artifacts, we simply run the risk of using the same term with two different meanings. For biological entities, the copy relation is a well-defined relation, warranting counterfactuals and based on the existence of well-known copying processes; for artifacts, the copy relation would not be a well-defined relation, will be unable to warrant counterfactuals, and will not be grounded in any clear copying process. Ultimately certain observed similarities among items of the same kind seem to be the only common features shared by biological kinds and artifact kinds.

11.4 Second Option

The second criterion proposed is

2. *o* is an object of an artifact kind K iff *o* is used for F.

Unlike in (1), the attribution of function according to use seems to rely mainly on the common pretheoretic habit of speaking of artifact function in this way. The main problem is that it is not clear if this practice perceives the attribution of functions as an attribution to single objects or to types of objects. Both possibilities must thus be taken into account. Henceforth I use *o* for single objects and *O* for types.

Usually what we mean by "o is used for F" is that o is normally used for F, but what does "normally" mean? In the debates over the notion of "normal," typically quantitative and qualitative interpretations are distinguished. In this case the most likely quantitative interpretation seems to be the following:

a. Most of the times that o is used, it is used for F.

A qualitative interpretation seems to be more difficult to give. A plausible possibility is the following semiqualitative criterion:

b. Most of the times that o is used by competent people, it is used for F.

Unfortunately both (a) and (b) are false or, more to the point, do not meet the common classification of artifacts into functional kinds.

We seem to be perfectly comfortable with the thought that something is still a chair (whatever "being a chair" means) even if it is used most of the time as a small stepladder. There is a certain intuitive resistance to considering such an object to be a stepladder rather than a chair. Analogously in the case of (b), it is perfectly conceivable to consider cases where we—supposedly competent users—will use the very same chair we are sitting on as a stepladder for years without forming the opinion that it is or has become a stepladder.

I take it to be symptomatic that we have linguistic instruments for distinguishing between instances of something being a chair and instances of something being used as a chair. There are circumstances in which we are willing both to assert that something is a chair and to deny that it is normally used as a chair, and yet other circumstances in which we deny that something is a chair even if it is commonly used as such. Indeed these linguistic facts can be viewed merely as cues for the thesis that we do not normally classify artifacts according to the function they perform or are used for.

Another possibility is to take o in (2) to range over types instead of tokens. This implies that we need some sort of paraphrase: a type is an abstract entity, so it does not seem to make much sense to say that O is literally used for F or that O is normally used for F. We can consider the following paraphrases:

c. Most of the times that tokens of O are used, they are used for F.

d. Most of the times that tokens of O are used by competent people, they are used for F.

These two solutions do endure. The critics moved to (a) and (b) because according to (c) and (d) a certain token is an artifact of a certain kind if it is a token of a type whose tokens are most frequently used for F. So it is perfectly possible for a single chair to be normally used as a stepladder but to nonetheless not be considered as a stepladder as long as other tokens of the same type—or at least most of them—are used as chairs. The problem is to specify the right types. It seems obvious that O in (c) and (d) has to be intended as a physical-structural type, but this interpretation excludes the following possibility: let us suppose that few items of a new type (physical-structural) of chair are produced but that none of them is ever used, as they are immediately put in a museum.

Fortunately the most intuitive way to understand (c) and (d) seems to be to imply that they have conditional forms of the following types:

c′. If tokens of O are used at all, then they are most often used for F.

d′. If tokens of O are used at all by competent users, they are most often used for F.

Thus according to (c′) and (d′) we can truly attribute the function F to types O, even when tokens of O are never used. But here we have two problems. First, it is not clear what exactly (c′) and (d′) mean with respect to a type whose tokens have never been used. How can we tell that (c′) or (d′) is true with respect to such a type? The second problem is that even if we could find the way to determine this (e.g., by establishing which is the most likely way an object with a certain structure could be used), we would have the undesirable result that all the nonused types of artifacts may have many more functions in common than our way of classifying them allows. That is, F in (c′) and (d′) ranges over all the functions that are compatible with the structure. It is clear that in these cases, however, we want to be able to say, for example, that the objects in question are chairs and that they have the same function that chairs typically have. The conditional interpretation thus fails to accommodate the fact that different types of unused artifacts may have different functions and may be classified accordingly as artifacts of different functional kinds.

If we interpret (c) and (d) according to a nonconditional form, (c) and (d) will fail to accommodate the fact that artifacts of a certain type may remain unused and nonetheless be artifacts of a certain functional kind.

11.5   Third Option

The third criterion is the attribution of function according to the intention of production, that is:

3. o is an object of artifact kind K iff o has been produced for F.

What does "an object has been produced for a certain function" mean?

According to the notion of "production" adopted here, production includes two phases: the execution or physical realization phase, implying some manipulation, and the designing phase. If an object is intentionally produced for F, the phase of the physical realization follows instructions posited during the design phase. The phase of designing is the phase of study, experimentation, and trial aimed at finding out which structural constraints an object must meet to exhibit certain physical dispositions. I do not want to enter the debate over the nature and main procedure of designing. I take this rough-and-ready characterization to be sufficient for my purposes. What is important is that the whole process of production must be, at least to some extent, successful. The success is evaluated with respect to two distinct factors. Concerning the conceptual design phase, the physical constraints posited must be adequate for the object to be able to perform the intended function, whereas the physical realization is successful when the realized object meets the structural requirements specified by the design. It is very difficult to individuate general constraints of success for both the design phase and the phase of physical realization, but there are no doubts that in many circumstances we actually do succeed in realizing successful designs and in producing objects according to such designs. I take it that this constitutes sufficient detail for the time being.

11.5.1 Two Related Problems

The phases of design and physical realization[7] can be intertwined in such a complicated manner that it can be very difficult to distinguish one from the other. On the one hand, the physical realization phase may not be a mere execution of intentions previously stated but it may also have creative aspects. On the other hand, the phase of designing may involve automatic procedures thanks to past experiences, either personal or collective, that do not require intentionality. These considerations have led to the belief that these two phases are nothing but mere abstractions. The conclusion may be drawn that it is not possible to isolate designers' intentions for the attribution of functions, which is why it is meaningless to attribute function according to design intentions.

A further problem concerns the attribution of functions to the component parts of artifacts. A possible way of producing an artifact *o* for a function F is by means of trial and error, and in such circumstances the designer may ignore how *o* performs the intended function F—that is, how the different parts of *o* contribute to the overall activity of *o*. The strategy of adopting an intentional criterion for the attribution of function to the parts of artifacts is therefore not a satisfactory strategy.

I reply to both criticisms as follows. Here I am not trying to formulate a general theory of function attribution to artifacts, so I do not think it is a real problem if the proposed principle does not capture some function attribution. My aim is to find out which criterion for function attribution may allow for a classification of artifacts that is compatible with a realist position. I agree that according to the present proposal if an object is produced without a full awareness of the goal, this object will not be an item of an artifact kind. Maybe it can still be considered an artifact according to some notions of "artifact" but such notions are not of interest here. I do not presuppose that all the objects we consider to be artifacts are objects of the same nature and hence objects that belong to the same type of kinds. What is of interest here is those objects that populate our everyday lives, objects that we judge to be—and that actually are—intentionally produced for a certain function. On the basis of this general presumption, we modulate our behavior, our choices of means, and our laws.

11.6   A Function for Artifact Kinds

Up until now, F has been treated like a black box. I argue that the naïve view that artifact kinds are functional kinds seems to be easier to defend if we adopt a criterion for function attribution that is based on the intentions of the producers of the object. It is important to reassert that I do not want to deny the possibility that sometimes we do attribute function to some version of (1) and (2). Nonetheless, I pursue the strategy of trying to account for artifact kinds as functional kinds by adopting a type (3) criterion that is less problematic and can meet certain strong intuitions we share on artifacts.

In detail, my approach to artifact kinds is (a) to adopt the idea that artifact kinds are functional kinds: *o* is an object of artifact kind K iff *o* has the *function* F; (b) to give an account of a function attribution for the categorization of artifacts of the following type: *o* has the function F iff *o* has been intentionally produced for F; and (c) to take F in the present formulation as a placeholder for a certain triple. The triple I propose is the following:

(T)   ⟨Input-Output, System of Interaction, Object Structure⟩ or ⟨I-O, S-I, O-S⟩

Hence (3) becomes

3′. *o* is an object of artifact kind K iff *o* has been produced for ⟨I-O, S-I, O-S⟩

My proposal is based on the idea that the intentions of artifact designers share the same general structure. Intuitively the production of an object of a certain artifact kind is always aimed at the realization of an object with certain dispositions. These dispositions are specified according to (T). (T) determines the kind.

(T) includes elements that in some contexts can be referred to as the function of an object. It is necessary to bear in mind that *function* can be used in many different ways and that in every context it is necessary to be clear about the use of the term that is adopted.

The first element of the triple is Input-Output. It corresponds to the most classical way of referring to functions, a way that is common in mathematics as well as in the notion of "function" applied to objects. A mathematical function is something that maps arguments to values. In a similar way the function of an object is, in the input-output sense, the disposition of an object to realize a certain final state given certain initial conditions. Commonly when we say that the function of a knife is to cut something, that the function of keys is to open and close locks, and so on, we use this characterization of function. It is worth observing that ordinarily when we refer to functional kinds we adopt an input-output notion of "function"—most of the time leaving implicit the input conditions.

The System of Interaction specifies the conditions in which the object is expected to realize a certain output starting from a given input—the output being some final status or activity concerning the system or some of its parts. The System of Interaction specifies all the relevant interactions with other objects and is particularly important for specifying the notion of use of an artifact. An artifact is conceived by its author and also by users as something that, used in a certain way, can bring about a certain result. Moreover, the System of Interaction is also relevant for assessing the functioning of the object. Among the classical requirements for a theory of function, there is what is known as the "malfunctioning requirement." We say, for example, that a chair is a malfunctioning chair if it is so unstable that a human being with a normal constitution can hardly sit on it, at least if the chair is being used in the right conditions. Let us, for example, consider the case of a chair standing on an uneven floor. Even if such a chair would equally prove to be unstable, we would be more cautious about calling it a malfunctioning chair. The reason lies in the fact that even if most of the time malfunctioning judgments seem to involve only an input-output characterization of function, they also actually presuppose that the object is properly used. We cannot say that a misused object is malfunctioning—strictly speaking, we simply do not know.[8] The System of Interaction provides the correct conditions of use for the artifact.

A further point is that the right conditions of use do not always involve an intentional notion of "use." For certain artifacts, it is not clear if we can say that their proper use is intentional, while for other artifacts even if there is an intentional use it is not clear if it is the intentional use that is relevant to the assessment of their functional properties. For example, a guardrail is something that has to be installed in the right way to perform its function, and this can be regarded as an intentional use. Still the *proper* function of the guardrail is to prevent vehicles—specified for ranges of velocity, weight, trajectory of impact, and so forth—from going off the road. The conditions in which the guardrail is expected to perform its function of stopping vehicles that hit it do not entail intentional use. When someone loses control of his or her car the properly functioning guardrail must be able to keep the car on the road: there is no intentional use.

For those claiming that an artifact is an object produced for a certain use, the problem is to adopt a notion of "use" that is sufficiently broad to include cases of nonintentional interaction between the user and the artifact, as in the case of the guardrail. Cases of artifacts produced for nonintentional interaction with the user are even more frequent at the artifact-component level: practically every mechanical or electric device is made of components that have been built to perform a function, but not for intentional use, at least not in the same sense in which chairs and tables are used. The System of Interaction specifies the right conditions for the object to perform the function and to be used.

The third element is the Object Structure, which must include a specification of all the components that are relevant from the functional point of view, a specification of materials and dimensions.

A criticism that can be leveled against my proposal is that a functional characterization of a kind is functional to the extent that it does not take into account the details of the physical realization but typically is of the input-output form, or even a simple output form. When we say that the function of a heart is to pump blood or that the function of a knife is to cut, we are adopting an output style. In the present proposal, the functional characterization includes instead the physical description of the structure of the object functionally characterized. I do not want to stick to the idea that artifact kinds must be purely functional in the sense here intended; what has been proposed here can be regarded as a functional-structural characterization.

The notion of "function" mentioned in the attribution criterion proposed here includes a specification of the object's structural type, and for this reason it largely avoids the problems related to multiple realizability. What I suggest is that artifact kinds are not in fact specified through those rough functional descriptions traditionally related to the meaning of ordinary artifact terms but through the specific functional-structural descriptions related to the design.

The mere input-output function is the one adopted in the characterization of kinds such as *chairs*, *watches*, *cars*, and so forth, typically mentioned in antirealist arguments. Both Millikan (1984) and Elder (1989) have already observed that kinds such as *car* or *chair* are simply too generic to be considered as real kinds and that more specific kinds might be better candidates for real artifact kinds. Elder proposed the example of the Eames desk chair 1957, Millikan the example of the 1969 Plymouth Valiant 100. The criticism could be further developed by saying that no matter how specific we are in the description of the kind, as long as we adopt a functional characterization, the possibility of multiple realizations remains, and this is precisely the difference with natural kinds. We can identify the essence of water with the molecular structure H_2O and perhaps the essence of a species with a certain genetic content, but for artifacts we either specify a functional kind, thus having the problem of multiple realizations, or we specify a structural type, thus abandoning the functional characterization of artifact kinds.

I reply as follows. First, it is worth observing that there is no straightforward way of concluding that multiple realizability holds for every functional kind. It is possible that a certain function might be realized in only one way, but of course this cannot be a conclusive argument.

Second, I contest the idea that significant multiple realizability is really possible for artifact kinds, because what happens during the design phase of artifact production is that a certain function is *related* to a certain physical structure. The type of artifact defined by a certain design is the result of such a connection. It seems to be rather arbitrary to decide to stop the characterization of artifact kinds at a general functional level, as the nature of the artifact precisely forms the specific connection between a certain input-output function and a certain structure. For practical reasons related to generic use, we may be satisfied with a general characterization of function, but for more specific uses—for example, when we need to substitute a component—such a general functional characterization is no longer adequate. We need to appeal to a finer-grain structural description of the artifact, and that is what the author's design provides.

The same criticism might be leveled in a different way. Someone could insist that we are not entitled to say that a characterization including a specification of a structure is functional and thus that in the end I fail to provide an authentic functional characterization of artifact kinds. These criticisms might be correct.

We can be driven by considerations concerning some expressions of use to regard physical structure as something extraneous to function, something belonging to the device that performs the function and not to the function itself. We say that an object with structure S performs function F, hence it seems that the function can be identified independently of S. But if for similar reasons we can say that a pen and a pencil have the same function even if they perform it in a different way, it also seems perfectly correct to say that a pen and a pencil have different functions of the same type; the same can be said of a bicycle and a motorbike or a Ferrari and a Fiat 500. The fact that we can describe function at a very high level of abstraction does not imply that descriptions mentioning both the input-output information and a physical characterization of the structure cannot be regarded as a more detailed characterization of the function itself. Thus criticisms of this type seem more allied to terminological rather than substantial points. The idea is not new in the literature. Think, for example, of the famous approach proposed by David Marr in *Vision* (1982). According to Marr, to understand a certain function or process, it is necessary to analyze it at three different levels: the level of *what* and *why*, the level of *representation* and *algorithm*, and finally the level of *physical realization*. According to him, the levels of the what description—the input-output level—and the physical structure are of necessity involved in the comprehension of the same phenomenon, that of the performing of a certain function.

An item of a historical-structural, that is to say, nonfunctional kind, would be described in words like "something that has been built with such and such a structure under certain conditions." I claim that the condition of the origin and the structure are not sufficient for the categorization of artifacts, which is why mere historical-structural kinds would not be appropriate for the classification of artifacts. As has just been observed, talking of function does not involve the exclusion of considerations concerning structure. Instead it allows one to choose the appropriate level of structural description; one can describe a function at a very general level or at a less general level by introducing more details concerning the structure. For this reason, the mere introduction of a reference to the structure in the criteria for the classification of artifacts is not sufficient to conclude that these criteria are not functional criteria.

11.7   Final Remarks

Some antirealists argue that artifact terms behave like abbreviations for functional descriptions, and that for this reason they cannot refer to real kinds but only to nominal kinds. I propose the two following theses:

I. Real artifact kinds are those specified by the author's design—that is, by the specific connection between a function and a structure;
II. The vernacular artifact kind terms correspond mainly to the Input-Output characterization of the object—that is, to a generic functional description.

Common artifact terms, then, simply name clusters of real artifact kinds sharing the same general functional characterization (input-output or merely output), and for the most part we lack adequate terms for the real artifact kinds. The present proposal can thus, on the one hand, provide an explanation for the fact that common artifact terms are normally considered just as names of nominal kinds, while, on the other hand, it can render this fact compatible with the existence of real artifact kinds, explaining the connection between such nominal kinds and real kinds. The general terms in everyday language may or may not refer to a general kind that can be a real kind. The problem is similar to that concerning the biological taxa that are of a higher order than species. The present proposal is neither meant to defend the reality of artifact kinds as ordinarily intended nor to reject it.

Finally, it is important to note that the artifact kinds described here are not to be confused with structural types. The fact that an object falls under a certain artifact kind has to do with its history of production and not merely its actual structure.

Notes

1. The distinction between strong and weak realism is drawn by Devitt (1997: 17–18).

2. These are the traditional examples of the nonnatural kinds to be found in philosophical literature. Indeed it is possible that kinds like *bachelor* or *widow* turn out to be theoretically relevant for a certain scientific discipline, for example, for certain sociological theory. In the literature, the burden of proof is normally laid on those who want to demonstrate that kinds of nonnatural science are natural kinds as well. This is a biased position I am not endorsing here.

3. It is possible to roughly distinguish between a reductionist approach and an eliminativist approach. Claiming that the kind *tables*, for example, is not a real kind and hence that tables are not real objects does not lead to the idea that there are no objects that can be truly said to be tables. According to reductionists, such as Devitt (1997) and Wiggins (2001), there are real physical objects classified as tables. Being a table is a property like being the first child born in 2006 or being a Christmas tree. That is, being a table is a real property that is accidental with respect to the nature of the objects having that property. Eliminativists, such as van Inwagen (1990) and Merriks (2001), also deny the existence of those material objects identified as tables, chairs, etc., hence according to these authors there is nothing that is strictly speaking a table—there are only particles and properties of particles and relations holding them together.

4. In the philosophical literature on function Millikan and Preston try to apply a selective criterion to function attribution even to artifacts. But the main input to such an approach comes either from sociological literature or

from those such as Dawkins (1986), who uphold the thesis that any design attribution—thus also any function attribution—is to be justified on the basis of selective processes. See also the original approach adopted by Elder (2004).

5. Here some clarification is needed in order to avoid possible misinterpretation of (FK) and (FA). These schemes are grounded in the idea that there is a one-to-one relation between functions attributed to objects according to a certain criterion and artifact kinds. For example, something that is a bicycle has a certain function, and something that is a hammer has another function. If our theory of function attribution allows for something to have both the function of a bicycle and the function of a hammer, then we must decide if either this object is an instantiation of a particular exotic kind of artifact, the *bicy-hammer*, or if it is an instantiation of two different kinds of artifact—that is, that it is both a bicycle and a hammer. It is necessary to pay attention here to the fact that for an object to be considered the instantiation of an artifact kind, it is necessary that the right relation exist between that object and the function. For example, if we adopt a criterion such that the function of an object is the one that it has been designed for, even if it is possible for something that has the function of a bicycle to be used as a hammer in certain peculiar circumstances, this is not sufficient for that object to be both a bicycle and a hammer.

6. The counterfactuals that need to be granted are of the following type: if a is a copy of b, with respect to a range of inheritable determinates $d_1, d_2, \ldots d_n$ of a determinable property D—for example the properties *brown, blond, red*, etc. with respect to the determinable *hair color*—then it is true that if b, the parents in the biological case, had a property d_n, a, the descendant in the biological case, would have a certain property d_m—where d_n and d_m can be different properties. The general idea is that the transmission of characters obeys some law like generalizations that support counterfactuals of this type. See Millikan (1984).

7. Designing does not necessarily lead to a physical entity like a drawing—a design can also be a mere mental plan.

8. Here I am not trying to base a conceptual distinction on linguistic uses—"malfunctioning" can actually also be used to refer to a chair standing on an uneven floor. I just want to stress that even if we label both a chair with a broken leg and a chair standing on an uneven floor as "malfunctioning," we nonetheless consider them to be malfunctioning in a somewhat different way.

References

Bloom, P. (1996). Intention, history, and artifact concepts. *Cognition, 60:* 1–29.

Bloom, P. (1998). Theories of artifact categorization. *Cognition, 66:* 87–93.

Cartwright, H. M. (1970). Quantities. *Philosophical Review, 79:* 25–42.

Cummins, R. (1975). Functional analysis. *The Journal of Philosophy, 72:* 741–765.

Cummins, R. (1983). *The Nature of Psychological Explanation.* Cambridge, Mass.: The MIT Press.

Cummins, R. (2002). Neo-Teleology. In: *Functions: New Essays in the Philosophy of Psychology and Biology* (Ariew, A., Cummins. R., and Perlman, M., eds.), 157–172. Oxford: Oxford University Press.

Darden, L., and Cain, J. A. (1989). Selection type theories. *Philosophy of Science, 56:* 106–129.

Dawkins, R. (1986). *The Blind Watchmaker.* New York: Norton.

Dennett, D. C. (1990). The interpretation of text, people and other artifacts. *Philosophy and Phenomenological Research, 50:* 177–194.

Devitt, M. (1997). *Realism and Truth, 2nd ed.* Princeton: Princeton University Press.

Dipert, R. (1993). *Artifacts, Art Works and Agency.* Philadelphia: Temple University Press.

Elder, C. L. (1989). Realism, naturalism, and culturally generated kinds. *The Philosophical Quarterly, 39:* 425–444.

Elder, C. L. (1995). A different kind of natural kind. *Australasian Journal of Philosophy, 73:* 516–531.

Elder, C. L. (1996). On the reality of medium-sized objects. *Philosophical Studies, 83:* 191–211.

Elder, C. L. (2004). *Real Natures and Familiar Objects.* Cambridge, Mass.: The MIT Press).

Hilpinen, R. (1992). On artifacts and works of arts. *Theoria, 58:* 58–82.

Hilpinen, R. (1993). Authors and Artifacts. *Proceedings of Aristotelian Society, 93:* 155–178.

Hoffman, J., and Rosenkrantz, G. S. (1997). *Substance: Its Nature and Existence.* London: Routledge.

Kornblith, H. (1980). Referring to artifacts. *Philosophical Review, 89:* 109–114.

Kripke, S. (1980). *Naming and Necessity.* Cambridge, Mass.: Harvard University Press.

Losonsky, M. (1990). The nature of artifacts. *Philosophy, 65:* 81–88.

Malt, B. C., and Johnson, E. C. (1998). Artifact category membership and the intentional-historical theory. *Cognition, 66:* 79–85.

Marr, D. (1982). *Vision: A Computational Investigation into the Human Representation and Processing of Visual Information.* New York: Freeman.

Merricks, T. (2001). *Objects and Persons.* Oxford: Oxford University Press.

Millikan, R. G. (1984). *Language, Thought and Other Biological Categories.* Cambridge, Mass.: The MIT Press.

Millikan, R. G. (1993). *White Queen Psychology and Other Essays for Alice.* Cambridge, Mass.: The MIT Press.

Millikan, R. G. (2000). *On Clear and Confused Ideas: An Essay on Substance Concepts.* Cambridge: Cambridge University Press.

Millikan, R. G. (2002). Biofunctions: Two paradigms. In: *Functions: New Essays in the Philosophy of Psychology and Biology* (Ariew, A., Cummins, R., and Perlman, M., eds.), 113–143. Oxford: Oxford University Press.

Millikan, R. G. (2004). *Varieties of Meaning: The 2002 Jean Nicod Lectures.* Cambridge, Mass.: The MIT Press.

Preston, B. (1998). Why is a wing like a spoon? A pluralist theory of function. *Journal of Philosophy, 95:* 215–254.

Putnam, H. (1970). Is semantics possible? In: *Language, Belief, and Metaphysics* (Kiefer, H. E., and Munitz, M. K., eds.), 50–63. Albany: New York Press.

Schwartz, S. P. (1977). *Naming, Necessity, and Natural Kinds.* Ithaka: Cornell University Press.

Schwartz, S. P. (1978). Putnam on artifacts. *Philosophical Review, 87:* 566–574.

Schwartz, S. P. (1980). Natural kinds and nominal kinds. *Mind, 89:* 182–195.

Simon, H. A. (1969). *The Science of the Artificial.* Cambridge, Mass.: The MIT Press.

Thomasson, A. (1999). *Fiction and Metaphysics.* Cambridge: Cambridge University Press.

Thomasson, A. (2003). Realism and human kinds. *Philosophy and Phenomenological Research, 67:* 580–609.

van Inwagen, P. (1990). *Material Beings.* Ithaca: Cornell University Press.

Wiggins, D. (1980). *Sameness and Substance.* Cambridge, Mass.: Harvard University Press.

Wiggins, D. (2001). *Sameness and Substance Renewed.* Cambridge, Mass.: Cambridge University Press.

Wright, L. (1973). Functions. *The Philosophical Review, 82:* 139–168.

12 A Device-Oriented Definition of Functions of Artifacts and Its Perspectives

Yoshinobu Kitamura and Riichiro Mizoguchi

12.1 Introduction

Functionality is a key aspect of technical artifacts and biological organisms. This chapter discusses definitions of the functionality of technical artifacts from the viewpoint of engineering design and ontological engineering. In engineering, much research on functionality has been conducted in areas such as functional representation (e.g., Chandrasekaran, Goel, and Iwasaki 1993; Chittaro et al. 1993), engineering design (e.g., Hubka and Eder 1988; Umeda et al. 1996; Hirtz et al. 2002), and value engineering (e.g., Miles 1961). Such research aims at establishing a *modeling framework* for computer models of artifacts from the teleological viewpoint (*functional models*), which can be used in engineering activities such as design and diagnosis. There have been many fundamental discussions on the functionality of artifacts, although there is no common definition of *functions* (Chandrasekaran and Josephson 2000; Hubka and Eder 2001; Stone and Chakrabarti 2005).

Such discussions in the field of engineering are motivated by the requirements placed on functional models, such as the consistency, reusability, and composability of the pieces that make up models. In practical situations engineers tend to describe functional models in an ad hoc way. Functions can be captured in different ways in different domains. To satisfy the above requirements, a *prescriptive definition* of functions is needed. Such a definition aims at giving authors of the functional models a conceptual schema and guidelines to restrict the viewpoint for describing the functions of target artifacts.

In this chapter, we first discuss our device-oriented definition of functions (Kitamura et al. 2002, 2006; Kitamura, Koji, and Mizoguchi 2006) as a definition of functions for engineering. Our definition of functions is prescriptive from the viewpoint of engineering devices; by intention, it describes only one kind of function. An important element of the prescriptive definition is the definition of the *behavior* of devices as a basis of functions. Then we show our definition of functions as a *role* played by a behavior (in that sense) under a context of use. We categorize functions into "component functions" and "external functions" according to the context of use. According to the essentiality to identity of the device performing the function, accidental functions are distinguished from essential functions. Then we examine the difference between our definition of technical functions and the definition of biological functions of Johansson and colleagues (2005).

Next we discuss other viewpoints for representing functions. We think that what function is being performed by an artifact depends on an agent's viewpoint (perspective). We discuss such *capturing perspectives* and show a variety of functions of technical artifacts based on them. We form categories of functions other than our device-centered definition and show some ontological distinctions, which are aspects for categorization of function definitions.

12.2   Device-Oriented Definition of Functions

This section discusses our definition of the function of technical artifacts from the device viewpoint.

12.2.1   Device as a System Structure

Two fundamental tasks for conceptualizing technical artifacts are determining what is a primitive in the model and how the primitives form the whole. Here we adopt the device-centered viewpoint from qualitative reasoning research in artificial intelligence (e.g., de Kleer and Brown 1984). The device-centered ontology has been widely adopted for performing engineering tasks including design, such as the German-style design methodology (Pahl and Beitz 1996). In the device ontology, a device has *ports*, through which it is connected to other devices. A *device* consists of other devices of smaller grain size, which form a whole-part hierarchy. A device operates on other things we call "operands," and thus it changes their physical states. The operand is something that flows through the device via ports and is affected by the device. Examples of operands include fluid, energy, motion, force, and information.

12.2.2   Function and Behavior

Before we give clear definitions of functionality, the concept of a "behavior" must be clarified as a basis of function. A distinction between behavior and function has been noted in philosophy and in the field of qualitative reasoning differently. Here we define *behavior* of a device as the change in the attribute value of an operand from the value at the input port of a device to the value at the output of the device. For example, the increase in the temperature of steam as it goes through a super-heater is a behavior of a heater. Such a description of the behavior of an artifact is constant with respect to the artifacts' situation and/or their context. By definition, it is also independent of the intentions of the designers and users. In this sense we say that "behavior is not subjective but objective."[1]

Unlike a behavior, a *function* is related to a *context of use* (i.e., teleological) and hence is context-dependent. A behavior can perform different functions according to contexts of

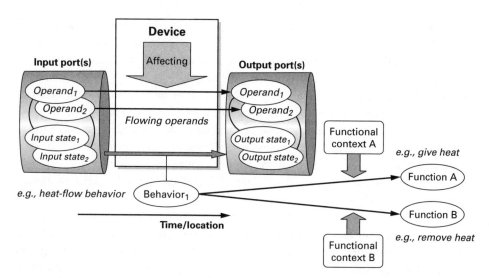

Figure 12.1
Device, operands, behavior, and function.

use. For example, a heat exchanger can be used as a heater in a boiler in a power plant or a radiator connected to an engine in a car.[2] The behavior is the same in any context; that is, heat flows from the warmer fluid to the colder one. As shown in figure 12.1, the functions of the heater and the radiator can be to give heat and to remove heat, respectively. This difference between the functions is dependent on the embedded system. Moreover, a function can be performed (realized) by different behaviors. A function is associated with specific constraints on some parts of behaviors to realize the function. A behavior can perform multiple functions simultaneously. Thus a function here is rather separated from a device.

On the basis of this discussion, we define a *(base-)function* as follows (Kitamura et al. 2006):

A (base-)function performed by a device is a role played by a behavior of the device to achieve a specific goal under a context of use, based on a certain capacity inherent in the device.

By "role" we mean such a concept that an entity plays in a specific *context* and cannot be defined without mentioning external concepts (Sunagawa et al. 2006). A role is *anti-rigid* (i.e., contingent with respect to identity), *dynamic* (i.e., temporary and multiple), and *founded* (i.e., is an extrinsic property defined with reference to an external concept) (Masolo et al. 2004). Accordingly a function can be defined as a role. First, a function (and a behavior as its basis) is *founded*, since a function of a device affects an entity other than the device itself (operand) and causes temporal changes to it. Second, the property that a behavior has to perform a certain function is anti-rigid, since a behavior can perform different functions according to contexts without loss of its identity. Third, a function can be performed (realized) by different behaviors. A behavior can perform multiple functions simultaneously. Thus a function is dynamic and multiple.

We say that "a behavior plays a function role." If a device performs a behavior and the behavior plays a *function role* in a context, then the device plays a *function-performer role* in the context. For example, the heat-exchange behavior plays the heat-removing function role, and a heat exchanger plays the function-performer role of removing heat as a radiator. The heat-exchange behavior implies changes in the temperatures of both fluids. When its function is recognized, one of the fluid flows becomes of further interest.

Note that the difference between function and behavior depends on their context. For example, "to change (increase) temperature" is a function of an electric heater, though the temperature-changing function is similar to the heat-exchange behavior (i.e., changes in the temperatures of both fluid flows) of the heat exchanger. In this case the behavioral model of the heater includes changes both in temperature and electricity. Thus another possible function of the electric heater is "to consume electricity." The temperature-changing function of the heater is dependent on the goal, whereas the heat-exchange behavior of the heat exchanger is independent of any goal.

Furthermore, note that a device's function here refers to local behaviors in the device (which we call a "local function"), though it is dependent on context. Thus its variety is restricted by the device's behavior. On the other hand, a *conjunct function* refers to behaviors of other components, the embedded system, or users. For example, the heat exchanger mentioned earlier can perform conjunct functions such as controlling the temperature of a room or preventing an engine from overheating. The conjunct functions cannot be exhaustively enumerated in nature. We use local functions as the base-functions here in order to realize composability of device models.

12.2.3   Functional Context

The *context of use* represents teleological goals to be achieved by the function (which we call the "functional context"). If the device is a whole product, its functional context (and its goal) is determined by how it is used by users externally (which we call the "external function context"). A function in an external function context (which we call the "external function") is one that is intended by a user. Some external functions are also intended by the designer, as discussed in section 12.2.5.

On the other hand, a function of a component embedded in a system contributes to achieving the system's function. Thus its functional context (which we call the "system function context") is determined by a functional structure in which the system's function is achieved by a sequence of (relatively finer-grained) component functions. We call this type of function with a system function context a "component function." The relationship is hierarchical and is called "is achieved by," as discussed in section 12.2.7.

In the appendix to our chapter, we try to show different notions of "context" in the literature by discussing other notions of context in contrast to our notion of context, and then locating our notion in those other notions.

12.2.4   Capacity to Perform a Function

A *function* in our definition is a role played by a behavior, which is performed or realized by a device. Such a performance is based on the device's *capacity* to perform the function, which is a feature (or property) of a device. For example, the heat-exchange behavior as a basis of the heat-giving function mentioned before can be realized through high thermal conductivity between channels of fluids. A device with such physical properties could be regarded as a thing that *has a capacity to perform the heat-giving function*. The capacity to perform a function is potential and inherent in a device. It can be induced and performed when an appropriate context and appropriate inputs are given to the device. The performed function is restricted by the capacity to perform the function.

In the engineering literature, Hubka and Eder (1988) define *functions* as follows: "The function is a property of the technical system, and describes its ability to fulfill a purpose, namely to convert an input measure into a required output measure under precisely given condition." In their definition, a *purpose* represents *intended effects* as output effects, while a *function* is the actual ability for an internal task of the technical systems. Here *purpose* and *function* roughly correspond to the *function* and *capacity to perform a function* in our terminology, respectively.

In philosophy, a *function* is typically a special feature of artifacts or biological organs. In particular, "[b]iological functions are typically taken as objective non-relational properties" (Vermaas this volume). In causal-role function analysis (Cummins 1975) and ICE theory (Vermaas and Houkes 2006), a technical function is regarded as a special kind of capacity (or disposition) to be ascribed to an artifact, as quoted in the appendix of this chapter.

A *function* in our definition is a role of a behavior, which is rather independent of an artifact, in contrast to the capacity, which is dependent on an artifact. In our definition, the existence of a function as a role has two states according to its fulfillment by a behavior. When a user wants a function before actual use, the function partially exists as a required function in a supposed context of use. By "partially-existing function" we mean that an instance of a function as a role without a behavior (a role-player) and an artifact (a performer) is neither realized nor performed but exists just as a thing required by the user within the specific context of use. Thus when a real artifact performs a behavior that plays the function in the duration of use by a user in the specified context, the function is performed and then fully exists.[3] Hence the existence of such a partial function is dependent on neither a behavior as a player of the function role nor a device as a performer but a context of use. From the engineering viewpoint, a behavior and (physical features of) an artifact are specific ways of realizing the required function. A function can be realized by different behaviors (and artifacts) in different ways. Thus a function should be independent of its realization. This engineering requirement justifies our definition of functions as being detached from artifacts. Biological functions, however, can be inherent to organs, as discussed in section 12.2.6.

Boorse (2002) makes a similar distinction in terms of a "weak function statement" and a "strong function statement." The former is "[an artifact] x performs the function Z in the [goal] G-ing of [a given system] S at t iff at t, the Z-ing of X is a causal contribution to G." The latter is such as "the function of X is Z" and "X has the function Z." Our definition of *function* is a kind of weak function statement. The strong function statement seems to regard a function as a feature dependent on an artifact. We describe such a statement as a statement about the *capacity to perform a function*. The existence of a function in such a sense is potential and hidden until it is induced by a user. People usually suppose essential functions (discussed in the next section) as the functions that an artifact has.

The actual performance of functions in our definition is similar to *functioning* in Johansson and colleagues (2005) and Dipert (2006) in the sense that it can be realized in temporal physical space. In Johansson and colleagues (2005), functionings are "processes, subject to a division into temporal parts" as a SPAN entity. Dipert (2006) points out that "there is a simple difference between an object having a function, and an object's functioning in some way. . . . I will call what functioning does its activity." *Functioning* in our definition, however, is for a behavior to play a role and then to make a function as a role a full existence.

12.2.5   Essential Functions and Accidental Functions

A device can perform some behavior(s) and behavior can perform some function(s). At least one of these functions is intended by a designer of an artifact (we call this an "essential function"). An engineering artifact is designed and manufactured in order to have a certain *capacity to perform* its essential function. Thus the essential function provides the artifact's identity. The names of many artifacts are derived from their essential functions.

On the other hand, a user can use a device differently from the use intended by the designer. In such a case we recognize that the device performs an incidental function (we call this an "accidental function") induced by the use. For example, a screwdriver performs a screwing function as its essential function. A user can use it for hitting (exerting linear force on) a nail as an accidental function. Such a kind of usage is the realization of one of its possible functions based on the device's capacities. This user-induced performance of an accidental function requires the screwdriver's capacity to perform the hitting function based on its physical feature, that is, a flat hard surface. The screwing function could be performed by a key as its accidental function. The hitting function is an essential function of a hammer.

Such capacities to perform accidental functions induced by accidental use, in principle, cannot be enumerated completely before used in reality. Note that the variety of accidental functions as local functions (discussed in section 12.2.2) are limited to those related to possible behaviors, unlike possible (unlimited) conjunct functions such as the screwdriver's accidental function using a nail for two pieces of wood.

The distinction between essential and accidental functions discussed thus far pertains to external functions, which are determined by a user. We can also consider distinctions among the component functions of components embedded in a system. A component is designed and manufactured for a specific function, which is its essential function. When a component is integrated into a system, a designer of the system normally uses a component to achieve its essential function. However, it is possible that the system's designer uses a component to fulfill a function different from its essential function. For example, a slurry containing diamond powder is manufactured for improving cutting efficiency. However, in a cutting device a slurry can also be used for cooling the cutting blade. In this case the slurry performs a cooling function as an accidental function.

Such accidental use is distinguished from a (proper) function in some philosophical writings (e.g., Wright 1973; Perlman 2004; Preston this volume). Such accidental use is called "function-as." We regard an accidental effect as a function if it is intended by a user in a specific context or if it is recognized in a context of a system's goal. If not, it is a behavior. One of our justifications for our use of the term *function* here is the effect-oriented definition from the engineering viewpoint. As actual effects, an essential function and an accidental function can have the same intended effect for users in our engineering sense. The second justification is to detach a function from a function performer such as a required function without a function performer, as discussed in section 12.2.4. A required function can be fulfilled as either an essential function or an accidental function. The third justification is that our function statements are mainly "weak function statements" (Boorse 2002), in contrast to "strong function statements" for which the proper functions are discussed. In fact Boorse does not distinguish a function's performance by accident in his "weak function statement."

Preston, in this volume, discusses "unintended proper functions" and "phantom functions" as difficulties of the qualification of intended use. Many examples for the former can be found with respect to social, economic, or political uses. The latter have no physical effects. Our definition of base-functions is based on physical and actual behavior, which excludes those cases (as a prescriptive definition).

12.2.6   A Brief Comparison with Biological Functions

The characteristics of the functions of technical artifacts are very different from those of biological organs. A biological organ performs its component function in a biological system under fixed context(s). For example, a component function of a heart is to increase the pressure of flowing fluid (blood) in the system context of the circulatory system. This is the same function as a pump. The component function of the circulatory (sub)system is to transport substances.

As pointed out in Johansson and colleagues (2005), a biological organ can perform multiple functions in a system. Those functions sometimes contribute to the functions of different subsystems (e.g., the oropharynx's functions; cf. Johansson et al. 2005). However, their system contexts are fixed to the organ and do not change. Thus a component function of a biological organ is rather tightly associated with the organ (unlike the component functions of engineering artifacts). In a fixed context a component function in a system contributes to the achievement of a goal of the system. Such contribution is shared with technical artifacts.

As external functions, a biological organ can perform several functions according to the given use contexts in the same manner as artifacts. For example, a nose can perform the external function of supporting glasses. We also view that a heart can perform a sound-making (heart sounds) function as an external function under the context of medical diagnosis whereby a medical doctor listens to heart sounds.[4]

The essentiality (i.e., whether it is an essential or accidental function) of such external functions of biological organs or organisms is out of the scope of our discussion here. As pointed out in McLaughlin (2001: 144), "the organisms are not attributed functions . . . only their traits or parts have functions."

In Johansson and colleagues (2005), a function of a biological organ is defined as "a disposition to act in a certain way to contribute to the realization of [a . . .] larger function on the part of that whole organism which is its host." This definition shares with our definition the "goal-oriented"-ness of a function and the agent's inherent property to perform a function. This definition, however, excludes *accidental functions* and says a function "inheres" in the entity (called "function bearer"), which can be regarded as a reflection of the characteristics of functions of biological organs as discussed earlier in this section. Thus it roughly corresponds to the "capacity to perform a function" in section 12.2.4, though a disposition (proneness) is different from a capacity. In fact Johansson and colleagues (2005) distinguish function from functioning. Our definition for engineering devices aims at a conceptualization of function separated from a function bearer, as discussed in section 12.2.5. It comes from an engineering requirement, that is, a function's independence of realization. The definition by Johansson and colleagues (2005) also includes the reliability of performing the function.

12.2.7 Relations Among Functions and the Way of Function Achievement

We believe that a clear understanding of relationships among functions contributes to a clear definition of *function*. We can distinguish "part-of" (called "is-achieved-by") from "is-a" relations. As pointed out in section 12.2.3 and further discussed in the appendix, in a system context a function of a system is (or can be) achieved by a series of finer-grained functions of components. (We call these functions a "goal-function" and "method functions," respectively). Figure 12.2 shows an example of a functional model of an artifact using such an "is-achieved-by" relation, a so-called function decomposition tree. In the literature in engineering, similar relation has been captured as a "degree of complexity" (Hubka and Eder 1988) and as a function decomposition (Pahl and Beitz 1996). This is-achieved-by relation shows a part of Cummins's causal role of the components for capacity of a containing system (Cummins 1975).

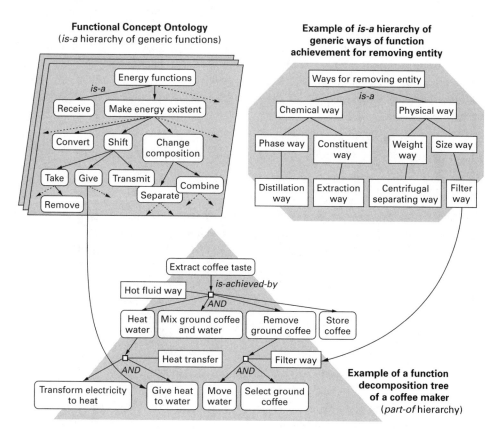

Figure 12.2
Relations among functions.

It is important for defining *function* to discriminate functions from "way of function achievement." A way of function achievement for a function shows background knowledge of the is-achieved-by relation such as physical principles and theories (Kitamura et al. 2002). The way of achievement helps us detach "how to achieve" (way) from "what is intended to achieve" (function). For example, "to weld something" should be decomposed into the "joining function" and "fusion way." This increases the generality and capacity of a functional model in that it accepts a wide range of ways to perform a function, such as using a bolt and nut to achieve the same goal. A feature of function decomposition can also be found as a "means" in Malmqvist (1997). We identify is-a relations between generic *ways of function achievement*. Figure 12.2 shows a portion of the is-a relations hierarchy of generic ways of function achievement for removing entities. The filter way used in the coffeemaker is a subtype of the "size way" and the "physical way." Such a knowledge base can be used by engineers to explore alternative ways to achieve a required function (Kitamura et al. 2006; Kitamura, Koji, and Mizoguchi 2006).

On the other hand, an is-a relation among generic functions (also called "a-kind-of," "categorization," "subsumption," etc.) shows abstraction of "what to achieve." We developed an ontology of generic types of functions (called "a functional concept ontology") with is-a relations (Kitamura et al. 2002). Figure 12.2 shows a portion of the is-a hierarchy of the generic types.

12.3   Perspectives for Capturing Functions and Categories of Functions

12.3.1   Overview and Approach

The definition of *function* in section 12.2 is strictly from the device-centered viewpoint, which is intended to prescribe guidelines for functional modeling of artifacts. Other types of functions, however, still remain to be investigated. Toward a more general account, this section discusses rather descriptive definitions of other kinds of functions.

To categorize functions, we generalize the device-oriented basic model discussed in section 12.2.2 into a generalized basic model. In the generalized model, a physical entity (agent) affects a target thing (operand) at a location. Thus a state of the operand changes in a time interval. The change, called an *effect,* is represented as a combination of the initial state s_1 at the start time point t_1 and the final state s_2 at the final time point t_2. Such an effect is a generalized type of the device-oriented behavior discussed in section 12.2.2.

The change (effect) plays a *goal-oriented role* under a *goal-oriented context.* The goal-oriented context consists of a *teleological context* and an *effect context.* The teleological context is related to the user's intention, as discussed in the following section. The effect context specifies the focused-on area in the model for capturing a role. By specifying the

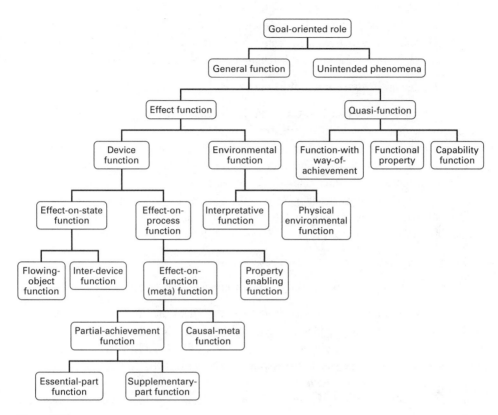

Figure 12.3
Categories of functions.

effect context, we can define subtypes of the goal-oriented role as a categorization of more general functions, as shown in figure 12.3. Note that figure 12.3 shows only an is-a hierarchy for the sake of readability, because some distinctions are independent of each other. These differences represent different *perspectives* for capturing functions. When a behavior is captured from a different perspective it is recognized as a different function. To some extent it might be possible to account for the perspectivity of function ascriptions. However, this is beyond the scope of this chapter.

12.3.2 Teleological Context

The *teleological context* is a generalized type of the functional context discussed in section 12.2.3 and is categorized into a function context and an *unintended context*. The function context represents that the effect (behavior) is "intended" by a user. With the function context, we can define a *general function* as a subtype of the goal-oriented role, as shown in figure 12.3. On the other hand, *unintended phenomena* such as faults can be defined a kind of the goal-oriented role under the unintended context. The causal process of faults can be represented as a goal-oriented achievement structure for (quasi-)goals. This is used for failure mechanism modeling (Koji, Kitamura, and Mizoguchi 2005).

12.3.3 Device and Environmental Functions

The *general function* discussed in section 12.3.2 is categorized into the *effect function* and the *quasi-function* (we will discuss this in section 12.3.6) as shown in figure 12.3. The effect function is based on an effect, which represents temporal changes in attributes of an external thing other than the agent. Its agent is either a device or a function.

The effect function is further categorized into a *device function* and an *environmental function*. The former represents changes in entities within the system boundary. The latter includes changes outside of the system boundary, especially those related to users or user actions. For example, an electric fan performs the following two functions:

"An electric fan moves air." (Device function)

"An electric fan cools the human body." (Environmental function)

In the latter, the cool-down effect by the moving air as an output of the electric fan device is on the human body and thus outside of the system boundary. The *environmental function* has two subtypes; a *physical environmental function* and an *interpretative function*. The former means physical changes in the system like the cooling function of the fan, while the latter sets up one of necessary conditions for cognitive interpretation. For example, an analog clock has the following two functions. The latter requires someone's cognitive interpretation.

"A clock rotates hands (in the specific and constant rate)." (Device function)

"A clock informs about time." (Interpretative function)

Chandrasekaran and Josephson (2000) discuss a similar kind of function, called an "environment function," as an effect on the environment (the surrounding world of the device). Some researchers distinguish *purpose* from *function* (e.g., Chittaro et al. 1993; Rosenman and Gero 1998), whereby the purpose represents a human-intended goal in a similar sense to this environmental function or interpretative function. We are extending our framework to include user actions as well (van der Vegte et al. 2004).

Dipert (2006) points out goal-dependence as follows: ". . . even components of artifacts have multiple and quite distinct functions. . . . Instead, they have a function with respect to an overall goal of some agent and with respect to a level of description." He shows some examples of a button on a DVD player, which we categorize into environmental functions. These include the designer's purposes and purposes of individuals higher in the corporate hierarchy as well as the user's purpose.

12.3.4 Effect on State or Process

According to the target thing changed by the effect, the *device function* is categorized into an *effect-on-state function* and an *effect-on-process function*, as shown in figure 12.3. The effect-on-state function refers to changes in physical attributes of a target thing. Its subtypes are; the *flowing-object* function, which corresponds to a *base-function* discussed in section 12.2.2, and the *inter-device function*, which refers to changes of another device. The latter's example is a rod's function "to push cam." The *cam* is another device, which is not considered as objects flowing through the rod.

The effect-on-process function is based on an effect on a process or change. A behavior as the basis of the device function can be regarded as a kind of a process. It has a subtype, the *effect-on-function function*, whereby a function plays a specific role for another function. The effect-on-function function is further categorized into a *causal-meta function* and *partial-achievement function*. The former corresponds to *meta-functions* such as ToDrive and ToEnable (Kitamura et al. 2002), which are collaborative roles played by a base-function for another base-function. The latter is performed by a *method function* for a *goal function* in the "is-achieved-by" relation discussed in section 12.2.7. It is categorized into an *essential-part function* and a *supplementary-part function* according to whether the contribution is mandatory or not. The distinction between primary and secondary functions (Pahl and Beitz 1996) is similar to this.

12.3.5 Negative Goal and Kinds of Time Interval

There are some categories of functions that are based on adding more descriptors to the categories discussed thus far. A negative function (as an antonym of *positive function*) is one of them, which has a goal to prevent a specific state (or a process) from occurring. The following two functions of a paperweight have a negative goal and a normal goal, respectively.

"A paperweight prevents a piece of paper from moving." (Negative function)

"A paperweight exerts vertical force on a piece of paper." (Positive function)

The effects are represented as changes in the values of attributes in a time interval. The same effect can imply different meanings for different kinds of time intervals. For example, an increasing-temperature function of a heater can imply the following:

"A heater increases the temperature of the air at a specific location in a room." (Absolute-functioning time function)

"A heater increases the temperature of the air at the output port to higher than that at the input port." (Flowing-object-functioning time function)

"In comparison with the original design, the redesign increases the highest temperature of the air." (Designing time function)

The first one refers to the changes of the operand's states at a specific absolute location in a time interval. The second one refers to the change while the operand flows from the input port to the output port, specified relative to the device. The last one is a quasi-function. During the deployment of our framework (Kitamura, Koji, and Mizoguchi 2006), we found that many engineers described such entities. As another example, a function of diamond powder in a cutting machine may be described as follows:

"The diamond powder increases the friction coefficient of the cutting blade." (Designing time function)

When the cutting machine is functioning, the friction coefficient of the cutting blade is high. The increase refers not to a change in the functioning time but to a comparison with the design case without the diamond powder.

12.3.6 Quasi-Functions

We recognize the following kinds of quasi-functions. Although the authors do not consider them as functions, it is found that a quasi-function is occasionally confused with a function. A *function-with-way-of-achievement* implies a specific way of achieving the function, as well as the function. Examples include washing, shearing, adhering (e.g., glue adheres A to B), linking (Hirtz et al. 2002), and "transportation by sea" (Hubka and Eder 1988), as well as welding, as mentioned earlier. Because the meaning of such functions is impure due to the additional meaning of how to achieve the functions, we regard them as quasi-functions.

A *functional property*[5] is mainly found in the materials science domain where a material whose function is dependent on its electronic, optical, or magnetic properties is called a *functional material* (EPSRC 2005). This is (usually implicitly) based on the *property enabling function*, which is based on an effect on a behavior (and/or a function) realized thanks to a physical *property*. Here a property is a conceptualization of "having (a range of) a value of an attribute." Because a property inheres in a physical entity, we regard this effect as a quasi-function. An example is "The high conductivity property of a conducting wire enables its electricity-conducting behavior (and function)" (property enabling [quasi-]function). This high-conductivity property is a necessary condition to perform the conducting behavior (and function). Although such a relationship exists in all realizations of behavior, if the property is a key factor for realizing the behavior, it might be captured as such a quasi-function. Moreover, because there is a direct relationship between the conductive property and the conducting function, the conductive property is called a *functional property*. Such a functional property provides part of the justification for the device's *capacity to perform a function* discussed in section 12.2.4. In addition to the *property of an agent* such as the high-conductivity property, the *property of an operand* can enable a behavior and/or a function: "The combustible property of oil enables the burning behavior of a boiler (and a heat-generation function)" (operand's property enabling [quasi-] function). Especially for human actions, we can say "a property of a *tool* enables human actions" such as "the sharp edge of a knife enables a human's cutting action" (user enabling [quasi-]function). This can be defined as a *user enabling function,* as a subtype of *environmental functions.*

A *capability function* represents that an entity can perform an activity that has no effect on others. For example, we may say, "Humans have a walking function" (capability function). The *property enabling function* and *capacity to perform a function* are similar, but they are based on the effect on other external entities. Last, we should note that the functions with designing time discussed in section 12.3.5 are quasi-functions as well.

12.4   Concluding Remarks

We discuss device-oriented definitions of functions of technical artifacts and some general categories of functions from different perspectives for capturing functions. Table 12.1 shows a summary of some of the distinctions discussed in this chapter, though it does not include detailed descriptions about some of the distinctions in section 12.3. From the viewpoint of engineering, we emphasize *accidental functions* based on contexts of accidental use and that our definition of function is separate from a device (*function as role of behavior*).

Table 12.1
Summary of Distinctions of Functions

Distinction	Distinguished types of functions (F denotes "function")	Section
locality	local F and conjunct F	12.2.2
context determiner	external F and component F	12.2.3
inherence	F as role of behavior and capacity to perform a function	12.2.4
essentiality	essential F and accidental F	12.2.5
generality (type)	generic F (types of F) and instances of F	12.2.7
achievement	goal F and method F	12.2.7
system boundary	device F and environmental F	12.3.3
effect	effect F and quasi-F	12.3.3, 12.3.6
device's operand	base-F and (causal) meta-F	12.3.4
operand (meta effect)	effect-on-state F and effect-on-process F (meta F)	12.3.4
negation of goal	positive F and negative F	12.3.5
—	other functions in figure 12.3	12.3

We are currently investigating the use of the generic categories of function shown in figure 12.3 as a "reference ontology" (its computational implementation was reported in (Kitamura, Takafuji and Mizoguchi 2007)) for translation of functional vocabularies (currently between Functional Basis (Hirtz et al. 2002) and ours) (Ookubo et al. 2007). Such a generic reference ontology of function, we believe, can clarify the ontological differences between functional vocabularies and thus improve the interoperability of functional knowledge. Furthermore, aiming at a computational basis with clearer semantics for discussion on the ontological nature of functions, we are also investigating a computational model of functions using the standardized ontology representation language OWL (Kitamura 2008).

Acknowledgments

The authors are most grateful to Professor Barry Smith of University at Buffalo for his extensive discussions that helped us to clarify our definition of *function* in section 12.2. The authors thank the anonymous reviewer(s) for their valuable comments. The authors would also like to thank Yusuke Koji, Kouji Kozaki, Munehiko Sasajima, Eiichi Sunagawa, Naoya Washio, and Masanori Ookubo for their contributions to this work.

Appendix: A Brief Comparison with Other Notions of Context of Function

Perlman points out the importance of "context" (this volume). He mainly discusses environmental contexts for functioning. As he points out, Cummins also emphasizes a function's dependence on an "analytical context" in the causal-role functional analysis as follows: "To ascribe a function to something is to ascribe a capacity to it which is singled out by its role in an analysis of some capacity of a containing system" (Cummins 1975: 765). In this definition, a function of a component is based on the component's causal role for a capacity of a containing system. In our definition, a *component function* is a role of a behavior for operands under the context of the system in which the component is embedded. In this context, there are causal relations among the components' functions and an is-achieved-by relation among the components' functions and the system's function.

Our *external function context* is based on the designer's or user's intention. In this sense our definition can be regarded as a kind of "goal-contribution theory" based on intentional states in the "recent past backward-looking reductionist category" in Perlman's categorization of definitions of functions (Perlman 2004). Among others, Boorse defines a *function* as a role as follows: an "artifact function [is] . . . an object's role in a human goal-directed activity" (2002: 68).

Vermaas and Houkes (2006) emphasize that functions are relative to use plans and human beliefs as follows:

A technical function of an artefact can be roughly described as the role the artefact plays in a use plan for the artefact that is justified and communicated to prospective users. . . . [F]unctions are features that are ascribed by agents to artefacts relative to use plans, human beliefs and actions, and bodies of evidence.

Here a function is a role played by an artifact in a use plan. Our *external context* depends on such a "use plan," though the functions in the ICE theory are based on an agent's beliefs on capacity and contribution, which are implicit in our definition. For a function of a component, the "bracketed" ICE definition of component's functions clearly captures its dependence on composition in a system configuration (Vermaas 2006). We revisit these definitions in section 12.2.4.

The *component function* is similar to the constituent function (Johansson 2006) as well:

A sufficient and necessary condition for something's being a constituent function is the following: F is a constituent function borne by B if and only if: (a) there is a functional whole A; (b) B is both a spatial part and a subunit of A; (c) B F's in relation to some other entities (X, Y, Z) that are relevant for A.

The embedded system in our terminology is precisely *a functional whole*. We try to describe an engineering model of the *relation* and the *relevance* mentioned in condition (c) in section 12.2.7.

Garbacz (2005) points out that a function is a *state of affairs*, which represents a connection between objects and processes. Our definition tries to define the connection from the device-oriented point of view.

Notes

1. Note that a behavior description depends on *modeling assumptions* made by a modeler. Behavior is independent of the intentions of designers and users.

2. You could consider just a cooler instead of a radiator.

3. Our discussion is mainly concerned with this full existence of functions that is being played by real behavior. The discussion here on the existence of the function required by a user, at first glance, seems to be applicable to a function required by a designer in the early phase in the engineering design process. However, there is room for further investigation because, unlike the use context, the function in the design phase seems to be insufficiently specified for the existence of a function's instance. We are currently investigating those issues on the existence of a function.

4. Note that this sound-making function is only *an external function* and not a *component function* of the heart for biological organisms. Many philosophers reject this function as a function of a heart (e.g., Wright 1973; Dipert 2006). We would also reject it under their definitions. As discussed in sections 12.2.4 and 12.2.5, our justification is to treat "a function as a role of behavior" from an engineering point of view. Moreover, the typical discussion of the rejection is about component and/or essential functions of the heart, which we do not discuss here. Cummins recognizes that this function "sounds wrong in

some context-free sense," though he suggests this function as a (possible) psychological function (1975: 762).

5. The term *functional* here is intended to represent neither a mathematical dependence relation nor attributes of functions, but function-oriented properties. The functional property is used as an antonym of the mechanical or structural property.

References

Boorse, C. (2002). A rebuttal on functions. In: *Functions: New Essays in the Philosophy of Psychology and Biology* (Ariew, A., Cummins. R., and Perlman, M., eds.), 63–112. Oxford: Oxford University Press.

Chandrasekaran, B., Goel, A. K., and Iwasaki, Y. (1993). Functional representation as design rationale. *Computer, 26:* 48–56.

Chandrasekaran, B., and Josephson, J. R. (2000). Function in device representation. *Engineering with Computers, 16:* 162–177.

Chittaro, L., Guida, G., Tasso, C., and Toppano, E. (1993). Functional and teleological knowledge in the multi-modeling approach for reasoning about physical systems: A case study in diagnosis. *IEEE Transactions on Systems, Man, and Cybernetics, 23:* 1718–1751.

Cummins, R. (1975). Functional analysis. *The Journal of Philosophy, 72:* 741–765.

de Kleer, J., and Brown, J. S. (1984). A qualitative physics based on confluences. *Artificial Intelligence, 24:* 7–83.

Dipert, R. R. (2006). The metaphysical grammar of "function" and the unification of artifactual and natural function. Presented at the 15th Altenberg Workshop in Theoretical Biology: Comparative Philosophy of Technical Artifacts and Biological Organisms. Altenberg, Austria: Konrad Lorenz Institute for Evolution and Cognition Research, September 21–24, 2006.

EPSRC (Engineering and Physical Sciences Research Council). (2005). http://www.epsrc.ac.uk/ResearchFunding/Programmes/Materials/ResearchPortfolio/EngineeringFunctionalMaterials.htm

Forbus, K. D. (1984). Qualitative process theory. *Artificial Intelligence, 24:* 85–168.

Garbacz, P. (2005). Towards a standard taxonomy of artifact functions. In: *Proceedings of the First Workshop FOMI 2005—Formal Ontologies Meet Industry.* CD-ROM.

Hirtz, J., Stone, R. B., McAdams, D. A., Szykman, S., and Wood, K. L. (2002). A functional basis for engineering design: Reconciling and evolving previous efforts. *Research in Engineering Design, 13:* 65–82.

Hubka, V., and Eder, W. E. (1988). *Theory of Technical Systems.* Berlin: Springer.

Hubka, V., and Eder, W. E. (2001). Functions revisited. *Proceedings of ICED 01.* CD-ROM.

Johansson, I. (2006). The constituent function analysis of functions. In: *Science—A Challenge to Philosophy?* (Koskinen, H. J., Pihlstrom, S., and Vilkko, R., eds.), 35–45, Frankfurt am Main: Peter Lang.

Johansson, I., Smith, B., Munn, K., Tsikolia, N., Elsner, K., Ernst, D., and Siebert, D. (2005). Functional anatomy: A taxonomic proposal. *Acta Biotheoretica, 53:* 153–166.

Kitamura, Y., Koji, Y., and Mizoguchi, R. (2006). An ontological model of device function: Industrial deployment and lessons learned. *Applied Ontology, 1:* 237–262.

Kitamura, Y., Sano, T., Namba, K., and Mizoguchi, R. (2002). A functional concept ontology and its application to automatic identification of functional structures. *Advanced Engineering Informatics, 16:* 145–163.

Kitamura, Y., Takafuji, S., and Mizoguchi, R. (2007). Towards a reference ontology for functional knowledge interoperability. In *Proceedings of ASME IDETC/CIE 2007,* DETC2006-35373.

Kitamura, Y., Washio, N., Koji, Y., and Mizoguchi, R. (2006). An ontology-based annotation framework for representing the functionality of engineering devices. In: *Proceedings of ASME IDETC/CIE 2006,* DETC2006-99131.

Kitamura, Y. (2008). Roles and function models in OWL and a comparison of functions of artifacts and biological organs. *A technical report of the Stanford Center for Biomedical Informatics Research, Stanford University*, to appear.

Koji, Y., Kitamura, Y., and Mizoguchi, R. (2005). Ontology-based transformation from an extended functional model to FMEA. In: *Proceedings of ICED 05:* 264.81.

Malmqvist, J. (1997). Improved function-means trees by inclusion of design history information. *Journal of Engineering Design, 8:* 107–117.

Masolo, C., Vieu, L., Bottazzi, E., Catenacci, C., Ferrario, R., Gengami, A., and Guarino, N. (2004). Social roles and their descriptions. In: *Proceedings of the 9th International Conference on the Principles of Knowledge Representation and Reasoning (KR2004),* 267–277.

McLaughlin, P. (2001). *What Functions Explain: Functional Explanation and Self-Reproducing Systems.* Cambridge, Mass.: Cambridge University Press.

Miles, L. D. (1961). *Techniques of Value Analysis and Engineering.* New York: McGraw-Hill.

Ookubo, M., Koji, Y., Sasajima, M., Kitamura, Y., and Mizoguchi, R. (2007) Towards interoperability between functional taxonomies using an ontology-based mapping, In: *Proceedings of ICED 07:* 154.

Pahl, G., and Beitz, W. (1996). *Engineering Design—A Systematic Approach.* London: Springer.

Perlman, M. (2004). The modern philosophical resurrection of teleology. *The Monist, 87:* 3–51.

Rosenman, M. A., and Gero, J. S. (1998). Purpose and function in design: From the socio-cultural to the technophysical. *Design Studies, 19:* 161–186.

Stone, R. B., and Chakrabarti, A. (eds.). (2005). Special Issues: Engineering applications of representations of function. *Journal of Artificial Intelligence for Engineering Design, Analysis and Manufacturing (AI EDAM) 19*(2 and 3).

Sunagawa, E., Kozaki, K., Kitamura, Y., and Mizoguchi, R. (2006). Role organization model in Hozo. In: *Proceedings of the 15th International Conference on Knowledge Engineering and Knowledge Management (EKAW 2006).* LNCS 4248, 67–81. Berlin: Springer.

Umeda, Y., Ishii, M., Yoshioka, M., Shimomura, Y., and Tomiyama, T. (1996). Supporting conceptual design based on the function-behavior-state modeler. *Artificial Intelligence for Engineering Design, Analysis and Manufacturing, 10:* 275–288.

van der Vegte, W. F., Kitamura, Y., Koji, Y., and Mizoguchi, R. (2004). Coping with unintended behavior of users and products: Ontological modeling of product functionality and use. In: *Proceedings of CIE 2004:* ASME 2004 DETC 2004. DETC2004-57720.

Vermaas, P. E. (2006). The physical connection: Engineering function ascriptions to technical artefacts and their components. *Studies in History and Philosophy of Science, 37:* 62–75.

Vermaas, P. E., and Houkes, W. (2006). Technical functions: A drawbridge between the intentional and structural natures of technical artefacts. *Studies in History and Philosophy of Science, 37:* 5–18.

Wright, L. (1973). Functions. *Philosophical Review, 82:* 139–168.

V EVOLUTIONARY PERSPECTIVES

Evolution is the process by which biological function bearers come into being, through the random variation of traits that are already present and the differential—or selective—survival of only some of the organisms. The concept is often transferred to the cultural history of technical artifacts, which can also be described as a process of more or less random modification and selective retention of certain modifications. The final part of this volume is therefore dedicated to analyses of organisms and artifacts from an evolutionary perspective. These analyses do not discuss whether evolution is also the source of the normativity of functions, as do the contributions to part II and the first contributions to part III. They rather consider some implications of the view that the functional organization of an entity is a result of evolutionary processes, inquire into the transfer of evolutionary concepts between the domains of biology and technology, and as far as the technological case is at stake, analyze the relation between evolution and intentionality.

Houkes assesses two different approaches of transferring evolutionary concepts to the explanation of technical artifacts. He argues that what he calls the "conflict image" of selected and intended functions is an oversimplification and that the boundary between the biological and technological domains is an open border rather than an iron curtain. His examples are evolutionary archaeology and evolutionary design in electronics. On the one hand, he criticizes the viewing of these approaches as genuine applications of the whole framework of evolutionary theory, while on the other hand, he takes the application of a limited set of evolutionary concepts in these fields seriously and queries how they can be combined with the intentionalist account that he sees still present and even dominant. He shows that evolutionary notions are introduced to solve specific problems in establishing artifact lineages and in design heuristics. He also demonstrates that in none of the cases are serious attempts made to find counterparts of more evolutionary notions than those that serve this immediate purpose. Intentionality is merely supplemented by evolutionary concepts or redescribed in terms of evolution. According to Houkes, this result shows that scientific practice does not allow for a clear distinction to be drawn between evolutionary and intentional approaches. There is a vast and variegated border area between the domains. This result, drawn as it is from scientific practice, leaves open the possibility that disciplines make eclectic use of any approach that turns out to be fruitfully applicable while the conceptual gap may persist.

Lewens puts forward the view that applying evolutionary accounts when explaining technological innovation is all too obvious to be denied but that the way in which evolutionary theory is applied to the cultural field is usually not the most fruitful way. What is generally acknowledged is the wide range of factors responsible for variation and selection in the field of engineering. However, Lewens delineates another important aspect of evolutionary approaches that requires closer consideration since it is acknowledged to be one of Darwin's important ideas but is not yet fully accepted as a basis for the explaining of technical innovation. It is the notion of population thinking as an alternative to typological thinking. He shows that population thinking in the version put forward by Richerson and Boyd (2005) is a most promising approach that may help to understand creative intentional processes in evolutionary terms. Considering theories of cultural change that ascribe fitness values to memes in the way that evolutionary biologists ascribe these values to organisms, Lewens demonstrates that even with these theories, which are based on typology and are not in line with Richerson and Boyd's approach, the inclusion of population level factors is indispensable in explaining evolutionary processes.

Krohs takes the example of the evolution of modular systems and compares biological accounts of the evolution of modules with what is known about the evolution of modular systems in engineering. A main line of argument used in attempts to explain the evolution of biological modularity is to explain modularization as a process of adaptation. Krohs argues that these explanations are incomplete. Though they explain the evolution of modules, most of these explanations cannot account for the fact that modularity is also restricted, that the borders of modules are fuzzy, and that a high degree of nonmodularity is present in many organisms. He proposes amending the arguments by considering not only the benefits but also the various kinds of costs, known from technological modularity, that modularity also entails for an organism. He finally discusses the problem of mapping functions on a modular structure, given that the structural and functional decomposition of biochemical and gene regulatory networks do not usually yield coinciding modular boundaries, and that almost any particular function within a molecular network is to be regarded not as being ascribable to a single component or module, but rather as being distributed over a large part of the whole network.

Kroes deals with the topic of emergent properties of technical systems. Emergence in the epistemic sense is the occurrence of unpredictable, novel systemic properties. He asks how this unpredictability affects the control paradigm in engineering. According to this paradigm, engineering practice aims at complete control of the behavior of a system by controlling the behavior of its constituent parts. It is precisely this that seems to be threatened by emergent behavior. He maintains that three issues related to the occurrence of emergent features in technical systems are particularly important in engineering practice: 1) emergent causal powers, 2) the tension between emergent features and functional decomposition, and 3) the unexpectedness and/or unpredictability of emergent features. To understand better the implications of emergence in engineering practice, he looks at whether, and in what respect, the functions of simple, stand-alone technical artifacts, such as everyday household appliances, can be regarded as emergent. He argues that the occurrence of epistemically emergent features in technical artifacts is not necessarily a threat to the control paradigm in engineering, since these features may be predictable and controllable on the basis of inductively established regularities. Emergence, though not the control paradigm, is also a central topic in the philosophy of biology, as is initially pointed out by Kroes. It is thus important to see that his results on what kinds of emergence are relevant in the field of technology fit in nicely with results obtained with respect to biological systems (Boogerd et al. 2005).

References

Boogerd, F. C., Bruggeman, F. J., Richardson, R. C., Stephan, A., and Westerhoff, H. V. (2005). Emergence and its place in nature: A case study of biochemical networks. *Synthese, 145:* 131–164.

Richerson, P., and Boyd, R. (2005). *Not by Genes Alone: How Culture Transformed Human Evolution.* Chicago: University of Chicago Press.

13 The Open Border: Two Cases of Concept Transfer from Organisms to Artifacts

Wybo Houkes

13.1 Introduction: The Conflict Image

Conceptually, organisms and artifacts have a long-standing but troubled relationship. This chapter is about one relatively recent aspect of this relationship, namely the transfer of evolutionary concepts and models, developed for the domain of organisms, to the domain of artifacts. More specifically I study whether this episode fits an image on which any such transfer from biology to technology would create a conflict between the conventional, intentionalist description of artifacts and the selectionist framework of evolutionary theory. On the basis of two case studies, I argue that there is no such conflict: the transfer of concepts and the relation between the intentionalist and selectionist frameworks are considerably more peaceful and subject to more varied and complicated interests and constraints.

Let me start by sketching what I call the "conflict image." On this image, artifacts cannot be described completely without making heavy use of teleological and intentional terminology: artifacts are *designed* to serve some particular *purpose,* and they are consequently *used* as *means* to some *end.* By contrast, the domain of organisms is selectionist: biological items are described as *reproduced* through genetic mechanisms, modified by *blind variation,* and subject to purposeless *natural selection.* The exact structure and status of these descriptions is of course quite different: the intentionalist description is a relatively loose set of "folk" notions, whereas the selectionist description is arguably a well-developed and robust scientific theory. I denote both descriptions loosely as "frameworks." Most relevant here is that on the conflict image the frameworks are incompatible: intentionality and natural selection do not mix, and notions used in one framework cannot be transposed to the other. The frameworks are separated by an iron curtain.

The possibilities for conflict do not end there, for one might maintain a strict separation of domains, parallel to that between frameworks. Consequently one ought to purge biology of all intentionalist (or even all teleological) elements, or proclaim such elements to be either metaphorical or misguided. Or one may argue that the role of intentionality in the construction of artifacts and sociocultural entities almost automatically frustrates the application of evolutionary theory. An alternative to this separatist image is to seek a unification by applying either the intentionalist or the selectionist framework to both domains and abolishing the other. Although this image does not involve an iron curtain between domains, it does maintain a strict separation of frameworks: applying the selectionist framework to artifacts would automatically mean that the intentionalist framework is abolished. In either the separatist or the unionist scenario, peaceful coexistence is ruled out on both sides of the border.

This conflict image of frameworks and domains is partly made of straw.[1] Yet if one analyzes the relation between artifacts and organisms by concentrating on a single concept, one might easily—and perhaps unwittingly—support one of these images. Thus if one concentrates on the notion of "selection," "design," or "function" alone, the results of the analysis are likely to be that the notion, as applied to artifacts and to organisms, a) means different things (e.g., "intentional" versus "natural" selection), so that any conceptual transfer trades on an ambiguity, or is merely metaphorical; or b) means the same thing, so that the domains are effectively unified.[2]

Part of what might make the conflict image attractive to philosophers is that it ascribes them a clear task: they may, or even should, monitor any conceptual traffic between the domains of artifacts and organisms. If one prefers separatism, there are ample opportunities for smuggling, that is, illicit transfer of concepts and models between the domains. Indeed if such transfer does not involve harmless souvenirs from the other domain (i.e., metaphors), it is very likely to amount to smuggling (i.e., making a category mistake). Alternatively if one endorses unionism, traffic between the domains may ultimately amount to a conceptual invasion in which philosophers can act as spies, pointing out sources of resistance (i.e., key concepts still in need of reduction). In either case the iron curtain image offers plenty of job opportunities for philosophers.

I argue, however, that the actual transfer of concepts from the domain of organisms to that of artifacts does not fit these projects of separation and unification. On them one would expect this transfer to involve either the promise of a unification of both realms or grand delusions and gross ambiguities—and perhaps both. In practice, however, one finds that intentionality and natural selection may coexist in various ways within the domain of technical artifacts.

Before my argument gets underway, however, let me briefly address an objection that it is redundant. As anyone who studies the interrelations between artifacts and organisms knows, so this objection goes, there is a vast and variegated border area between the domains, which might be regarded as contested terrain. This no-man's-land is populated by transgenic mice, nature-identical flavors, and restored landscapes, to give a few examples. Moreover, new items are continuously added to the border area through activities such as breeding and genetic manipulation. Given this border area of objects and activities, refuting the conflict image seems more like kicking a dead horse than like burning a straw man.

Yet although there is no denying that there is a no-man's-land between both domains, it is not so obvious that its mere existence undermines the conflict image. The contested area could, after all, be clearly divided between the domain of artifacts and that of organisms through stipulation. Such a stipulation might involve some arbitrariness, but proponents of the conflict image could maintain that at least one division carves nature (or rather nature-culture) at its joints.

This is an implausible response, but what makes it implausible shows that there is a need for further discussion, for the objection did not rest only on mere existence of a contested area but also on the fact of its growth, through activities such as breeding and genetic manipulation. These activities in turn might well have characteristics in common with the intentional selection of traditional design and with natural selection: a breeder, for instance, intentionally selects for some traits but has to rely partly on genetic reproduction. In genetic manipulation, the possibilities of intentional selection appear to have increased—"at the price of" natural selection. Thus a mixture of intentionalist and selectionist notions seems appropriate for describing both activities.

Again this observation is probably correct, but its very formulation shows that there is more to be said. Which mixture, if any, is appropriate for describing both activities? Are intentional and natural selection indeed communicating vessels, meaning that an increase in one causes a decrease in the other? Do breeding and genetic manipulation, or the description of these activities, involve the application of concepts that were earlier used exclusively for other domains or activities? And to what extent do intentionalist and selectionist notions coexist in descriptions of breeding and genetic manipulation? Some answers to these questions surely would undermine the conflict image, but it is also possible to give highly segregationist answers. Only a close review of actual theories or descriptions would show which answers are correct and whether the conflict image is really undermined by the existence of oncomice and breeding. Without such a review, the contested-area objection is only based on intuitions—albeit highly plausible ones.

In this chapter I provide such a review, not of theories of breeding and genetic manipulation, which are activities that have a very wide scope and for which there are, to the best of my knowledge, no concise theories or models available. Instead I have selected two more determinate and narrow fields of inquiry and activity. In sections 13.2 and 13.3 I discuss two types of artifact-oriented research in which evolutionary theory has seen very recent use: evolutionary design, especially in electronics, and evolutionary archaeology. These discussions are like journalistic border reports. Given the rapid development and lack of consensus within these fields, it is virtually impossible to get an accurate overview of the situation and to tell where promises end and results begin. I present work in both fields as concisely and straightforwardly as possible, but this does involve considerable reconstruction. Still, it can be shown that the application of evolutionary concepts and models is, in both cases, problem-driven and open-ended, that neither application is fruitfully analyzed as merely metaphorical, and that neither leads to abolishing the intentionalist framework; in both cases, the conceptual border between organisms and artifacts is open, and intentionality and selection coexist. Finally, in section 13.4, I present an argument against the conflict images of this section, based on similarities between both cases; I offer an alternative, open-border image; and I describe briefly which tasks this image leaves for philosophical analysis.

13.2   First Case: Evolutionary Design in Electronics

Over the past decade electrical engineers have become increasingly interested in the possibilities of designing circuitry through processes that resemble those that produce biological items. Attempts to construct such design processes are known under such names as "hardware evolution," "bio-inspired systems," and—the name I adopt here—"evolutionary design" (ED). The idea of "growing designs" that are capable of adaptive self-reproduction is certainly much older than the 1990s, but in the latter half of the decade it has rapidly developed beyond the visionary stage: there are now many research teams and several conference series devoted to ED, and there is significant interest from industry.[3] Simultaneously there is a growing trend among biologists to simulate evolutionary processes, for instance, by building computer models. Some of the results in these artificial-life programs are presented at the same conferences and published in the same journals as those of the engineering research just mentioned.

13.2.1   Aim and Approach

ED is motivated by the hope of designing circuitry quickly, innovatively, and without continuous designer interference. The guiding idea in the field is that through defining a set of eligible components, a procedure or algorithm for constructing circuits from these components, and a fitness measure for evaluating the constructed circuits, the construction and evaluation of circuits can be fully automated.[4] Moreover, researchers hope that the results might outperform traditional design solutions; they believe that ED processes may explore fruitful portions of the component or circuit "design space" that are ignored by human designers.[5] The following hypotheses are representative of this goal:

(H1) Conventional design methods can only work within constrained regions of design space. Most of the whole design space is never considered. (H2) Evolutionary algorithms can explore some of the regions in design space that are beyond the scope of conventional methods. In principle, this raises the possibility that designs can be found that are in some sense better. (Thompson, Layzell, and Zebulum 1999: 167)

Given this ambition of supplementing or outperforming conventional methods, most researchers in this field attempt to develop ED processes without preprogramming them for success through, for example, carefully selecting the design space of components, defining a very strict fitness measure, or removing unwanted elements from the space of circuits before recombination. After all, such preprogramming might introduce conventional design methods and solutions through the back door and thus fail to fulfill ED's promise.

The challenges facing such a hands-off approach are understandably enormous and it is impossible to predict whether this field will bear fruit in the sense described earlier. But even if, from a practical point of view, ED would be a failed growth on the tree of engineering science, the problems currently identified in the field and some of the methods used to solve these problems are interesting for my present purpose.

13.2.2 Current State of Development

One key problem in ED is that of scalability. Finding a viable solution to a circuitry design problem becomes harder with increasing complexity of the circuitry: designing a functional two-bit adder is much easier than designing an evolvable motherboard. Thus many evolutionary designs for complex electronics fail by leading to solutions that are vastly inferior to traditional designs. In principle, this problem could be solved by limiting the search for a functional design to a small portion of the available design space—then, in many cases, evolutionary processes yield successful, larger-scale designs. However, this is typically regarded as defeating ED's purpose of developing functional and innovative solutions.

A large number of contributions to conference proceedings and of journal papers address the scalability problem, so there appears to be consensus on the problem. This cannot be said for the solution, and few proposals amount to more than promissory notes. Still, one intriguing proposal is to increase ED's scalability and problem-solving capacity by making the ED process more similar to natural evolution. The idea behind these proposals appears to be that since nature has managed to develop solutions to design problems that are even more complex than those facing engineers, evolutionary electronics can benefit from imitating nature (Bentley 1999). This does not mean that engineers seek to imitate natural objects, for example, by making a silicon brain; instead they are interested in modeling and mimicking the mechanisms by which they think that nature has overcome the scalability problem.

In practice this imitation has at least two levels of intricacy. First, researchers have noted that complex natural objects typically have various structural features that improve scalability, such as modularity and iteration. Attempts have been made to develop ED processes that make use of these structural features; one example is the Cellular Encoding approach (e.g., Koza et al. 1999). In some cases these features are more or less programmed into the process, roughly speaking by including iteration and modularity rules in a tree of developmental stages where both the configuration of the artifacts to be designed and the rules for changing these configurations are represented by a "genetic" code. It can be shown, however, that such so-called static or explicit approaches at best provide very partial

solutions to the scalability problem: even for relatively simple problems, such as constructing patterns of tessellating tiles, the number of rules needed increases dramatically, with
no improvement in the fitness of the solutions, with increasing complexity (Bentley and
Kumar 1999). Consequently some researchers have introduced a second level of nature
imitation. They do not just employ a distinction between circuit phenotypes and an underlying code but they also seek to imitate the embryogenesis of organisms, that is, the way in
which items develop through the interaction of the genetic code with a constantly changing
environment. This approach does not make use of a large and intricate set of rules—not
even of a set that may evolve during the design process. Instead a random set of starting
configurations are used in combination with simple rules, which are activated or ignored
depending on the state of the environment (i.e., the intermediate product of the design
process), applied in parallel instead of sequentially, and which—perhaps most important—can be changed or supplemented during the evolutionary process; extra rules may
be added without the intervention of the designer, or the activation conditions of existing
rules may be changed. In this so-called implicit approach, more of the design process is
put beyond the control of the human engineer. Some promising, albeit very preliminary,
results have been reported: an implicit embryogenesis appears to solve complex problems
more quickly, more reliably, and more diversely than explicit approaches (Bentley and
Kumar 1999; Kumar and Bentley 2003a; Gordon and Bentley 2002, 2005).

13.2.3 Preliminary Assessment

What is, for my purposes, most salient about this research is that some researchers in ED
are trying to overcome a specific problem, that of scalability, by means of an increasingly
intricate transfer of selectionist concepts and models to the domain of artifacts. Thus
electrical engineers do not transfer concepts and model mechanisms because there is no
argument not to do so, or because they envisage a unification of biology and engineering
science. Instead they have noted that there is at least a structural similarity between a
problem in the domain of artifacts and one in the domain of organisms. This scalability
problem is the driving force behind the interdomain transfer of concepts and the modeling
of mechanisms.

The researchers themselves most often describe their efforts as "biologically inspired"
or based on "metaphors."[6] If this were true, organisms would hold no privileged position
over other objects that may be a source of inspiration—circuit designers might just as well
have looked at the clouds. Yet the choice for this particular source is far from arbitrary.
The reason for looking at organisms rather than clouds is, as said, the similarity of the
scalability problems encountered by nature and electrical engineers. Moreover, the interest
of biologists in ED programs is almost comparable with the interest of engineers in genetic
mechanisms.[7] This is explained straightforwardly by assuming that the prospects of ED
programs increase the more accurately nature's strategies for solving the scalability

problem are modeled. Since, however, our current knowledge of nature's strategies is incomplete, this search for accuracy within ED may also increase our knowledge of biological mechanisms. This largely tacit accuracy aim[8] therefore explains the convergence of interests that is manifested in the conference proceedings and journals referred to at the start of this section; regarding the conceptual transfer as metaphorical cannot do justice to this phenomenon.

Finally, the conceptual transfer in ED is to some extent open-ended. Suppose the current transfer does not resolve the scalability problem in ED. Then if biologists would arrive at a better understanding of genetic and embryogenic mechanisms, perhaps using additional concepts we may safely predict that ED researchers will adopt these concepts in their quest for innovative circuitry designs. Thus we have a reason to assume that conceptual transfer from biology to ED might continue in the future. Still, I have not found any electrical engineer who claims that the selectionist framework might, at some time, completely replace the intentionalist framework as a description of design processes: transfer is open-ended but it does not amount to a conceptual invasion.

13.3 Second Case: Evolutionary Archaeology

Archaeology involves both the excavation and interpretation of the material remains of earlier civilizations. For the latter goal, a wide variety of methods have been and currently are in use, ranging from the hermeneutic methods of structuralist and postmodernist archaeology to the positivism of processualist archaeology.[9] One relatively recent addition to this spectrum is evolutionary archaeology (EA). Musings about extending the theory of evolution to archaeology and anthropology probably precede Darwin, and have led to many different models and theories. Some of the better-known and controversial theories in this vein may be sociobiology, Dawkins's (1982) speculations about the extended phenotype, and Boyd and Richerson's (1988, 2005) dual-inheritance model of cultural evolution. EA may be continuous with or even dependent on some of these various approaches but I shall not consider this connection here. Whatever the dependence, EA has more specific goals and methods.[10]

13.3.1 Aim and Approach

Proponents of EA emphasize that their work is based on an explicit, even stipulative, distinction between so-called stylistic and functional attributes of artifacts, proposed by Robert Dunnell in the 1970s. The guiding idea behind this dichotomy is that whereas functional features quickly gain and then maintain prominence in the archaeological record, stylistic features are more variable; they become popular, stay in fashion for a while, and are then replaced by other styles. Thus styles are useful for constructing historical traditions within the archaeological record.[11]

This dichotomy is supplemented with a hypothesis about the underlying processes and with an explicit explanatory goal. First, functional attributes are understood as fitness-conferring adaptations, subject to natural selection. Second, styles are assumed to be "selectively neutral" (Dunnell 1978: 199) results of cultural transmission; the idea is that toolmakers transmitted stylistic features to one another, occasionally varying on the theme, producing old-fashioned or avant-garde arrowheads, and so forth.[12] Thus a combination of two mechanisms, natural selection and cultural transmission, is used to reconstruct archaeological traditions or lineages, and to explain them.[13]

The primary goal of EA is, however, not classificatory but explanatory. Specifically the application of evolutionary theory is supposed to circumvent the need for intentionalist or mentalist explanations of the archaeological record. In EA, artifact traditions are constructed and explained, not by reconstructing the intentions or mentality of the producers of ancient artifacts, but by directly constructing artifact lineages and identifying styles and cultural transmission. The reason is quite simple: "Individuals do make decisions, but evidence for these decisions cannot be recovered by archaeologists" (Flannery 1967: 122); and more recently, "Although we endorse the notion that new variants are intentionally created at least some of the time . . . we have yet to determine how such intentions are to be identified analytically in the archaeological record" (Lyman and O'Brien 2000: 41).[14] Thus researchers in EA seek to combine evolutionary concepts and models with a commitment to designer intentions: the former are regarded as potential explanatory replacements of the latter.

13.3.2   Current State of Development

Very roughly, this is the conceptual framework of EA. To fulfill their promise to avoid intentions in explaining the archaeological record, advocates of EA need to demonstrate that it is actually possible to construct a lineage of phylogenetically related artifacts and to explain this lineage as a product of cultural transmission. This is far from easy: we know from personal experience, ethnographical studies, and the historical record that there are many different processes that may result in similarities and differences among artifacts of consecutive generations. Thus an underdetermination problem arises. Suppose that we find two slightly different projectile points, A and B, in adjacent layers of some excavation site, and that dating methods show that they are approximately one generation apart in age.[15] It makes sense to place these artifacts in a lineage and to presume that B was produced by person P, who (directly or indirectly) learned his trade from Q, the producer of A. This may explain the similarity between A and B. For their differences, however, a large variety of explanations is available: P might have come up with a functionally equivalent stylistic variation, he might accidentally have produced an unfaithful copy, he might have used a slightly different production process, or he might have adapted the arrowhead to a perceived change in environments. Alternatively, B might have been robbed from a neighboring tribe, who produced this functionally equivalent but stylistically different artifact. Additional information about the goals, environment, and way of life of P and Q may partly resolve this problem, but in the absence of such information the problem seems insurmountable: even the distinction between functional and stylistic features, which is a cornerstone of EA approaches, can hardly be made on the basis of evidence.[16]

Advocates of EA have recently and tentatively started to seek solutions. One type of solution is to use cladistic methods. Some researchers have noted that there is a similarity, or even an isomorphism, between the frequency diagrams used by some archaeologists to monitor the diversity within a class of tools and the clade-diversity diagrams used by some biologists and paleontologists to monitor the diversity within a class of organisms (Lyman and O'Brien 2000: 48–50). Furthermore these researchers (e.g., Lyman 2001: 77) have noted that the problem of distinguishing functional and stylistic features—natural selection and cultural transmission—bears a strong resemblance to the biological problem of distinguishing analogies from homologies—characteristics that are morphologically similar because of convergent evolution or because of common ancestry, respectively.[17] They note that biologists are able, at least in principle, to make this distinction by using cladistic methods, in which items are classified in a nested hierarchy of branches, that is, a lineage is constructed.[18] Based on these similarities, researchers have attempted to transfer these methods from biology to archaeology. Some preliminary results have been obtained, for example, cladograms of projectile points found in the southeastern United States and of ceramics from the lower Mississippi Valley (Lyman and O'Brien 2000; O'Brien, Darwent, and Lyman 2001; O'Brien et al. 2002; O'Brien and Lyman 2002).

It is unclear whether these results are more than a happy coincidence. What is more certain is that cladistics cannot be the panacea of EA: even if there are no conceptual obstacles, cladograms are unstable if the set of data and characteristics used to construct them is small (a typical situation in archaeology). And the problems do not end here, for constructing a cladogram is only a first step for evolutionary archaeologists.[19] Suppose that, after entering a set of artifact characteristics, the data-processing software comes up with a single conjecture of their phylogenetic relations—what O'Brien and Lyman claim to be the case for the projectile points. And suppose that we establish that all artifacts are functionally equivalent—as seems to be assumed by calling all the classified items "projectile points." Even on these assumptions, the underdetermination problem described earlier remains unsolved. The reason is that the phylogenetic relation is still compatible with a large number of very different cultural-transmission processes: faulty copying, faithful transmission with a purely stylistic variation, faithful transmission with adaptation to a perceived change in the environment, protoindustrial espionage, and so forth. This problem becomes manifest when Lyman and O'Brien discuss an evolutionary model for vacuum-tube radios in the early twentieth century, following a study by Michael Schiffer (1996). Although they show that Schiffer's results can be reproduced by using cladograms and clade-diversity diagrams, Lyman and O'Brien admit that Schiffer's ability to explain his results in terms of stimulated variation and other transmission processes is due largely to the availability of historical data on the use and production of vacuum-tube radios. In the absence of such data, explaining changes in the archaeological record "is fraught with analytical difficulties" (Lyman and O'Brien 2000: 55). Yet if information on the use and production of artifacts is available, there seems little need to avoid traditional intentionalist reconstructions by means of evolutionary models.

Thus the goals of EA may be clear, but the promise to avoid intentionalist reconstructions remains largely unfulfilled—and it is difficult to see how it can be fulfilled. Some techniques are available for the first, classificatory step, but models and techniques for taking the crucial second, explanatory step seem lacking. At least some researchers in the discipline are aware of this lack, and they seem ready to apply further (semi)evolutionary techniques to solve their "analytical difficulties."

13.4   The Open Border

In the previous sections, I report on two fields that deal with artifacts, but that use concepts and mechanisms drawn from the selectionist framework of evolutionary biology. The aims of these fields are different—one studies primitive artifacts and aims at classification and explanation; the other involves state-of-the-art technology and aims at effective and innovative design. The differences in aim are partly reflected in the conceptual transfer involved. Still, there are marked similarities: transfer in EA is as problem-oriented, nonmetaphorical, and open-ended as in ED. To sum up, selectionist concepts and models in EA serve specifically to avoid appealing to the intentions in explanations of the archaeological record, structural similarities between classifications of organisms and artifacts are used to support the transfer of cladistic methods, and researchers take an active interest in transferring more concepts and models.

Although researchers in both fields frequently describe their own research as involving "metaphors," they actually attempt a faithful and, to some extent, incremental application of evolutionary concepts and mechanisms. This does not mean that they are interested in applying the full selectionist framework to their field. Instead they take care to apply the concepts and techniques transferred from biology accurately in order to solve some specific problem, and they take an active interest in transferring more concepts and techniques to solve additional problems. Such transfer is, moreover, often supported by considerations of structural similarity. Thus research in both disciplines is not characterized by a flash of inspiration followed by a short-lived transfer from biology to technology. Instead researchers in EA and ED actively search to increase conceptual and methodological similarities between artifacts and organisms and to import selectionist concepts and mechanisms on the basis of these similarities—if this suits their explanatory, descriptive, or constructive purposes.

Moreover, in transferring concepts, researchers seek to establish a particular mode of coexistence of the intentionalist and selectionist frameworks. Neither in EA nor in ED does this coexistence amount to a facile eclecticism or a tangle of ambiguities: researchers seem aware that the success of their field depends on realizing the coexistence purpose, and their research is shaped by this awareness. Summing up, the goal of ED is to supplement traditional design methods with evolutionary algorithms that explore uncharted portions of design space; I have not found any claims that ED might replace traditional design, that is, that it might lead to more efficient or effective designs in the well-traveled portions of design space. That evolutionary processes should supplement traditional, intentionalist methods is a driving force behind the development of ED: In section 13.2.2, I have described how various procedures are evaluated and improved with regard to their capacity for yielding innovative designs. In EA, coexistence is constructed along different lines. Researchers admit that in principle the intentionalist framework can be used to describe and explain the archaeological record, but they claim that lack of information prevents them from using it in practice. Therefore evolutionary concepts and models are chosen as substitutes. Again this mode of coexistence shapes research: if it turns out that evolutionary models can do their explanatory work only by reintroducing intentionalist elements, researchers admit that there are difficulties, and additional transfer is seen as advisable. In practice these difficulties may turn out to be insurmountable—but that would be a success criterion derived from the explanatory-substitution mode of coexistence. Thus the two modes of coexistence both lead to constraints on satisfactory results within the program and act as heuristics for modifying the program.

This leads to my argument against the conflict image presented in section 13.1—specifically, against the separatist and unionist image of the relation between organisms and artifacts. On these images, a problem-oriented, nonmetaphorical, and open-ended transfer of concepts between the domain of organisms and that of artifacts is either ruled out or vastly lacking in ambition. Neither option is, in my opinion, attractive.

On the first, separatist judgment, researchers that indulge in the type of transfer described earlier should be charged with smuggling by the philosophical border patrol. This should not lead to symbolic sanctions: if the classificatory, explanatory, and constructive resources employed in EA and ED belong in the realm of organisms, transferring them to a qualitatively different realm can only yield results by accident. Both programs should fail. It may be possible to suspend this judgment as long as the techniques employed and the concepts transferred are sufficiently general to avoid ambiguities. Insofar as cladistics involves only a statistical analysis of the resemblances within a set of objects[20] and embryogenesis involves no more than some translation mechanism, this may be correct. However, ED and EA are bound to exhaust this tolerance rather quickly, given the researchers' willingness to employ structural similarities between organisms and artifacts and to transfer additional concepts. As soon as the border patrol runs out of ways to euphemize current or future smuggling as harmless verbal interchange or metaphor coining, the programs will be outlawed.

Thus the separatist iron curtain image puts tight and, ultimately, uninteresting constraints on the viability of both evolutionary research programs: they are either nonstarters or bound to exhaust philosophical patience quite soon. Moreover, this judgment may be passed on the basis of conceptual analysis—and creative reformulation—alone. This consequence seems unacceptable. Whether ED and EA are successful programs should be a specific and (partly) empirical, not a general and a priori, matter. As philosophers have learned the hard way, a priori and general limitations on scientific research typically expire long before the programs they try to constrain. Kant's admonition that neither chemistry nor psychology could ever become a scientific discipline comes to mind (Friedman 1992: ch. 5.III), as do neo-Kantian resistance against general relativity theory and the nagging complaint that the social sciences lack universal laws.

On a unionist image, the prospects for EA and ED are hardly better than on a separatist image, for unionism entails that researchers in both programs are underachievers: instead of seeking a particular mode of coexistence, they should seek to replace the intentionalist framework with the selectionist. This has the minimal effect of increasing the burden resting on both programs far beyond their expected carrying capacity: as I have described, researchers in both programs are already struggling to realize their more modest goals. What is more, unionism might undermine EA: if the intentionalist mechanism of cultural transmission would be replaced with natural selection, the central dichotomy between functional and stylistic features would disappear.

The conceptual transfer in EA and ED fits neither the separatist nor the unionist image presented in the introduction. The relation between the domains of organisms and artifacts must be understood in a different way.

Let me end by describing one such way. On this open-border image, there is no general limitation on the transfer of concepts from the domain of organisms to that of artifacts. Instead such transfer is a decidedly pragmatic affair. Researchers may attempt to apply elements of the framework used to describe the other domain, provided that there is a well-determined need for such a transfer and that there are similarities among the domains that support a nonmetaphorical transfer. Thus conceptual transfer is more than an heuristic process; it may play a role in descriptions and explanations in the domain of artifacts—and not just in the search for such descriptions and explanations. The success of this transfer is not determined by general principles, but there are constraints on transfer set by the specific project. To stay with the original metaphor, transferring concepts and models from the domain of organisms to that of artifacts involves neither smuggling nor an attempt at conquest, but resembles immigration for economic reasons.

Although there ought to be no a priori regulations for this type of transfer, the open-border image does not lead to philosophical quietism: even if one acknowledges that there is no need for separatist border patrol or unionist espionage, philosophers may be actively involved in open-border transfer in various ways.[21] First, I have described how considerations of structural similarity between the domains motivate the transfer of concepts and models, and I have elevated the existence of these similarities to a criterion of evaluation. Philosophers might study these similarities, or claims about them in scientific research programs, and pass judgment on which episodes of transfer are warranted by them and which are either metaphorical or misguided. This type of evaluation is not straightforward; it calls for a detailed and well-informed analysis of rapidly evolving fields. The same goes for the second type of task, which is to study whether the intentionalist and selectionist frameworks are, in the end, incompatible in the domain of artifacts. Despite the argument given in this section, this incompatibility aspect of the conflict image may still be defensible. Although both the fields of EA and ED involve the transfer of concepts and models from the domain of organisms to that of artifacts, neither seems to involve transfer between frameworks. In a sense, intentions and natural selection are kept apart in both fields, albeit within the domain of artifacts. Thus one might still argue that the frameworks are incompatible and that coexistence can or should not lead to any real conceptual interchange. Evaluating this multicultural image would require a detailed study of the frameworks used in EA and ED, and it would require philosophers to consider multiple concepts, chart their relations, and keep track of changes within this conceptual framework used in a domain or field. Most challenging, it requires the development of tools for analyzing the relations between intentionalist and selectionist concepts in the domain of artifacts. The situation at the open border between organisms and artifacts leaves philosophers little choice but to face these two tasks.

Notes

1. The iron curtain image may shape debates on the relation between artifacts and organisms, and the selectionist and teleological frameworks—such as the Intelligent Design controversy, (anti)adaptationism in biology, and the ongoing search for a memetic mechanism for cultural evolution. In all these cases there is a tendency either to a) keep the two frameworks carefully apart and to rule out conceptual incursions, or b) extend one framework to another domain in its entirety.

2. One way to avoid the iron curtain image in "single-conceptual-analysis" projects is to develop a single *nonintentionalist and nonselectionist* concept and apply it to both domains. Ulrich Krohs's (2004) theory of functions may provide the only example of this type of analysis.

3. For example, International Conference on Evolvable Systems (ICES; roughly biennial since 1996; proceedings published in Springer's *Lecture Notes on Computer Science* series), NASA/DoD conference on Evolvable Hardware (EH; annual since 1999), and Genetic and Evolutionary Computing Conference (GECCO; biennial 1989–2000, annual since 2000; selected papers published in *Natural Computing* and *Genetic Programming and Evolvable Machines*).

4. The interested reader may consult the introductory chapters of Thompson (1998) or Layzell (2001) for a schematic impression of the actual methods involved in ED.

5. Recently some researchers have argued that there might also be design problems that can be solved *only* by ED methods (Thompson 2002).

6. In the description of one leading journal, one finds the following description of the field: "Characteristic for man-designed computing inspired by nature is the metaphorical use of concepts, principles and mechanisms underlying natural systems" (source: online description of the international journal *Natural Computing* on SpringerLink).

7. See, e.g., many of the essays in Kumar and Bentley (2003b).

8. The accuracy aim is sometimes made explicit, e.g.: "[This model of development for evolutionary design] is intended *to model biological development very closely* in order to discover the key components of development and their potential for computer science" (Kumar and Bentley 2003a: 57; emphasis added). Yet this quote is taken from a paper that carries "biologically inspired" in its title, and "biologically plausible" in the reference on Peter Bentley's Web site!

9. See Renfrew and Bahn (2004), especially ch. 12, for an overview of various archaeological methods.

10. In the following, I mainly rely on one particular line of work in EA, that of R. Lee Lyman, Michael O'Brien, and various cooperators.

11. Some researchers admit that functional features may also be useful for these purposes (e.g., O'Brien and Leonard 2001: 5–6).

12. "[A]rtifacts are stylistically similar as a result of *cultural* transmission" (Lyman and O'Brien 2000: 44).

13. For example, "The Darwinian mechanisms of selection and transmission . . . provide exactly what culture historians were looking for: the tools to begin explaining cultural lineages" (O'Brien and Lyman 2002: 35); "Only with explicit adoption of the tenets of Darwinian evolutionary theory has it become clear *why* historical types behave the way they do" (Lyman and O'Brien 2000: 47).

14. This nonintentionalism occasionally turns into anti-intentionalism, e.g.: "It is increasingly common to explain human outcomes in terms of the intentions of the agents involved. Unfortunately, this leads to a vitalistic explanation of little merit. . . . [T]here is a significant discrepancy between intentions and outcomes. Every pre-historic farmer who ever put hoe or digging stick to earth intended success. Many failed. To explain the success of the successful in terms of their intentions is absurd. They were successful not because of their intentions but because of the particular variant they generated, the vagaries of chance and the operation of natural selection" (O'Brien and Leonard 2001: 26). This argument is puzzling, for it changes the explanandum of EA from "human outcomes" to *successful* outcomes. Furthermore, this argument addresses functional features and selection rather than stylistic features and cultural transmission, which are central to the explanatory project of EA. In cultural transmission, (partial) failure seems just as important for explaining stylistic variation as success is for explaining stylistic continuity.

15. The error margins of dating methods in archaeology actually may be too large to make reliable statements of this kind.

16. Early papers in EA suggest that functional features are easily distinguished within the archaeological record because they would show a "directional increase." This is now widely admitted to be false: the frequency of stylistic features may show the same directionality (O'Brien and Leonard 2001: 8–9). Still, most researchers in EA maintain that styles "should behave randomly in relation to the selective environment" (Hurt and Rakita 2001: xxvi).

17. This rough characterization of homologies and analogies is similar to that given by Mayr (2001).

18. An introduction to cladistics, the problems in constructing reliable cladograms, and the advantages of cladistics over other methods of biological classification can be found in Ridley (1996, chs. 14 and 17).

19. Constructing a cladogram is not even a very important first step in EA. Following the so-called Ford-Spaulding debate of the 1950s, most (evolutionary) archaeologists do not believe in objective artifact kinds; instead they maintain that classification depends on the interests of the archaeologists. Thus cladograms do not show real artifact kinds but are ways of classifying artifacts such that cultural transmission can be studied.

20. Lyman and O'Brien sometimes praise cladistics for this neutrality, e.g.: "It depends solely on heritable continuity, irrespective of the mode of transmission" (O'Brien and Lyman 2002: 30).

21. The tasks described here resemble those described by Lewens (2004) in evaluating the applicability of the "artefact model" in biology.

References

Bentley, P. J. (ed.). (1999). *Evolutionary Design by Computers.* San Francisco: Morgan-Kaufmann.

Bentley, P. J., and Kumar, S. (1999). Three ways to grow designs. In: *Proceeding of the Genetic and Evolutionary Computation Conference (GECCO '99),* 35–43.

Boyd, R., and Richerson, P. J. (1988). *Culture and the Evolutionary Process.* Chicago: University of Chicago Press.

Boyd, R., and Richerson, P. J. (2005). *The Origin and Evolution of Cultures.* New York: Oxford University Press.

Dawkins, R. (1982). *The Extended Phenotype.* Oxford: Oxford University Press.

Dunnell, R. C. (1978). Style and function: A fundamental dichotomy. *American Antiquity, 43:* 192–202.

Flannery, K. V. (1967). Culture history versus cultural progress: A debate in American archaeology. *Scientific American, 217:* 119–122.

Friedman, M. (1992). *Kant and the Exact Sciences.* Cambridge, Mass.: Harvard University Press.

Gordon, T. W., and Bentley, P. J. (2002). Towards development in evolvable hardware. In: *Proceedings of the 2002 NASA/DoD Conference on Evolvable Hardware,* 241–250.

Gordon, T. W., and Bentley, P. J. (2005). Development brings scalability to hardware evolution. In: *Proceedings of the 2005 NASA/DoD Conference on Evolvable Hardware,* 272–279.

Hurt, T. D., and Rakita, G. F. M. (eds.). (2001). *Style and Function: Conceptual Issues in Evolutionary Archaeology.* Westport: Bergin & Garvey.

Koza, J., Bennett, F. H. I., Andre, D., and Keane, M. A. (1999). *Genetic Programming III.* San Francisco: Morgan-Kaufmann.

Krohs, U. (2004). *Eine Theorie biologischer Theorien: Status und Gehalt von Funktionsaussagen und informationstheoretischen Modellen.* Springer: Berlin.

Kumar, S., and Bentley, P. J. (2003a). Biologically inspired evolutionary development. In: *Proceedings of the 5th International Conference on Evolvable Systems: From Biology to Hardware* (Tyrrell, A., Haddow, P., and Torresen, J., eds.), 57–68. Springer: Berlin.

Kumar, S., and Bentley, P. J. (eds.). (2003b). *On Growth, Form, and Computers.* London: Academic Press.

Layzell, P. (2001). *Hardware Evolution: On the Nature of Artificially Evolved Electronic Circuits.* Unpublished Ph.D. thesis. Available at http://www.informatics.sussex.ac.uk/users/adrianth/lazwebpag/web/Publications/THESIS/ssxdphil.pdf

Lewens, T. (2004). *Organisms and Artifacts: Design in Nature and Elsewhere.* Cambridge, Mass.: The MIT Press.

Lyman, R. L. (2001). Culture-historical and biological approaches to identifying homologous traits. In: *Style and Function: Conceptual Issues in Evolutionary Archaeology* (Hurt, T. D., and Rakita, G. F. M., eds.), 69–89. Westport: Bergin & Garvey.

Lyman, R. L., and O'Brien, M. J. (2000). Measuring and explaining change in artifact variation with clade-diversity diagrams. *Journal of Anthropological Archaeology, 19:* 39–74.

Mayr, E. (2001). *What Evolution Is.* New York: Basic Books.

O'Brien, M. J., Darwent, J., and Lyman, R. L. (2001). Cladistics is useful for reconstructing archaeological phylogenies. *Journal of Archaeological Science, 28:* 1115–1136.

O'Brien, M. J., and Leonard, R. D. (2001). Style and function: An introduction. In: *Style and Function: Conceptual Issues in Evolutionary Archaeology* (Hurt, T. D., and Rakita, G. F. M., eds.), 1–24. Westport: Bergin & Garvey.

O'Brien, M. J., and Lyman, R. L. (2002). Evolutionary archaeology: Current status and future prospects. *Evolutionary Anthropology, 11:* 26–36.

O'Brien, M. J., Lyman, R. L., Saab, Y., Saab, E., Darwent, J., and Glover, D. S. (2002). Two issues in archaeological phylogenetics: Taxon construction and outgroup selection. *Journal of Theoretical Biology, 215:* 133–150.

Renfrew, C., and Bahn, P. G. (2004). *Archaeology: Theories, Methods and Practice,* 4th ed. London: Thames & Hudson.

Ridley, M. (1996). *Evolution,* 2nd ed. Cambridge, Mass.: Blackwell Science.

Schiffer, M. B. (1996). Some relations between behavioral and evolutionary archaeologies. *American Antiquity, 61:* 643–662.

Thompson, A. (1998). *Hardware Evolution: Automatic Design of Electronic Circuits in Reconfigurable Hardware by Artificial Evolution.* Berlin: Springer.

Thompson, A. (2002). Notes on design through artificial evolution: Opportunities and algorithms. Presented at Adaptive Computing in Design and Manufacture V. http://www.informatics.sussex.ac.uk/users/adrianth/acdm2002/paper.pdf.

Thompson, A., Layzell, P., and Zebulum, R. S. (1999). Explorations in design space: Unconventional electronics design through artificial evolution. *IEEE Transactions on Evolutionary Computing, 3:* 167–196.

14 Innovation and Population

Tim Lewens

14.1 The Problem for Evolutionary Theories of Technology Change

According to one standard story, Darwin's achievement is twofold (see, e.g., Sober 2003: 267; Waters 2003: 117–118). First, Darwin offers a view of the *pattern* of biological change; namely descent with modification. Second, he gives us a *mechanism* for how a significant proportion of that change—including, especially, adaptive change—occurs. That mechanism is natural selection. Against an intellectual background that sees each species as specially created by a beneficent intelligence, these claims are radical, for they deny both the pattern and the process underlying special creation.

Evolutionary models of technological innovation and change come in many forms. Some of the best-known works in this area include books by Basalla (1988) and Mokyr (1990), and a collection of articles on the subject edited by Ziman (2000). To illustrate some of the suspicions one might have about the value of evolutionary models when applied in this domain, let us try to apply each of Darwin's two insights to technological innovation and technological change. First, what does it mean to defend an evolutionary view of the pattern of technological change? On the face of it this involves only the claim that novel artifacts are not produced ex nihilo without influence from previous artifact generations. Admittedly the recommendation always to look for ancestral forms of either whole artifacts or their parts may be a useful heuristic on occasions, but the problem here is that "great man" theories of innovation, and appeals to individual genius, have been out of fashion for a very long time. Historians of technology will not be surprised to learn that most of the great material innovations we may wish to study (Watt's steam engine, the Wright brothers' airplane) have ancestors that at least partially resemble them.

Second, what of the view that innovation proceeds by a form of natural selection? Once again this threatens to be an underwhelming assertion, at least if we construe natural selection in a loose enough manner that permits us to swat away potential counterexamples founded on disanalogies between the organic and technological realms (Lewens 2002). Darwin himself used artificial selection to illustrate the principle of natural selection. In a sense, then, natural selection was already copied from the realm of artifacts. According to many standard presentations, selection explanations account for adaptation by telling us that a variety of forms are produced; the ones that fit local demands well are retained for further modification, and gradually a well-adapted system is built up. If this is how selection explanations work, then the selectionist view seems once again to be true for technological innovation, but rather too obviously true to provide much insight to the student of technical change. (For an elaboration of this argument, see Lewens 2002; Lewens 2004: ch. 8.)

14.2   Two Responses to the Basic Problem

There are many responses we could give to the mean-spirited antievolutionary argument sketched in the previous section. One response consists of pointing out that we should not expect too much from evolutionary theories of technology change. For example, an evolutionary view of technical innovation should neglect neither variation nor selection. This means that evolutionary views of innovation will credit a wide range of factors with explanatory value. On the variation side, these factors will include the current state of technical know-how, the nature of dominant design heuristics, and the material basis of artifact manufacture, all of which help to determine what alternatives are available for selection to act upon. On the selection side, these factors will need to include the competitive environment a given technology finds itself in, as well as consumers' conscious and unconscious desiderata for technical artifacts. Since the latter can, in principle at least, take many forms, evolutionary views are likely to lead to skepticism of any grand theory that sticks its neck out regarding the general determinants of artifact success. Certainly we should not expect the most useful, or the best-designed, or the cheapest, artifacts always to be the ones that succeed.

Depending on one's standpoint, then, it may be a strength rather than a weakness of the evolutionary view that it offers few clues in itself regarding which factors are most important in explaining technological change. Evolutionary theories provide useful standpoints from which to articulate the rashness of monistic, or "deterministic," theories of technology change, regardless of whether such monistic theories locate determining power in technologies themselves or in the societies that produce and make use of them.

Note that what is distinctive about evolutionary views here is precisely their lack of distinctiveness: they tend to partially endorse many aspects of competing theories of technological change. Consonant with this, the culmination of a series of discussions of evolutionary theories of technical innovation by John Ziman (2000) and collaborators is the modest claim that evolution provides a unifying "paradigm of rationality" for describing technology change (Ziman 2000: 313). I take this to mean that evolutionary theories provide a framework in which contentious issues familiar to historians and economists regarding the relative importance in technological explanation of social, economic, psychological, material, and other factors can be articulated and discussed, without this evolutionary framework itself offering adjudication regarding these issues.

In this chapter I focus on a different way of responding to skepticism about the ability of evolutionary models to explain technical innovation. This response denies that Darwin's great contributions number just two. One of the many seminal claims that we owe to Ernst Mayr is the view that Darwin "replaced typological thinking with population thinking" (1976: 27). According to Mayr, population thinking is Darwin's third great contribution to biology. Whatever "population thinking" is, it is supposed to be distinct both from the hypothesis of evolution and from the principle of natural selection. Hence if we are unimpressed by the contribution that common descent and natural selection can make to the study of technology, we might be more impressed by the contribution made by population thinking.

14.3   Mayr's Population Thinking

Evolutionary anthropologists Robert Boyd and Peter Richerson have championed the view that population thinking is the key to an evolutionary understanding of culture (e.g., Boyd and Richerson 2000; Richerson and Boyd 2005). Although their work looks at cultural evolution in general, they count technology as part of culture (2005: 29), and some of their discussions focus on technological change. They explain their stance succinctly at the beginning of an important recent work:

Eminent biologist Ernst Mayr has argued that "population thinking" was Charles Darwin's key contribution to biology. . . . Population thinking is the core of the theory of culture we defend in this book. (Richerson and Boyd, 2005: 5)

I argue in this section that "population thinking," in Mayr's core sense of that term, does not in fact offer much of interest to the evolutionary theorist of technological change. From section 14.5 onward I argue that Boyd and Richerson's rather different brand of population thinking is far more promising.

Mayr defines *population thinking* by contrast with typological thinking (Mayr 1976). The typological thinker believes there is some small number of stable "types" or "forms," which explain the observed patterns of diversity in the biological world. The "vertebrate archetype" of Richard Owen, for example, was an effort to represent a common structural plan, modified to various degrees in particular species, which underlies all vertebrates. We can think of "types" as explanatory posits: some forms are seen rarely or not at all because there is no corresponding type. Others are seen frequently because they are variations on an underlying type (see Sober 1980; the following presentation is adapted from Lewens 2006: ch. 3).

Moving away from biology for a moment, it seems appropriate to offer something like a typological explanation when we try to understand why some crystal structures are seen frequently while others are not seen at all. We can take reference to "types" here to be shorthand for sets of physical facts that make some crystalline forms stable, others unstable. Perhaps we can think of organic types in a similar way. The typologist claims that only a few basic organic configurations are stable. These stable configurations then explain the diversity of forms manifested by individual organisms.

Darwin says that species are formed from natural selection acting on slight variation. His position demands then that these small variations, if they can be added up to produce new species, are themselves stable. Hence the reason why we do not observe forms that are intermediate between existing species cannot be that these forms are unstable. This presents Darwin with a dilemma. On the face of things, if he is right about common ancestry and the stability of slight variations, there should be no gaps between existing organic forms. On the other hand, Darwin learns from typological thinkers such as Geoffroy St. Hilaire and Owen that this is not the pattern we observe. So Darwin needs to give a nontypological explanation for apparently typological phenomena.

Ron Amundson has argued persuasively that Darwin's response is to reinterpret Owen's archetypes as ancestors: the diverse vertebrate species appear to be variations on a common theme not because they are manifestations of a single timeless ground plan but because they have retained the characteristics of a common ancestor (Darwin 1985 [1859]: 416; see also Amundson [2005]: ch. 4). But Darwin's way of thinking about shared history does not guarantee that the world contains species that are what he calls "tolerably well-defined objects" (Darwin 1985 [1859]: 210). We still need some explanation for the coherence of species, and the gaps between them.

One of Darwin's primary explanatory tools for discharging this task is an offshoot of the more general principle of natural selection, which Darwin calls the "principle of divergence of character." Darwin had learned from Adam Smith that competition will be most intense between individuals in the same line of business. Darwin argues that in the economy of nature, no less than in human affairs, competitive advantage will come to those who can open new markets, and find new ways of making a living:

the more diversified the descendants from any one species become in structure, constitution, and habits, by so much will they be better enabled to seize on many and widely diversified places in the polity of nature, and so be enabled to increase in numbers. (Darwin 1985 [1859]: 156)

Over time, generalists are squeezed out, and diverse specialists come to predominate. By coupling principles such as this one to his hypothesis of common ancestry Darwin is able to explain the existence of discrete species while also accounting for their underlying commonalities, and he is able to do so in a nontypological way. This much is good news for Mayr. What is not such good news for Mayr is that the primary resources Darwin uses to replace typological explanation are natural selection and the "tree of life" hypothesis.

This makes it hard to characterize "population thinking" as a third conceptual innovation wholly distinct from Darwin's better-known ideas.

If this is right, then this form of "population thinking" is unlikely to offer us any new set of conceptual resources not already implicit in the ideas of natural selection and common ancestry. So if we are already unimpressed by the contribution these two ideas can make to understanding technological change, we should not expect much additional insight to follow from using this form of population thinking. To see this, remember that on Darwin's view species are "tolerably well-defined objects" in virtue of the corralling forces of local environments, which discipline the tendencies of individuals to vary, and thereby maintain coherence over time at the level of the population. This population-level coherence is achieved in spite of differences constantly being introduced among individuals, not because of something shared by all individuals.

What would it mean to apply this way of thinking to a technological lineage? It would involve explaining technical trajectories in terms of the corralling forces of local market environments instead of in terms of the shared internal properties of token artifacts. The problem is not that this is an inappropriate way of explaining technical change. The problem is that it is too obviously an appropriate way. We are quite used to thinking that the absence of some kinds of artifact and the concentrated presence of others owes itself primarily to discontinuities in market demands. There are no chocolate teapots, not because they are impossible to construct, but because teapots made from chocolate would be useless. To the extent that Darwin's population thinking is novel, he does not so much devise a new way of thinking as show how the general modes of explanation available for the form of artifacts can be plausibly applied to the explanation of organic form.

14.4 A Place for Typological Thinking

"Typological" styles of explanation are not, as I have been discussing them, genuinely incompatible with "populational" styles of explanation. It is possible to explain the population-level coherence of an organic lineage by appealing to a combination of characteristic biasing forces affecting the range of variation that can be produced in that lineage, and by the winnowing effects of local environmental forces. This mixed stance is the one held by many workers in contemporary evolutionary developmental biology (often abbreviated to "EvoDevo"). It is, for example, precisely the stance expressed in Wallace Arthur's "biased embryos" program (Arthur 2004).

If population thinking is too obviously applicable to technical change, we might think a "typological" style of explanation is the place to look for surprising results in the history of technology. Typological styles of explanation would seek to discover characteristic forms of bias on the variants that arise in technical lineages. This may mean looking to the constraining or facilitating properties of commonly found construction materials, but it may also mean looking to characteristic patterns of thought among innovators. The "heuristics and biases" literature (e.g., Gilovich, Griffin, and Kahneman 2002; Kahneman, Slovic, and Tversky 1982), when used to shed light on the psychological dispositions that affect what sorts of artifacts are considered for use and development, would constitute a typological explanation of this second sort.

Just as evolutionary developmental biologists have argued that standard models of evolutionary change ignore the relevance of the concrete details of organisms and instead focus exclusively on selection pressures for the explanation of organic change over time, so an EvoDevo school in technology change might find fault with standard evolutionary models for ignoring the relevance of the concrete details of artifact production and focusing instead on selection pressures exerted by users (or other selectors) in explaining technological evolution. Models that marry heuristics and biases with market-based selection mechanisms would be the technical analogues of the biased embryos program in evolutionary developmental biology. So-called evolutionary economics seeks to broker just such a marriage (Nelson and Winter 1982; see also MacKenzie 1996).

14.5   Boyd and Richerson's Population Thinking

Mayr uses "population thinking" as a label for many different, albeit related, forms of thinking. The core sense that I outline in section 14.3 is not the only form of population thinking praised by Mayr. If Boyd and Richerson feel that population thinking is the key to an informative evolutionary theory of cultural change, then perhaps what they mean by "population thinking" corresponds to one of these alternative forms.

One of the curious things about population thinking is that its advocates often seem undecided on whether it is primarily about populations at all. Mayr says, "Averages are mere statistical abstractions; only the individuals of which the populations are composed have reality . . ." (Mayr 1976: 29). Of course this does not commit Mayr to denying that populations exist, but it credits only individuals with "reality" and hints at least at a skepticism regarding the explanatory importance of population-level properties (such as averages). In an important article, Elliott Sober takes issue with Mayr regarding this point. Sober suggests that the true importance of population thinking lies in crediting population-level properties with explanatory efficacy (Sober 1980). It appears that for Mayr, however, we should explain population-level phenomena in terms of the properties of individual organisms and their interactions.

Boyd and Richerson's population thinking expresses their belief that we should understand culture in terms of the combined effects of the interactions of the individuals that make up cultural groups. This explains, once again, why their "population thinking" stresses the importance of keeping track of individuals and their properties:

Remember that the essential feature of Darwin's theory of evolution is population thinking. . . . All of the large-scale features of life—its beautiful adaptations and its intricate historical patterns—can be explained by the events in individual lives that cause some genetic variants to spread and others to diminish. (Richerson and Boyd 2005: 59)

Population thinking, on this view, is really nothing more than what we might call "aggregative thinking." It is the kind of thinking one engages in when one explains the behavior of a unit composed of varied parts in terms of the properties of those parts and their interactions. An approach of this sort seeks to explain population-level phenomena in terms of individual-level, rather than population-level, properties:

The processes that cause . . . cultural change arise in the everyday lives of individuals as people acquire and use cultural information. . . . In the short run, a population-level theory of culture has to explain the net effect of such processes on the distribution of beliefs and values in a population during the previous generation. Over the longer run, the theory explains how these processes, repeated generation after generation, account for observed patterns of cultural variation. The heart of this book is an account of how the population-level consequences of imitation and teaching work. (Richerson and Boyd 2005: 6)

So population thinking for Richerson and Boyd is all about explaining how population-level patterns emerge from the collective behavior of the diverse individuals that make up the population. Some of their broad methodological statements might appear to exclude attributing causal powers to populations in their own right; however, population-level properties can feed back to the individual. An example from a traditional organic selection model might be when the reproductive success of an individual with trait T depends on the frequency of T in the population. An example from a cultural evolutionary model might be when the chance of an individual coming to believe that P depends on the proportion of people in the population at large who believe that P.

A similar population-based methodology features in the work of evolutionary economists Richard Nelson and Sidney Winter. They have much in common with so-called behavioral economists (see, once again, Kahneman, Slovic, and Tversky 1982), specifically regarding the rejection of the rationality assumptions of classical economics. However, they explain how they differ from behavioral economists in the following way:

We diverge from the behavioral theorists in our interest in building an explicit theory of industry behavior, as contrasted with individual firm behavior. This means on the one hand that our characterizations of individual firms are much simpler and more stylized than those employed by the behavioral theorists, and on the other hand that our models contain a considerable amount of apparatus linking together the behavior of collections of firms. (Nelson and Winter 1982: 36)

It is because of the potentially counterintuitive nature of the aggregation of individual-level events that population thinking of this sort has value. This general moral underlies the work of Boyd and Richerson as well as Nelson and Winter. Consider this simple illustrative example, which Richerson and Boyd draw from the domain of technological change

(2005: 70). If we assume that all individuals in a population have a psychology that disposes them to find frequently encountered technologies especially attractive, then we can predict that various phenomena will emerge at the level of a population made up of such individuals. Specifically we can predict that the rate of uptake of a new technology measured across the population as a whole will increase over time. To show why this is the case requires some (admittedly very elementary) mathematical thinking. As the frequency of individuals using the technique increases, so the attractiveness of the technology to a typical individual increases, and the chance of a new individual adopting the technique increases. Elementary population thinking enables us to explain why technology adoption in a population follows the so-called S-curve.

14.6   Sober's Challenge

Elliott Sober is somewhat skeptical of the value of models of cultural evolution (Sober 1992). Sober says that sociologists are interested by and large in questions about what makes one technology, for example, more attractive to the typical individual than another. One wants to know not just whether Rollerblades™ are more attractive to users than roller skates, but why they are more attractive. Evolutionary models of cultural change rarely promise answers to these questions. Evolutionary models do, on the other hand, give us rules for determining what will happen at the population level once we have determined which technology is the more attractive one. But Sober's complaint is that in many cases this sort of calculation is too obvious to be of much value. Once we know that Rollerblades are more attractive than roller skates, we can infer that Rollerblades will replace roller skates.

Boyd and Richerson respond by saying that Sober's objection assumes that "we are all good intuitive population thinkers" (Richerson and Boyd 2005: 97). Sober assumes that it is obvious how individual-level dispositions to prefer one cultural variant to another will combine to yield population-level phenomena. We should concede to Boyd and Richerson that, on occasions, naïve population thinking might let us down. It is perhaps not immediately obvious that a shared psychological disposition to adopt the most frequently encountered technique will lead to an S-shaped curve describing the adoption of new technology across the population. Other ways of determining how individual-level psychological dispositions will play out at the population level are even less intuitive.

Boyd and Richerson's appeal to population thinking in defense of cultural evolutionary models is legitimate. Indeed the same defense was put forward by Sober in the very paper they criticize:

So the question about the usefulness of these models of cultural evolution to the day-to-day research of social scientists comes to this: Are social scientists good at intuitive population thinking? If they are, then their explanations will not be undermined by precise models of cultural evolution. If they are not, then social scientists should correct their explanations (and the intuitions on which they rely) by studying these models. (Sober 1992: 492)

An example of an area where the intuition of a prominent social scientist has been challenged by this form of population thinking comes from Nelson and Winter's work in economics. Why think that firms act as efficient profit maximizers? Because, says Milton Friedman, if they did not do so, they would not have survived (Nelson and Winter 1982: 140). This form of intuitive evolutionary argument is no supplement for a population-level model, which asks under what circumstances only the efficient profit-maximizing firms will survive, and under what circumstances a population comprising a significant proportion of nonmaximizers could persist. Nelson and Winter claim that their more rigorous models show Friedman's argument to be flawed, or at least grossly oversimplified (Nelson and Winter 1982: 141).

14.7 Population and Innovation

Let me now turn more directly to the question of population thinking and innovation. The questions we need to keep in mind from here onward—the questions prompted by Sober's challenge—are not only whether population thinking yields hypotheses that are true but whether population thinking has enough heuristic value to yield hypotheses that one might not otherwise think to test. Simple demographic facts relating to such things as the size of a population can affect the likelihood of innovation being produced in that population. This of course is hardly a surprising outcome of formal modeling. The result was intuitive enough for Darwin to have noted it both in the case of organic evolution, and in the case of technical change. In *The Origin of Species* he remarks:

as variations manifestly useful or pleasing to man appear only occasionally, the chance of their appearance will be much increased by a large number of individuals being kept; and hence this comes to be of the highest importance to success. (Darwin 1985 [1859]: 41)

Darwin makes a related point in his discussion of technical innovation in *The Descent of Man*:

[I]f some one man in a tribe, more sagacious than the others, invented a new snare or weapon . . . the plainest self-interest, without the assistance of much reasoning power, would prompt the other members to imitate him; and all would thus profit. . . . If the new invention were an important one, the tribe would increase in number, spread, and supplant other tribes. . . . In a tribe thus rendered more numerous there would always be a rather greater chance of the birth of other superior and inventive members. (Darwin 2004 [1877]: 154)

If an important invention renders a tribe more numerous, the invention thereby increases the chances, merely by increasing the size of the tribe, of inventive members being born into that tribe, and producing yet more inventions. This result may be intuitive but this does not make it trivial, because it reminds us that there are explanations of different levels of innovative success in different nations, or among different ethnic groups, that need not appeal to differences in culture, environment, level of investment, or social institutions.

Jared Diamond's explanation for the higher rate of innovation in Europe compared to America is a little less intuitive, but it makes use of a similar populational perspective. As Richerson and Boyd (2005: 54) put it:

> Diamond argues that the greater size of the Eurasian continent, coupled with its east-west orientation, meant that it had more total innovations per unit time than smaller land masses, and that these innovations could easily spread throughout long east-west bands of ecologically similar territory. The Americas are not only smaller but are oriented north-south, making it difficult to diffuse useful cultivars, like maize from (say) temperate North America to temperate South America, or domesticated animals in the opposite direction. As a result, the set of adaptations necessary to support complex urbanized societies was assembled more slowly in the Americas.

In a similar vein, Boyd and Richerson explain the disappearance of important technologies on Tasmania by reference to declining population size alone. Drawing on the work of anthropologist Joseph Henrich, they suggest that the maintenance of technologies and the associated behaviors required to produce and operate them may require a population that is large enough for the rate of innovation to offset the degradation that results from error-prone imitation (Richerson and Boyd 2005: 138). Again these hypotheses are by no means obvious, and they arise from aggregative thinking, which prompts us to ask what number of individuals, with fallible imitative abilities and limited innovative abilities, is required to sustain complex technical know-how.

14.8 Population Thinking and Memes

The claim that Sober's challenge naïvely assumes we are good intuitive population thinkers works well as a defense of Boyd and Richerson's views. It works far less well as a defense of so-called memetic theories of cultural change. Here fitnesses are assigned directly to ideas, and sometimes to techniques or even artifacts, according to the expected growth rate of those entities over time within a population. Suppose, for example, we decide that Rollerblades are fitter than roller skates, meaning that the former have a higher long-run growth rate in the population than the latter. Once we have assigned a higher fitness to Rollerblades than to roller-skates it is hard to see how intuitive population thinking might fail us. Again the explanatory interest lies not in seeing how the consequences of these different fitnesses play out at the population level; it lies in seeing what makes Rollerblades fitter than roller-skates.

Boyd and Richerson are not defenders of memetics. They do not attempt to assign reproductive fitnesses directly to entities like ideas or artifacts. As we have seen, they are interested in seeing what effect individual psychological dispositions—especially dispositions to learn and imitate—have on the cultural makeup of human populations:

> A population-based theory of cultural change tells us how the details of individual psychology affect what kinds of skills, beliefs, and values that individuals acquire. (Richerson and Boyd 2005: 8)

In showing that this kind of thinking bears fruit, one does not thereby show that it is worthwhile to take an evolutionary stance on individual cultural items (tools, ideas, or whatever) and the populations they form, of the type endorsed by memeticists. However, although we can now see that artifact fitnesses, understood as long-run growth rates, may illuminate rather little by themselves, we should not assume that an artifact's long-run growth rate is merely a simple function of how attractive it is to typical users, any more than an organic type's long-run growth rate is a simple function of its suitability to its local environment (Lewens 2006: ch. 7). Take the case of a particular recording of a song on a CD. What facts might make recordings of this song more likely to spread than others? In part, of course, we can point to facts that make the song catchy—psychological facts that make an individual who has the CD more likely to play it, and facts that make individuals who hear it more likely to buy a copy for themselves. But a song could score comparatively poorly on these characteristics and still spread faster than its competitors simply because it is ubiquitous. If a record company ensures that a melody is played through all available radio and TV networks, then even a recording that is comparatively uncatchy will quickly be purchased by millions. We cannot infer from the swift spread of a CD through a population that the song the CD has on it has features that make it likely to hop from mind to mind. The song may not be especially contagious or catchy at all; the song's producers may just be powerful enough to make it ubiquitous, hence more likely to be purchased than far catchier but more poorly funded competitors. The moral of this example is that if we choose to build a cultural evolutionary model that assigns fitnesses to technologies themselves, one will need to include population-level factors in addition to facts about typical individual psychology, among the determinants of artifact fitnesses. For some students of technology change, that may be a significant lesson in itself.

14.9 The Needham Question

We have already considered the fairly intuitive positive effects of large population size on technical innovation. Population thinking also prompts us to consider the less intuitive possibility that small population size might also have positive effects on innovation. Sewall Wright famously argued that drift can foster adaptation. Drift is more likely to occur in small populations than large ones. This means that in a small population, whichever variant is better suited to the local environment is more likely to be eliminated than it is if in a large population. Wright's language of "fitness landscapes" allows us to articulate the possibility that small populations can thereby "drift" toward the bottoms of small adaptive peaks, which allows them to scale even higher peaks through the action of selection. Wright argued, in other words, that a large population has a greater chance of getting stuck on some "local optimum" than a series of smaller populations, whose subdivisions allow for the exploration of alternative adaptive peaks through the action of drift. Once a high peak is found by a subgroup, its members will thrive and invade the other subgroups of the population. In this way, the overall population comes to occupy a high peak, in a way that would be less likely if selection were acting alone. (See Ridley 1996: 217–219, for an accessible presentation of Wright's ideas.)

This shifting balance model was criticized by R. A. Fisher, and is not widely accepted (Ridley 1996: 219). Even so, population thinking of this sort might lead one to complicate Diamond's explanation of innovative success by pointing to the potential trade-off between, on the one hand, societies that are large enough, conformist enough, and have the right norms of communication to enable a successful technology to spread rapidly and faithfully and, on the other hand, societies that are fragmented, prone to errors in communication, and tolerant enough to allow diverse experiments that will prevent convergence on local technological optima.

One could use this kind of model to fashion an answer to the so-called Needham Question (Needham 1975). This is the question sometimes asked by historians of science and technology of why China lagged behind the West in the period when Europe was enjoying great technical and scientific creativity. Perhaps Wright's shifting-balance model could inspire a novel populational answer, in terms of the subdivided nature of Europe compared to the national unity of China. Perhaps the fragmented nations of Europe permitted a hedging of innovative bets not possible in the more monolithic China, while international trade allowed successful techniques developed in one European nation to spill over into others.

This is indeed a hypothesis worth testing, but it does not show decisively that population thinking has heuristic value. David Hume used more intuitive forms of thinking to arrive at a very similar hypothesis some time ago. Hume claimed:

That nothing is more favourable to the rise of politeness and learning, than a number of neighbouring and independent states, connected together by commerce and policy. The emulation, which naturally arises among those neighbouring states, is an obvious source of improvement: But what I would chiefly insist on is the stop, which such limited territories give both to *power* and to *authority.* (Hume 1994 [1742]: 64)

Hume goes on to explain the contrast between Europe and China, kicking off with a diagnosis of what the Greeks got right:

Greece was a cluster of little principalities, which soon became republics; and being united both by their near neighbourhood, and by the ties of the same language and interest, they entered into the closest intercourse of commerce and learning. There concurred a happy climate, a soil not unfertile, and a most harmonious and comprehensive language; so that every circumstance among that people seemed to favour the rise of the arts and sciences. . . . Europe is at present a copy, at large, of what Greece was formerly a pattern in miniature. (Ibid.: 65)

And he finishes by telling us what the Chinese got wrong:

In China, there seems to be a pretty considerable stock of politeness and science, which, in the course of so many centuries, might naturally be expected to ripen into something more perfect and finished, than what has yet arisen from them. . . . But China is one vast empire, speaking one language, governed by one law, and sympathising in the same manners. The authority of any teacher, such as Confucius, was propagated easily from one corner of the empire to the other. None had courage to resist the torrent of popular opinion. (Ibid.: 66)

14.10   Population Thinking and Evolution

We should not be disheartened by our exploration of the Needham Question. It is true that on some occasions the conjectures we reach through population thinking could just as well be reached by informal reflection. But work by the likes of Richerson and Boyd (2005), Nelson and Winter (1982), and Philip Kitcher (1993), in relation to other aspects of innovation, suggests that the intuitive sketches of population explanations we arrive at informally (such as Hume's explanation for the success of innovation in Europe) can be tested and corrected by more formal populational modeling. This is the lesson of Nelson and Winter's skeptical evaluation of Friedman's defense of the assumption of profit-maximization. It is also the lesson of many of their more complex models, which try to ascertain, for example, under what circumstances imitation and innovation can coexist as research and development strategies, and under what circumstances imitators will drive out more effortful innovators, in a population of competing firms (Nelson and Winter 1982: pt. V).

It also seems plausible that formal modeling of this sort can throw up new hypotheses for empirical testing. Boyd and Richerson draw usefully on the statistical thinking that, although largely absent from Darwin's work, was central to the establishment of selection as an important factor in evolutionary change during the modern synthesis. Consider, for example, Fisher's remarks on particulate inheritance. He claimed that if blending inheritance were the dominant mode, then selection could only lead to permanent evolutionary change if mutation rates were very high. Otherwise the population would always tend to regress to the mean, regardless of how well individuals with advantageous mutations might do in virtue of them. Observed mutation rates are, as a matter of fact, comparatively low. Hence in the organic world, Fisher's population thinking shows that for selection to be efficacious, inheritance must be particulate. Now this does not show (as memeticists may be inclined to assume) that for selection to lead to permanent changes to artifacts, the resources that underpin technological inheritance must be particulate also; rather it prompts Boyd and Richerson to ask whether mutation rates are high enough in this domain for blending inheritance to permit cumulative change (Richerson and Boyd (2005): 88–90). Questions of this sort are perhaps especially easy to access from the perspective of populational evolutionary models.

In a similar vein, population thinking prompts Kitcher to test whether a group of scientists who pursue truth in a disinterested manner are in fact instantiating an optimal strategy for attaining the truth. One of his formal models suggests that a group of scientists who care only about getting at the truth may be less efficient at generating new knowledge than a group of scientists who also care about taking the credit for making a new discovery (Kitcher 1993: 308–314). Roughly speaking, this is because those who care about getting to the truth will tend to behave in a uniform fashion: If received wisdom suggests that a particular prominent scientist's views are along the right lines, then they will all borrow those views. If the prominent scientist's views are not well-regarded, then they will all ignore them. Kitcher's "sullied scientists," who also care about their own reputations, have more of an incentive to pursue unfancied, or unfashionable, avenues of research, which may lead to their being seen to make an important discovery that goes against the grain of the community. But this also means that a community of more egotistical scientists will tend to pursue diverse avenues of research—it will not put all its eggs in one basket. And this, in turn, can increase the chances, from the community perspective, of making important breakthroughs.

Kitcher's model is recognizably "populational." Like Boyd and Richerson, he seeks to show how the properties of a population—in this case, a scientific community—depend in counterintuitive ways on the properties of the entities that make it up—in this case, scientists. This is another instance of the form of population thinking that I have been referring to as "aggregative thinking." At a stretch, one might also describe Kitcher's populational models as "evolutionary," simply because they try to explain the unfolding behavior over time of a group of interacting entities. But Kitcher's models rarely have any obvious analogue to natural selection, reproduction, replication, or drift. No such concepts feature in his populational attempts to understand the epistemic fortunes of communities of sullied and pure scientists. This suggests that "population thinking," understood as "aggregative thinking," is an important part of the evolutionary biologist's toolbox, but it is not a distinctively biological, or even a distinctively evolutionary, way of thinking. Even so, the application of these formal, populational modeling techniques may be among the most promising ways in which styles of thinking familiar to evolutionary biologists will shed light on the domain of technical innovation.[1]

Note

1. This chapter was first presented at the KLI Workshop on the Comparative Philosophy of Technical Artifacts and Biological Organisms in September 2006, and a much earlier ancestor of it was presented in Delft. I am grateful to both audiences for comments and criticism, and especially to Peter Kroes and Ulrich Krohs.

References

Amundson, R. (2005). *The Changing Role of the Embyro in Evolutionary Thought.* Cambridge: Cambridge University Press.

Arthur, W. (2004). *Biased Embryos and Evolution.* Cambridge: Cambridge University Press.

Basalla, G. (1988). *The Evolution of Technology.* Cambridge: Cambridge University Press.

Boyd, R., and Richerson, P. (2000). Memes: Universal acid or a better mousetrap? In: *Darwinizing Culture* (Aunger, R., ed.), 143–162. Oxford: Oxford University Press.

Darwin, C. (1985 [1859]). *The Origin of Species,* 1st ed. (Burrow, J. W., ed.). London: Penguin.

Darwin, C. (2004 [1877]). *The Descent of Man,* 2nd ed. (Desmond, A., and Moore, J., eds.). London: Penguin.

Dennett, D. (1995). *Darwin's Dangerous Idea: Evolution and the Meanings of Life.* New York: Norton.

Dobzhansky, T. (1951). *Genetics and the Origin of Species,* 3rd ed. New York: Columbia University Press.

Gilovich, T., Griffin, D., and Kahneman, D. (2002). *Heuristics and Biases: The Psychology of Intuitive Judgement.* Cambridge: Cambridge University Press.

Hume, D. (1994 [1742]). Of the rise and progress of the arts and sciences. In his: *Political Essays* (Haakonssen, K., ed). Cambridge: Cambridge University Press.

Kahneman, D., Slovic, P., and Tversky, A. (1982). *Judgement under Uncertainty: Heuristics and Biases.* Cambridge: Cambridge University Press.

Kitcher, P. (1993). *The Advancement of Science.* Oxford: Oxford University Press.

Lewens, T. (2002). Darwinnovation! *Studies in History and Philosophy of Science, 33:* 199–207.

Lewens, T. (2004). *Organisms and Artifacts: Design in Nature and Elsewhere.* Cambridge, Mass.: The MIT Press.

Lewens, T. (2006). *Darwin.* Routledge: London.

MacKenzie, D. (1996). Economic and sociological explanations of technological change. In his: *Knowing Machines: Essays on Technical Change.* Cambridge, Mass.: The MIT Press.

Mayr, E. (1976). Typological versus population thinking. In his: *Evolution and the Diversity of Life.* Cambridge, Mass.: Harvard University Press.

Mokyr, J. (1990). *The Lever of Riches: Technological Creativity and Economic Progress.* Oxford: Oxford University Press.

Needham, J. (1975). *The Grand Titration: Science and Society in East and West.* London: George Allen and Unwin.

Nelson, R., and Winter, S. (1982). *An Evolutionary Theory of Economic Change.* Cambridge, Mass.: Harvard University Press.

Richerson, P., and Boyd, R. (2005). *Not by Genes Alone: How Culture Transformed Human Evolution.* Chicago: University of Chicago Press.

Ridley, M. (1996). *Evolution,* 2nd ed. Oxford: Blackwell Science.

Sober, E. (1980). Evolution, population thinking, and essentialism. *Philosophy of Science, 47:* 350–383.

Sober, E. (1992). Models of cultural evolution. In: *Trees of Life: Essays in the Philosophy of Biology* (Griffiths, P., ed.), 17–39. Dordrecht: Kluwer.

Sober, E. (2003). Metaphysical and epistemological issues in modern Darwinian theory. In: *The Cambridge Companion to Darwin* (Hodge, J., and Radick, G., eds.), 267–287. Cambridge: Cambridge University Press.

Waters, K. (2003). The arguments in the *Origin of Species.* In: *The Cambridge Companion to Darwin* (Hodge, J., and Radick, G., eds.), 116–139. Cambridge: Cambridge University Press.

Ziman, J. (2000). *Technological Innovation as an Evolutionary Process.* Cambridge: Cambridge University Press.

15 The Cost of Modularity

Ulrich Krohs

15.1 Introduction

Biological organisms are complex systems, as are most modern technical artifacts. However, most of the entities of both classes are much less complex than the number of their components would allow for. This is in part due to their modular organization—the fact that they are not maximally integrated systems. Modularity means that the parts of the system are grouped in such a way that strong interactions occur within each group or module, but parts belonging to different modules interact only weakly (Simon 1969; Lewontin 1978; cf. also Alexander 1964). Since the maximum possible degree of complexity depends on the number of components of a system and on the number of interactions between these components (e.g., Simon 1969: 184), a limitation of interactions in a system that consists of partly independent subsystems reduces complexity. Exactly such limited interaction among subsystems occurs in systems organized in a modular way.

Modularity may be best known from industrially produced technical artifacts. The whole industry that produces integrated electronic circuit elements is built on the idea of grouping several parts together and using such integrated modules as components of larger systems. Similar design principles hold for washing machines, cars, and most obviously, stereo sets, which even require the user to plug together physically separated modules. Modularization is economically advantageous as it facilitates designing, constructing, and maintaining artifacts (Ulrich and Eppinger 2003; Pahl et al. 2007). But modular organization is not restricted to the realm of artifacts, in which economic principles rule. Biological evolution brought about modular systems long before technology did. Anatomists and physiologists have been aware of the modularity of organisms for a long time. Near-decomposability, the analytic equivalent of modularity,[1] does not apply only on the macroscopic level, where the organs of higher metazoans form clearly delineated structures that perform a limited set of functions. Metabolic and gene regulatory networks, cellular signaling systems, and developmental pathways are other instances of nearly decomposable systems (see section 15.2).[2]

A hot topic with respect to biological modularity is how to explain its evolution. I discuss arguments that have been put forward to explain the evolution of modularity, and confront them with the fact that modularity is not an all-or-nothing issue, but comes in degrees (sections 15.3 and 15.4). The evolutionary explanations do not offer any reason why biological organisms are less than maximally modularized, or why even secondary integration of modules has occurred, for example, in endosymbiogenesis. A comparison of the biological arguments with arguments about modularity in technology shows that an important aspect is missing in the biological considerations: they focus almost exclusively on the benefits of modularity, while in engineering it is acknowledged that a modular structure may also have disadvantages, such as a larger weight of a modular device as compared to an integrated one, or less flexibility in meeting other than standard requirements. I show that not only the benefits but also the costs must be considered to explain the evolution of modular organization of biological organisms (section 15.5). This is meant not as a biological argument about the evolution of modularity but as a contribution to the question which structure an argument about adaptive processes needs to have in order to be of explanatory value.

A second aim in this chapter concerns the aspect of functionality in discussions of biological and technological modularity. Function ascriptions are used to delineate modules in entities of both realms. However, the results are quite different in both cases. Technological functional modules largely coincide with structural subunits, while in biological systems this is often not the case (section 15.6). The mismatch between the results of structural and functional decomposition gives rise to claims in the field of systems biology that only the structural approach should count as yielding adequate results. I look into the divergence of structural and functional modules from a different perspective. I first discuss the question of whether functional modularity may be of any relevance in explaining the evolution of organisms, and whether or not cost considerations can help in explaining evolution in this case as well (section 15.7). My final concern with respect to the mismatch between functional and structural modules then is to draw some conclusions regarding the epistemic consequences of accepting the mismatch that occurs in particular systems (section 15.8).

15.2   Delineating Modules

In many cases modules can be delineated morphologically. The organs of animals, for example, are discernible structures, which show strong internal interactions and comparably weak and few interactions among each other. They also have distinct functions (pumping and detoxifying blood, mediating gas exchange, digesting food, etc.).

Morphological criteria, however, are not always applicable when decomposing a complex system. Think of the network of the metabolism of the cell. Here, the various components, for example, enzymes and metabolites, may occur in almost even distribution within the cell (though there will in fact be considerable heterogeneity and even compartmentalization). Such a network can nevertheless be decomposed into modules, either according to structural or to functional criteria. The older method is functional decomposition: cell metabolism is decomposed into capacities that are brought about by metabolic pathways such as glycolysis, the citric acid cycle, beta-oxidation, catabolic and anabolic pathways in amino acid metabolism, and so forth. The metabolic pathways, then, are regarded as functional units or modules. Some of these pathways may be regarded equally as structural modules in the sense introduced in section 15.1: the strength of internal interactions and the weakness of external interactions allows for their separation.[3] However, this does not hold true in general, since functional relations do not guarantee independency of the pathways in the sense of near-decomposability, as the many interactions among the pathways prove. This so-called crosstalk with other pathways often turns out to encompass stronger interactions to the metabolic surroundings than can be found within the pathway under consideration. The functional delineation of pathways within a metabolic network need not coincide with structural modules.[4] Consequently, functional analysis as a method to identify modules in biological systems is not undisputed and often regarded as biased (Rohwer, Schuster, and Westerhoff 1996; Koza et al. 2002; Friedman 2004; Papin, Reed, and Palsson 2004; for a discussion see Krohs and Callebaut 2007).

To avoid functional bias, and to end up with a picture that accounts for the structural near-decomposability of the system, metabolic networks have been delineated more recently according to the strength and relevance of static relations, as well as to dynamic interactions. Both together constitute the structure of a network. The importance of the dynamic dimension for the structural picture was emphasized by Simon (1969: 198):

the short-run behavior of each of the component subsystems [i.e., modules] is approximately independent of the short-run behavior of the other components; . . . in the long-run the behavior of any of the components depends in only an aggregate way on the behavior of the other components.

Though this delineation criterion looks as if it were straightforwardly applicable, its operationalization is difficult. Sophisticated mathematical methods had to be developed to allow for a breakdown of a network into structural modules that satisfy the criterion. They are neutral with respect to functional considerations and often seem to end up with a picture of the organization of a network that differs significantly from the results of functional decomposition.

Only the results of structural decomposition are generally regarded as being capable of delivering an authentic picture of a network (Bechtel and Richardson 1993; Schaffner 1998; Onami et al. 2002; Papin, Reed, and Palsson 2004; Palsson 2006).[5] I therefore stay with structural modules for the main part of my argument, returning to the possible relevance of functional modules in evolution in sections 15.6 and 15.7.

Since modularity includes a dynamic dimension, the concept is applicable to developmental processes as well. Characters of adult organisms do not pop up out of nothing; they develop during the ontogenesis of the organism. It was an interesting finding that developmental processes show near-decomposability as well. An example is the development of the hind limb of tetrapods. The hind limb is both a discrete structure and a developmental module with a unique and intrinsic set of patterning mechanisms (Raff 1996; Franz-Odendaal and Hall 2006). The character of the adult organisms is used to delineate the developmental module—the set of developmental pathways and resources that brings about this very character. This is again a functional delineation, and nothing guarantees that the interactions within this functional module are stronger than the interactions with processes involved in the development of other characters. But the structural delineation of developmental processes indicates that functional and structural modules seem to coincide in this case. However, biological evolution has not always delineated modules as nicely as engineers tend to do. Biological modularity (but also its technological counterpart) comes in degrees and nature's joints are sometimes fuzzy. So if structural modularization is the aim, developmental modules must be delineated not by their products but—like the modules of metabolic networks—by unbiased methods that identify semi-autonomous developmental pathways (Raff 1996), as is in fact successfully performed (e.g., Davidson et al. 2002; Davidson and Erwin 2006).[6]

15.3   Explanations for the Evolution of Modularity

Modular systems can evolve from different starting points by changes going in opposite directions: by parcellation of a highly integrated system, or by integration of existing systems (Callebaut 2005: 9). The mechanisms producing modularity are usually described as specialization of existent structures in the case of parcellation, and assembly processes in the case of integration (Simon 1969: 193). Since not every specialization needs to end up in parcellation, and components integrated by assembly may simultaneously specialize, evolutionary modularization processes need not belong to only one of the kinds. However, it is clear that both ways of modularization are relevant for biological evolution: 1) The eukaryotic cell evolved by integration of prokaryotic cells of different species. In particular, incorporated bacteria that already possessed a respiratory chain became mitochondria, and photosynthetic bacteria were modified into plastids (Sagan [Margulis] 1967; Margulis 1970, 1981; for the history of this idea see Khakhina 1992). 2) Parcellation by specialization of eukaryotic cells into different tissues and organs occurred during the further evolution of higher metazoa (animals and plants).

The mere description of the different ways by which modularity may or did evolve is not all that biologists aim for. The evolutionary processes also require causal explanation. Different explanations were proposed and discussed in the literature. I focus here on the most widely applied class of explanations and show that current arguments belonging to this class are incomplete and therefore not yet satisfying. In the following sections of the chapter I propose a way of completing the arguments. Arguments of the kind in question are adaptive explanations that refer to the evolutionary options that a modular structure opens up. In short, since modularity allows for evolutionary plasticity, it is regarded as favored by natural selection (e.g., Altenberg 1995; Galis 1999; Wagner and Altenberg 1996). Some authors even claim that the modular organization of metazoa is the result of selection *for* evolvability (e.g., Gerhart and Kirschner 1997).

Before discussing the argument behind this claim, and reframing it in a way that makes it more plausible than it appears now, I proceed to investigate its equivalent within the realm of technological evolution, where a sound argument in favor of selection for evolvability can be made when a latent premise is made explicit. Technical artifacts with a modular design can be modified by substituting or reassembling modules without much effort, as it is familiar from construction kits. Modularity thus allows to cover a huge design space easily. Not only substitution and reassembly but other modifications are possible as well. Baldwin and Clark (2003) give a long, but still incomplete, list of operations that can be performed on modules: splitting, substituting, augmenting, excluding, inverting, porting as well as replicating, combining and extending. Given this flexibility in modifying a modular design, it is obvious that new kinds of systems can evolve more easily from modular systems than from fully integrated ones. The latter requires complete redesigning to end up with another functional system, while the former can be modified stepwise, module-by-module, with a high probability that the intermediate forms are still working (stable).[7] Industry takes advantage of this: "Through widespread adoption of modular designs, the computer industry has dramatically increased its rate of innovation" (Baldwin and Clark 1997; see also Langlois and Robertson 1992). Design methodology relies heavily on a modular approach, exactly for the reason that this allows for quick evolution of products (e.g., Ulrich and Eppinger 2003; Pahl et al. 2007). Therefore, in the field of technology, modularity is present and favored because of the high evolvability it enables. It may well be that in some cases there is selection not only of evolvability but also selection for evolvability.

Accepting that one reason why modularity drives technical innovation is that modular design is often chosen because it allows further evolvability, it is tempting to draw the parallel between technological and biological evolution, resulting in the aforementioned claim that modularity has evolved *because of* the evolvability of modular systems. Modularity in fact opens up possibility spaces also for the evolution of biological organisms, and high evolvability is found with respect to many modular traits of organisms. So modularity allows for quick biological evolution and high evolvability (Schlosser and Wagner 2004; Callebaut and Rasskin-Gutman 2005). However, it is difficult to see how high evolvability could be a character that natural selection can act upon. Mutations, that is, undirected heritable modifications occurring in the offspring, do not usually increase the fitness of an individual.[8] It is much more plausible to regard evolvability as a by-product of other selective processes. Following Elliott Sober's distinction between "selection for" and "selection of" (Sober 1984: 97–102), there is merely selection *of* modularity and evolvability. As Wagner, Mezey, and Calabretta (2005) concisely put it, evolvability is selected only indirectly.

An argument similar to the one against selection for evolvability also holds with respect to selection for modularity due to a supposed increased evolvability of modular systems. This becomes clear from a comparison with the technological case. In engineering it is well known that modular structure, though enabling for quick modifications of the system within a specified range of related designs, may in the long run decrease the further evolvability of a system. Existing modules, in particular if they are used in parallel in different contexts, may pose constraints on the evolvability of a class of artifacts (Pahl et al. 2007: 509). A modular design is less flexible than individual design with respect to the adoption of changing requirements as soon as a certain design space covered by the modular systems is left. What can be learned from the technological example is that modularity does not increase evolvability in every case and each respect, but that any particular modular design also poses evolutionary constraints on the organism in question, mostly in cases where a particular module serves different roles in an organism. Any change needs to be compatible with all of these roles. This limits the range of viable variation and may well limit the evolvability of modular structures. Consequently selection of modular structures cannot be explained in general as selection of modularity due to increased evolvability.

Let me come back to arguments in favor of selection of modular systems that do not refer to a supposed increase in evolvability. What must still be expected from an adaptive explanation of modularity is an argument for why there is selection of modular structures—though not of modularity as such. In their discussion of evolutionary explanations of modularity, Wagner, Mezey, and Calabretta (2005) compile scenarios of eight evolutionary paths, with the proviso that the scenarios will have to be adjusted to new findings, since in most cases empirical data are still missing. Among these scenarios are the stepwise lowering of integration, occurrence of new modules after duplication of components so that the additional component may take a new function, or the occurrence of developmental modules by a sorting process that collects genes contributing to the development of the same character. Because empirical data are lacking, simulations are crucially important to support the relevance of the scenarios. Interestingly only some of the simulations the authors quote yield modular structures, and they do this only under a limited range of assumed conditions. The degree of modularization that is reached is low in most cases.[9] However, if one looks not to the mathematical implementation but to the conceptual background of the simulation models—which boils down to looking at the arguments discussed here as arguments to make plausible why modular structures evolved—there seems to be no reason for this resistance to full modularization. The conceptual framework that is applied supports an adaptive view on modularity. But it seems to know of the advantages of modularity only. The simulations, which show results with even less modularization than that found in actual biological organisms, reveal that modularity may have disadvantages under a broad range of conditions. Consequently the adaptive explanations of modularity as given by now are far from being satisfying.

In the case of modularization by duplication (Raff 1996; Calabretta et al. 2000), it is clear that modular*ity* or evolvabil*ity* are not the traits being selected for. Instead selection acts on the particular modules, which arose from the duplication process, or on individuals possessing these modules. So again not modularity but the secondary adaptation of specific modules requires evolutionary explanation (or better, as becomes clear in the next section, the specific degree of modularity). But in this case, too, the adaptive argument does not show why the process results to some degree in integration rather than in the highest possible independence of the modules.

15.4 The Desideratum of Explaining Nonmodularity

Integration, parcellation, or specialization of duplications, whichever way may have led in any particular case to a modular organization: according to arguments of the kind discussed, the modular structure results in a selective advantage over a fully integrated system. So one might tend to conclude that after ample time for evolution all structures of biological organisms should be modular. But this is not the case, and it was not expected. Even after biologists had become knowledgeable about modularity of metabolic networks, it was an important and unexpected *finding* that genetic networks are modularized as well (see Callebaut 2005). Moreover, modules are often much less clearly separated than the paradigmatic examples in the debate might suggest, and there seems to be no borderline between modular and integrated (sub)systems, but rather a continuity spanning the whole range that lies between the extremes. Thus the citric acid cycle is a functional unit that is not at all a structural module, as shown in section 15.6. What about the urea cycle? There the internal interactions may be larger than the external ones, but is the difference large enough for a structural delineation of a module? In β-oxidation of fatty acids this difference is larger, so in this case one might tend to talk about a structural module proper. So the urea cycle seems to be an intermediate case. The conclusion needs to be as follows: modularization comes in degrees—as does decomposability, its conceptually related analytic counterpart—and many networks show only intermediate, or even low degrees of modularization. Metabolic networks, for instance, although unanimously classified as modular, are found to be much less nicely decomposable according to structural criteria, and integrated much higher than straightforward mathematical methods can deal with (cf. Davidson et al. 2002; Palsson 2006).

The situation thus is the following: while modularity in biological systems comes in degrees, the present explanations of the evolution of modularity account for strong modularity only. They do not state any reason that could explain why the evolution of modular structures did not bring about modules that are separated more nicely than actually found. So the arguments about the evolution of modularity are incomplete.

There are several possibilities for supplementing the arguments. Some of these could be assumed to be tacitly accepted—that the time available for evolution was not long enough to bring about full modularization; that, by chance, the additional mutations required for further modularization did not occur; that the actual degree of modularity makes the organisms sufficiently adapted; or that, in the case of extreme separation, morphologically distinct modules would fall apart without being viable anymore. None of these explanations of the limitations of modularity is satisfying as long as no support by adequate data is provided, because all can be brought up as ad hoc arguments whenever needed. They have no specific explanatory power. (An exception may be the argument about the separation of morphological modules, but this is hardly applicable to modular networks.)

Though this critique is focused on parcellation arguments, it covers assembly arguments as well. It even gets additional strength from data collected on assembled systems. After an assembly has taken place, be it on the endosymbiotic pathway or by duplication, an evolutionary integration process takes place that *lowers* the degree of decomposability of the assembled system. So we have a process—usually described as an adaptive process under natural selection—by which additional interaction and interconnection of the modules is established. This integration seems to proceed much further than is physiologically required. Mitochondria may serve as an example. Genes are transferred from the bacteria that became mitochondria into the nucleus, which increases the integration of the modules. But the modularization argument does not explain why this integration (i.e., this lowering of the degree of modularity) goes much further in the case of mitochondria than in cases of endosymbiotic bacteria as found in certain flagellates. Even the mere existence of the adaptive part of the integrative path to modularity—the adaptive part being the modification that follows the integrative step—shows that both modularity *and* integration bring selective advantages. Schank and Wimsatt (2001) have pointed to such advantages of integration. However, the adaptive explanations of the evolution of modularity mentioned do not recognize this issue.[10]

Let me return to the argument that explains modularity by adaptive parcellation of an integrated system. Since it refers to selective advantages, it is quite clear that the incompleteness of modularization has to be explained by reference to selective disadvantages of a modular organization or to the advantages of integration as mentioned with respect to the integrative pathway. The consideration of the benefits of modularity ought to be supplemented by the cost side. Modularity has fitness-decreasing effects as well.

While the development of the demanded argument has to be left to biologists, philosophy can give, in addition to criticisms of the argument, some hints in which direction to proceed. In the following I therefore single out some kinds of possible costs of modularity by comparing the biological with the technological case, and ask for their consideration in biological explanations of the evolution of modularity.

15.5   Kinds of Costs of Modularity

While the cost side is missing in current explanations of the evolution of nearly decomposable systems, cost-benefit analysis was present from the very beginning of the modularity
debate. It was already part of Herbert Simon's famous watchmaker metaphor from 1962:
Two watchmakers, Hora and Tempus, both built their watches from 1,000 parts. Tempus's
watch was highly integrated, while Hora's watch was modular, consisting of stable subassemblies of 10 parts each. Both watchmakers were frequently disturbed by telephone calls.
To accept a call, they had to put down the assembly they are working on, which then fell
apart. Tempus hardly ever finished a watch and became poorer and poorer while Hora
prospered (Simon 1969: 188).

The reason for the different success of Hora and Tempus is that Tempus, on an interruption by a telephone call, loses the time for up to 999 assembly steps. Hora, in contrast,
loses on no single call more time than needed for 9 steps, because every tenth step yields
a stable assembly.[11] This is the benefit of modularity. But, in contrast to the biological
arguments mentioned, Simon also considers the costs. Hora needs more steps to finish a
watch: he has to complete 111 subassemblies and needs a total of 1,110 steps. Tempus
needs only 1,000 steps. So the cost of modularity is an additional 11 percent of work.[12]
What happens if we disregard, in contrast to Simon, the cost side and focus on the benefit
alone? We could then maximize the estimated benefit of modularity by making the modules
smaller and smaller. In the extreme, every module may consist of only 2 parts. This would
minimize the loss of work on interruptions by telephone calls. But there are in fact additional costs of such a strong modularization so that in the end it would not pay off. A
watchmaker—let his name be Minuta—who applied such extreme modular design needed
$1,000 + 500 + 250 + \ldots + 1 =$ about 1,999 steps to finish a watch.[13] So Minuta needed
almost twice as many different steps as Tempus to assemble a watch, which might be
worth his while only under extreme phone harassment.

Additional kinds of costs may be associated with modularity. I therefore describe different classes of costs as they can be derived from descriptions of modular design in
technical artifacts, and apply them to the field of biology.

1) The aforementioned costs of additional assembly steps that must be performed find
their equivalent in biology in the extended time required for ontogenesis. The ontogenesis
of a modularized organism, according to these considerations, needs more time than the
ontogenesis of a higher integrated organism.[14] The *costs of a longer ontogenetic process*
can be seen in higher energy requirements for development and a higher risk of dying
before offspring are produced.

2 and 3) Keeping a region of a network that is singled out as a module working entails
additional energetic effort and need for material. The module is nearly decoupled from the
rest of the network. In engineering, it is well known that a modular organization thus
causes higher weight and volume and material effort (Pahl et al. 2007: 509). This can be

made clear from a modular stereo set. 2) More housings are needed, one for each module. This causes additional *tara costs*. 3) Each module also requires its own power supply pack and so forth. This causes what can be called *autonomy costs*. Equivalents to costs of both kinds do occur in the biological case. 2) Organs, being macroscopic morphological modules, are wrapped into fascia. And gene clusters, forming the core of many developmental modules, have their regulatory sequences and additional structural markers as a kind of packaging. 3) Autonomy of organs requires individual vascular and nervous connections. On the molecular level, regulatory cassettes of transcriptional regulators may serve as an example. Such cassettes constitute a network containing several coactivator genes, and only these seem to mediate the autonomy of the cassette as a regulatory module (Kardon, Heanue, and Tabin 2004). It follows that in the biological case there are tara and autonomy costs as well.

4) In case of a failure of a modular technical artifact, the defect is localized only by checking whole modules; instead of repairing modules on the level of their components, they are simply exchanged when defective (White 1999: 475). The failure must not be localized within a module. This "diagnostic opaqueness" of modules in technology leads, on the one hand, to decreased diagnostic and exchange costs in terms of hours of work, and, on the other hand, to an increased requirement of material for repair and maintenance, since whole modules are discarded instead of only single defective parts. In biological systems, an equivalent can be found wherever whole, morphologically distinct modules are discarded in development and self-reproduction, especially where whole cells are sacrificed, for example, during the renewal of epithelia, and in many cases of apoptosis. So there are material and energetic *costs for module-wise replacement*.

To supplement adaptive arguments about modularization, costs of the identified kinds have to be taken into account. It of course requires empirical data to further specify these costs. It should be noted that with respect to biological organisms only phenomena that decrease fitness on a level relevant for selection may count as costs in arguments about evolutionary adaptation. This includes increases of the energy requirement of processes going on in the organism, but excludes "costs" of evolutionary processes themselves. This is in contrast to the case of technical artifacts where costs of the designing process also need to be taken into account (see section 15.7).

15.6   Mapping Functions on Modular Structures

Up to this point, only the evolution of structural modules has been discussed, since only modules of this kind are usually regarded as relevant for the organization of a network or other hierarchical system. However, as presented in section 15.2, another way of decomposing biological systems relies on functional criteria. The following two findings require a closer look at the functional view of biological networks, since they may either be relevant for the evolution of modularity, or they themselves may be in need of an evolutionary explanation: 1) functional modules do not in general coincide with structural ones in biological systems, and 2) functionality in metabolic and gene regulatory networks is not localized at particular components of the system but delocalized or distributed over entire subnetworks (Boogerd et al. 2005). The relevance of functional modularity for evolution needs to be investigated.

The concept of a biological function is notoriously problematic from the philosopher's point of view, so I need to explain which concept is to be applied. First, I should point out that "function" must not be equated with "dynamics" (see Krohs 2004: 41). As any complex physical entity, biological entities have a structure and display change in time, which is described as its dynamics. (Even being static is a kind of dynamic in this sense: a change of measure zero.)[15] "Function," then, is ascribed to an entity that exhibits its dynamics if these dynamics contribute to some capacity of the biological (or technical) system of which the entity is a component.[16] While "dynamics" denotes the processes an entity undergoes, "function" refers to the relation of this process to a more comprehensive process within a system of a certain kind. A function can thus be considered the contribution of an entity to a capacity of the system the entity is embedded in—with the caveat that the system and/or the capacity at all qualifies as being functionally organized. The dispute among philosophers about the concept of function is over how to specify these further conditions. In the following paragraph I indicate an explication of the concept of function that is adequate for use in systems biology.

In general, functions are ascribed to metabolic and other networks and to their substructures by physiological analysis, not by an analysis of adaptive processes in the evolution of the network. Consequently systems biologists do not refer to etiological functions, which are specified with respect to evolutionary processes, but to systemic functions, to be conceived as roles in a system (Boogerd et al. 2005). For the latter, merely the present contribution to a capacity or disposition to such a contribution is relevant. However, it is not satisfying to rely on Cummins's (1975) approach as an explication of the concept of function in question, since this approach does not allow to make any difference between function and dysfunction. But it is possible to modify the Cummins approach suitably by introducing reference to a norm for functionality and dysfunctionality. Such a norm already is present in the way systems (and other) biologists refer to their subject of inquiry: they do not describe the metabolic network of one individual of the species *E. coli* or *Drosophila melanogaster* but the type of network present in individuals belonging to these species. The type of the network is described as fixed by genetic and epigenetic factors, growth conditions (which exactly for the reason of sticking with one type need to be standardized when networks are to be investigated), and so forth. Not only the networks as wholes but their components are described as being of fixed types. The fixed types also fix the roles of the components within a network. Such fixed roles are conceived as functions. Any deviation from the function that corresponds to the fixed type may therefore be identified and classified as dysfunctional (Krohs 2004, 2008b).[17] This allows for the following explication of the concept of function:

A function is a contribution of a type-fixed component to a capacity of a type-fixed system (Krohs 2008b). "Contribution" is to be taken in a dispositional meaning, as in Cummins (1975). To ascribe a function, it is sufficient to single out the contribution of a component to the capacity according to the type fixation.

This explication of the concept of function reconstructs how biologists discern functionality from dysfunctionality without referring to the evolutionary history of an organism but merely by reference to its physiology and ontogeny.[18] When identifying modules with the components the explication refers to, function talk in approaches to functional modularization can be understood in this way. The explication is also applicable to function ascriptions in the technological realm, so it allows for a comparison of findings about functional modules in biology and technology.[19]

A look at functional modules of engineered systems casts some light on the difficulties we envisage with respect to functional modularity of biological networks, and also shows where the difficulties originate. With respect to technical artifacts, Pahl and colleagues (2007: 496) discern two kinds of modules: functional modules and production modules. Though conceptually different, production modules and functional modules usually coincide. The reason for this is to be found in the rational planning of the design process. Early on, the desired capacities of the system are specified and broken down into functional modules. Each single functional module is then designed as a separate production module (Pahl et al. 2007: 499–508). In the realized modular artifact, a production module, or an assembly of several such, becomes a structural module. Consequently the structural modules (henceforth "S-modules") coincide with the functional modules ("F-modules"). The only reason for this congruence, however, is that the S-modules are designed as realizations of F-modules. Such a rationale of the design process is missing in the biological case: nobody has designed biological systems to have a 1:1 S-module:F-module map. The modules have evolved by processes of adaptation, response to constraints, self-organization, and so on. Since we are confronted with the empirical findings of distributed functionality and overlapping functional modules anyway, it is unsurprising that F- and S-modules of biological networks are often found not to coincide. To the contrary, cases where F- and S-modules coincide require explanation.[20] In such cases one must identify external causes or internal constraints that "adjust" the system in the direction of such congruence of S- and F-modules.

As an example for a mismatch of F- and S-modules, consider the citric acid cycle. It is delineated functionally (see Krohs 2004: 173) and consequently must be taken as a metabolic F-module. It is not at the same time an S-module, for the following reasons: Each metabolic intermediate of the cycle is also involved in many anabolic and catabolic reactions not belonging to the cycle, the so-called anaplerotic and cataplerotic sequences that heavily influence and help regulate the size of the pools of each of the intermediates

(Kornberg 1965; Owen, Kalhan, and Hanson 2002). Moreover, the cofactors that are involved in the cycle occur in many other metabolic pathways as well. In contrast to such a plethora of external interactions, only two reactions integrate each intermediate into the cycle. So the external interactions of the cycle seem to be stronger than the internal ones, which is the opposite of what is required for structural modularity.[21] In biological systems, other than in engineered ones, functional and structural modules are not usually congruent.

15.7   A Lesson from the Function-Structure Map

In the case of technical artifacts, the congruence of F-modules and S-modules was seen to originate from design methodology. Unsurprisingly such congruence is not generally found in biological systems, since evolution cannot follow any methodology. Nevertheless, in many developmental and evolutionary modules, a unique attribution of function(s) to a structural module is in fact possible (Wagner, Mezey, and Calabretta 2005). In such cases biological F- and S-modules coincide. In developmental modules something similar is found to what is in the subjects of more classical physiological disciplines, where functional descriptions often map fairly reliably on structural descriptions (Krohs 2004). An evolutionary and developmental module—for example, one that gives rise to a signaling pathway—forms an S-module that is simultaneously an F-module. A characteristic of such modules is that they contribute, identically or with slight variations, to different capacities of an organism. It is precisely this multiple involvement that is regarded as one of the main benefits of biological modularity (Schlosser and Wagner 2004; Callebaut and Rasskin-Gutman 2005). But, in light of the result of the discussion here, the congruence of F- and S-modules needs an evolutionary explanation.

Functionality within networks depends on the properties of the network components and on the structure of the network. According to the explication of type-fixed structures given in section 15.6, any change of an F-module is the result of a change of the structure of the system. So the explanation of any congruence of F- and S-modules will be equivalent to an explanation of a change of the structure of a system until congruence of F- and S-modules is achieved. One may be tempted to conclude that this evolutionary change of structure creates evolutionary costs. In the case of technical artifacts such costs do arise during the design process and therefore contribute to the price of the product. They are outweighed by lower costs of construction, diagnosis, and maintenance. However, evolutionary "costs" are not comparable with costs of the four kinds that I have identified here. Their relevance differs between the biological and the technological case. In contrast to the costs of design processes in engineering, evolutionary "costs" do not influence the energy requirement, or any other measure for the costs of the ontogenesis or the life span processes of an organism. And evolutionary processes as such do not cause costs because nobody is waiting for a particular result. Time and effort (in terms of, e.g., lethal variations) required in evolution, though they may have drastic consequences on the species level, are simply irrelevant with respect to the evolved organism—in sharp contrast to the engineering case.

If the congruence of F- and S-modules is an outcome of evolutionary processes and thus contingent, a conclusion ought to be drawn with respect to the theoretical treatment of the mismatch found in large metabolic networks. The epistemic goal, then, should not be to identify the one and only valid method of decomposing a network and trying to show that the other way of decomposition distorts the picture. From the functional view, the structural picture looks similarly distorted, as does the functional map from a structural perspective. Bias is relative and can therefore be ascribed reciprocally. To discredit one method, one had to show that it misconceives the subject of inquiry instead of demonstrating a bias with respect to some other approach. But as long as functionality is considered as relevant in biology at all, one should allow for an integration of functional modularization into the biological account of a network. I admit that conceptual problems do at first occur with mismatching modules. However, if biological organisms are as they seem to be, and if physiology is still regarded as relevant to biology, then the task is to solve the conceptual problems and develop a more differentiated account of biological networks.

15.8   Conclusion

I show that insofar as they focus almost exclusively on the benefits or positive fitness effects of modularity, present explanations of the evolution of modularity of biological organisms are incomplete. They can neither account for modularity that originates in integrative rather than in parcellation processes nor do they explain that modularity comes in degrees. What is missing is the consideration of the cost of modularity, as it is known from the field of technology. I single out different kinds of costs: costs of a longer period of development, tara costs, autonomy costs, and costs for module-wise replacement. The list is not meant to be complete, but it must not include costs for evolutionary processes.

What then should an explanation of the evolution of modularity look like? It must demonstrate, for the particular case considered, that the balance of the costs and benefits of modularity lies on the side of modularity, and it must estimate the expectable degree of quasi-independence of the modules. This requires data that are neither available at present nor easy to collect. Nevertheless, without such effort, any explanation of the evolution of modularity by selective processes, even if it is based on sophisticated mathematical models from population genetics, is but an adaptive story: it shows that there *might* have been an adaptive evolutionary path leading to the modular organization observed, without ruling out the possibility that exactly the proposed evolutionary pathway was highly unlikely to occur because of fitness-decreasing "side effects." Without such a supplementation, the parcellation path to a modular structure cannot be regarded as being satisfactorily explained either.

Another issue is identified that demands an explanation in terms of evolution: the sometimes occurring congruence of functional and structural modules in metabolic and in many gene regulatory networks. Such congruence is a precondition for the multiple use of a functional module in an organism. It is shown that this congruence, though almost always present in technical artifacts due to their design methodology, is not trivially present in biological systems but only brought about by evolutionary processes. Here as well, a gap is diagnosed in the biological arguments about the evolution of modular systems. This latter gap is not to be closed by cost considerations. It poses, in contrast, epistemic costs: Where congruence is found, it needs to be explained. And where it is absent, an account of the system needs to be developed that is more differentiated than those currently available.

Notes

1. Simon consistently takes the analytical perspective and leaves open "whether we are able to understand the world because it is hierarchic [i.e., modular] or whether it appears hierarchic because those aspects of it which are not elude our understanding and observation." He gives reasons for supposing "that the former is at least half the truth—that evolving complexity would tend to be hierarchic—but it may not be the whole truth" (Simon 1969: 208).

2. I focus on biological modularity and its relation to modularity in engineering, and do not discuss the somewhat different and still unsettled issue of the modularity of mind. For a discussion of the latter issue see, e.g., Callebaut and Rasskin-Gutman (2005), García (2007), and Sarnecki (2007).

3. A recently proposed explication of two different concepts of functional modularity (García 2007) combines functional and structural criteria in each of the considered cases, functional integration and functional independence of modules. In each of the two cases this results in counting only those subsystems as modules that satisfy simultaneously the modularity conditions of functional and structural approaches. I discuss modules of this kind in section 15.7 but stay, when arguing about functional models, with the classical concept of functional decomposition, which largely disregards criteria of network structure.

4. I further explain this in section 15.6. In contrast to the biological case, the coincidence of functional and structural modules usually holds with respect to technical artifacts, as also discussed in section 15.6.

5. It was clear from the very beginning of the modularity debate (e.g., Simon 1969) that *any* analysis of a system in terms of modular components distorts the picture of the network (Krohs and Callebaut 2007).

6. *Evolutionary* modules, however, are sometimes even defined as subsystems that are both functional and developmental units (Brandon 2005; see also Schlosser 2005).

7. "[C]omplex systems will evolve from simple systems more rapidly if there are stable intermediate forms than if there are not" (Simon 1969: 196).

8. Similarly, it is doubtful whether a "gene for high mutability" could be selected in diploid organisms under usual selective regimes (Wagner, Mezey, and Calabretta 2005).

9. Fell (2007) supports the validity of this finding by comparison with similar results obtained from evolutionary approaches to engineering electric circuits by means of genetic algorithms (Koza et al. 1999; Bennett et al. 2000).

10. I do not claim that the authors of these arguments are not aware of such adaptive advantage of integration. The arguments, however, do not cover this issue.

11. With the given frequency of telephone calls that Simon assumes, Tempus loses on average twenty times as much work as Hora per interruption.

12. Simon's calculation is correct only if attaching the second to the first part of an assembly counts as two steps, not only one. This assumption holds if positioning a part—and not sticking it to another part—consumes most of the time of an assembly step.

13. The exact number of steps depends on the number of 3-part modules, which are unavoidable in this example. Minuta perhaps better redesigned his watch to contain 2-part modules only, ending up, e.g., with an arrangement of $2^{10} = 1,024$ parts. This of course posed additional costs for the new design process.

14. This in no way means that the latter would have been a possible product of evolution. Relevant in evolution are small differences among organisms with different but similar degrees of modularization, and this is never likely to result in the simultaneous appearance of highly differently modularized but otherwise similar organisms.

15. The abstract *formal* structure of a network, and "structure" is used in this sense in systems biology, is to be understood as embracing both the physical structure of an entity and its dynamics.

16. Cf. Kitamura and Mizoguchi this volume, figure 12.1.

17. The type-token relation is regarded as the weakest relation that introduces a norm suitable as a reference for functionality (see McLaughlin this volume).

18. Type fixation was of course brought about by evolution, but when applying the explication given here, only the present state and not its evolutionary history needs to be taken into account. Consequently we need not ask whether the component was type-fixed *in order to* make the contribution to the capacity (see also Krohs 2008a). So the explication is in accordance with evolutionary theory and any notion of evolutionary goal directedness is absent.

19. See also Franssen's discussion of the concept (this volume).

20. This demand holds also if only such F+S-modules are counted as modules proper, as is often proposed or presupposed by authors focusing on evolutionary modules (e.g., Brandon 2005; Schlosser 2005; Garciá 2007). In this case, F- and S-modules may be regarded as systems-biological theoretical terms. F+S-modules are discussed in section 15.7.

21. As a consequence of these structural peculiarities, many authors feel that the cycle needs to be emphasized in depictions of the metabolic network in order to make it visually discernable as a substructure (e.g., Alberts et al. 2002: 69).

References

Alberts, B., Johnson, A., Lewis, J., Raff, M., Roberts, K., and Walter, P. (2002). *Molecular Biology of the Cell.* New York: Garland.

Alexander, C. (1964). *Notes on the Synthesis of Form.* Cambridge, Mass.: The MIT Press.

Altenberg, L. (1995). The schema theorem and Price's theorem. In: *Foundations of Genetic Algorithms 3* (Whitley, D., and Vose, M. D., eds.), 23–49. Cambridge, Mass: The MIT Press.

Baldwin, C. Y., and Clark, K. B. (1997). Managing in the age of modularity. *Harvard Business Review, 75:* 84–93.

Baldwin, C. Y., and Clark, K. B. (2003). Commentary [on Baldwin and Clark, 1997]. In: *Managing in the Modular Age: Architectures, Networks and Organizations* (Garud, R., Kumaraswamy, A., and Langlois, R. N., eds.), 161–171. Oxford: Blackwell Publishers. Also published as: Modularity after the crash. Harvard NOM Research Paper No. 01-05 (2001).

Bechtel, W., and Richardson, R. C. (1993). *Discovering Complexity: Decomposition and Localization as Strategies in Scientific Research.* Princeton, N.J.: Princeton University Press.

Bennett, F. H. III, Koza, J. R., Yu, J., Mydlowec, W. (2000). Automatic synthesis, placement, and routing of an amplifier circuit by means of genetic programming. Lecture Notes in Computer Science, 1801: 1–10.

Boogerd, F. C., Bruggeman, F. J., Richardson, R. C., Stephan, A., and Westerhoff, H. V. (2005). Emergence and its place in nature: A case study of biochemical networks. *Synthese, 145:* 131–164.

Brandon, R. N. (2005). Evolutionary modules: Conceptual analyses and empirical hypotheses. In: *Modularity* (Callebaut, W., and Rasskin-Gutman, D., eds.), 51–60. Cambridge, Mass.: The MIT Press.

Calabretta, R. S., Nolfi, S., Parisi, D., and Wagner, G. P. (2000). Duplication of modules facilitates the evolution of functional specialization. *Artificial Life, 6:* 69–84.

Callebaut, W. (2005). The ubiquity of modularity. In: *Modularity* (Callebaut, W., and Rasskin-Gutman, D., eds.), 3–28. Cambridge, Mass.: The MIT Press.

Callebaut, W., and Rasskin-Gutman, D. (eds.). (2005). *Modularity: Understanding the Development and Evolution of Natural Complex Systems.* Cambridge, Mass.: The MIT Press.

Cummins, R. (1975). Functional analysis. *The Journal of Philosophy, 72:* 741–765.

Davidson, E. H., and Erwin, D. H. (2006). Gene regulatory networks and the evolution of animal body plans. *Science, 311:* 796–800.

Davidson, E. H., et al. (2002). A genomic regulatory network for development. *Science, 295:* 1669–1678.

Fell, D. A. (2007). How can we understand metabolism? In: *Systems Biology: Philosophical Foundations* (Boogerd, F. C., Bruggeman, F. J., Hofmeyr, J.-H. S., and Westerhoff, H. V., eds.), 87–101. Amsterdam: Elsevier.

Franz-Odendaal, T. A., and Hall, B. K. (2006). Modularity and sense organs in the blind cavefish, *Astyanax mexicanus. Evolution and Development, 8:* 94–100.

Friedman, N. (2004). Inferring cellular networks using probabilistic graphical models. *Science, 303:* 799–805.

Galis, F. (1999). Why do almost all mammals have seven cervical vertebrae? Developmental constraints, Hox genes, and cancer. *Journal of Experimental Zoology Part B: Molecular and Developmental Evolution, 285:* 19–26.

García, C. L. (2007). Cognitive modularity, biological modularity,and evolvability. *Biological Theory, 2:* 62–73.

Gerhart, J., and Kirschner, M. (1997). *Cells, Embryos, and Evolution.* Malden, Mass.: Blackwell Science.

Kardon, G., Heanue, T. A., and Tabin, C. J. (2004). The *Pax/Six/Eya/Dach* network in development. In: *Modularity in Development and Evolution* (Schlosser, G., and Wagner, G. P., eds.), 59–80. Chicago: University of Chicago Press.

Khakhina, L. N. (1992). *Concepts of Symbiogenesis* (Margulis, L., and McMenamin, M., eds.). New Haven: Yale University Press (1979 Russian edition, Leningrad).

Kornberg, H. L. (1965). Anaplerotic sequences in microbial metabolism. *Angewandte Chemie International Edition, 4:* 558–565.

Koza, J. R., Bennett, F. H. III, Andre, D., Keene, M. A. (1999). *Genetic Programming III: Darwinian Invention & Problem Solving.* San Fancisco: Morgan Kauffmann.

Koza, J. R., Mydlowec, W., Lanza, G., Yu, J., and Keane, M. A. (2002). Automated reverse engineering of metabolic pathways from observed data by means of genetic programming. In: *Foundations of Systems Biology* (Kitano, H., ed.), 95–121. Cambridge, Mass.: The MIT Press.

Krohs, U. (2004). Eine Theorie biologischer Theorien: Status und Gehalt von Funktionsaussagen und informationstheoretischen. *Modellen.* Berlin: Springer.

Krohs, U. (2008a). Co-designing social systems by designing technical artifacts: A conceptual approach. In: *Philosophy and Design: From Engineering to Architecture* (Vermaas, P. E., Kroes, P., Light, A., and Moore, S. A., eds.), 233–245. Dordrecht: Springer.

Krohs, U. (2008b). Functions as based on a concept of general design. Synthese: in press.

Krohs, U., and Callebaut, W. (2007). Data without models merging with models without data. In: *Systems Biology: Philosophical Foundations* (Boogerd, F. C., Bruggeman, F. J., Hofmeyr, J.-H. S., and Westerhoff, H. V., eds.), 181–213. Amsterdam: Elsevier.

Langlois, R. N., and Robertson, P. L. (1992). Networks and innovation in a modular system: Lessons from the microcomputer and stereo component industries. *Research Policy, 21:* 297–313.

Lewontin, R. C. (1978). Adaptation. *Scientific American*, *239:* 156–169.

Margulis, L. (1970). *Origin of Eukaryotic Cells.* New Haven: Yale University Press.

Margulis, L. (1981). *Symbiosis in Cell Evolution.* San Francisco: W. H. Freeman.

Onami, S., Kyoda, K. M., Morohashi, M., and Kitano, H. (2002). The DBRF method for inferring a gene network from large-scale steady-state gene expression data. In: *Foundations of Systems Biology* (Kitano, H., ed.), 59–75. Cambridge, Mass.: The MIT Press.

Owen, O. E., Kalhan, S. C., and Hanson, R. W. (2002). The key role of anaplerosis and cataplerosis for citric acid cycle function. *Journal of Biological Chemistry*, *277:* 30409–30412.

Pahl, G., Beitz, W., Feldhusen, J., and Grote, K. H. (2007). *Engineering Design: A Systematic Approach*, 3rd ed. London: Springer.

Palsson, B. Ø. (2006). *Systems Biology.* Cambridge, Mass.: Cambridge University Press.

Papin, J. A., Reed, J. L., and Palsson, B. O. (2004). Hierarchical thinking in network biology: The unbiased modularization of biochemical networks. *Trends in Biochemical Sciences*, *29:* 641–647.

Raff, R. A. (1996). *The Shape of Life: Genes, Development, and the Evolution of Animal Form.* Chicago: University of Chicago Press.

Rohwer, J. M., Schuster, S., and Westerhoff, H. V. (1996). How to recognize monofunctional units in a metabolic system. *Journal of Theoretical Biology*, *179:* 213–228.

Sagan, L. [Margulis, L.] (1967). On the origin of mitosing cells. *Journal of Theoretical Biology*, *14:* 225–274.

Sarnecki, J. (2007). Developmental objections to evolutionary modularity. *Biology and Philosophy*, *22:* 529–546.

Schaffner, K. F. (1998). Genes, behavior, and developmental emergentism: One process, indivisible? *Philosophy of Science*, *65:* 209–252.

Schank, J. C., and Wimsatt, W. C. (2001). Evolvability, adaptation and modularity. In: *Thinking About Evolution, Vol. 2: Historical, Philosophical, and Political Perspectives* (Sing, R. S., Krimbas, C. B., Paul, D. B., and Beatty, J., eds.), 322–335. New York: Cambridge University Press.

Schlosser, G. (2005). Amphibian variations: The role of modules in mosaic evolution. In: *Modularity* (Callebaut, W., and Rasskin-Gutman, D., eds.), 143–179. Cambridge, Mass.: The MIT Press.

Schlosser, G., and Wagner, G. P. (eds.). (2004). *Modularity in Development and Evolution.* Chicago: University of Chicago Press.

Simon, H. A. (1969 [1996]). *The Sciences of the Artificial*, 3rd ed. Cambridge, Mass.: The MIT Press.

Sober, E. (1984). *The Nature of Selection.* Cambridge, Mass.: The MIT Press.

Ulrich, K. T., and Eppinger, S. D. (2003). *Product Design and Development*, 3rd ed. Boston: McGraw Hill.

Wagner, G. P., and Altenberg, L. (1996). Perspective: Complex adaptations and the evolution of evolvability. *Evolution*, *50:* 967–976.

Wagner, G. P., Mezey, J., and Calabretta, R. (2005). Natural selection and the origin of modules. In: *Modularity* (Callebaut, W., and Rasskin-Gutman, D., eds.), 33–49. Cambridge, Mass.: The MIT Press.

White, K. P. (1999). Systems design. In: *Handbook of Systems Engineering and Management* (Sage, A. P., Rouse, W. B., eds.), 455–481. New York: Wiley.

16 Technical Artifacts, Engineering Practice, and Emergence

Peter Kroes

16.1 Introduction: Emergence as a Practical Problem

Mainly driven by an *intellectual* challenge, philosophers, biologists, and more recently also physicists have devoted a lot of attention to the analysis of what they call "emergent properties" and "emergent behavior or capacities."[1] Their aim is to understand how the overall properties of (complex) systems, composed of various parts, are related to the properties of their parts and their relations. Emergence is said to occur when certain properties appear in a system that are novel or unexpected and go beyond the properties of the parts of that system. Paradigmatic examples of emergent features studied in these fields are consciousness and the brain, life in biological organisms, and chaotic behavior of complex dynamical physical systems. Especially within the biological sciences, discussions about emergent behavior have a long history. Can features of living organisms such as homeostasis, plasticity, or adaptation be reduced without residue to the properties, behavior, and relations of their underlying parts or not, and if not, in what sense can these features be claimed to be emergent, that is, "more than the sum of the parts" (see, e.g., Feltz, Crommelinck, and Goujon 2006; Boogerd et al. 2005)? The issues about emergence in these various fields are confounded by profound ambiguities related to the notions of "reduction" and "emergence." With regard to the mind-body problem, various interpretations of the notions of reduction and emergence have resulted in a proliferation of positions in recent decades. In a review article, Van Gulick (2001) distinguishes at least ten varieties for each of the notions of reduction and of emergence, giving rise to a host of possible combinations to interpret the mind-body problem.

In recent times, the notion of "emergence" has also become a topic of debate in engineering circles (Buchli and Santini 2005; Deguet, Demazeau, and Magnin 2006; Johnson n.d.). The science of complex systems as well as engineering itself are developing fields in which emergent properties are seen as a defining feature of complexity.[2] Complex systems may exhibit processes of self-organization that are of particular interest for engineering when these processes lead to emergent systemic properties such as adaptivity, robustness, and self-repair. What is taken to be a less desirable feature of these emergent properties is that their occurrence is often unexpected and unpredictable. Engineers would not be engineers, however, if they did not try to harness these emergent properties for human benefit. Within an engineering context, the challenges that emergent phenomena pose are primarily not intellectual—they are *practical* in nature. Understanding emergent properties is one thing; manipulating them to solve practical problems is another. There certainly appears to be a need to manipulate emergent features within engineering practice. On the one hand, emergent phenomena in complex systems can have disastrous effects; blackouts in electric energy supply systems are often claimed to be emergent features of these systems.[3] On the other hand, they may be beneficial with respect to certain desirable properties of those systems; for instance, complex adaptive systems may be more robust to changing conditions in the environment.[4] Depending upon whether the emergent features are undesirable or desirable, engineers face the task of either trying to control or avoid the emergent features or design them into systems. This is a branch of engineering sometimes referred to as "emergence engineering" or "complexity engineering" (Potgieter 2004; Buchli and Santini 2005). However, is it at all possible to take emergence into account and design systems in a predictable and controllable way, given that these emergent properties are characterized as unexpected and unpredictable? Does the combination of engineering and emergence make sense at all?

Discussions of emergence within engineering practice suffer from the same problem as the ones mentioned here: there is much confusion about the meaning of the term *emergence* and consequently about its implications for engineering practice. I make no attempt to review the various interpretations given to emergence.[5] Instead let me point out three issues from the various discussions about emergence that strike me as being of particular importance for engineering practice.

1. The causal powers of emergent features.   Emergent features without causal powers of their own (often called "epiphenomena") appear prima facie not to be very interesting for engineering practice; they do not present new opportunities for causally influencing the physical environment. Why bother trying to predict and control phenomena that have no causal impact on the world? Emergent features with causal powers are more interesting in this respect but confront engineers with the problem of how to control these causal powers.

2. Emergence and functional decomposition.   To tackle complicated design problems, engineers deploy a divide-and-conquer strategy known as functional decomposition: the function of the system as a whole is divided into subfunctions (and so on) that are performed by the constituent parts of the system. This strategy does not seem to work for systems functions that are based on emergent properties, since these properties cannot be understood in terms of the properties of the parts of the system. As Johnson (n.d.: 5) remarks, "The idea that there are properties of systems that cannot intrinsically be understood in terms of lower level concepts seems entirely at odds with many contemporary approaches to engineering. For example, this would suggest that there are many risks that cannot be identified by following the functional decomposition that is implicit within techniques such as FMECA [Failure Modes, Effects, and Criticality Analysis]."

3. The unexpectedness and/or unpredictability of emergent features. If emergent features are unexpected or unpredictable, then it seems rather odd to try to design emergent properties into systems, no matter how desirable they may be. Is it at all possible to design such properties into a system and is it possible to prevent them from occurring? Of course engineers are accustomed to unexpected and unpredictable behavior associated with the things they design and produce. From a traditional engineering perspective, however, such behavior is to be prevented as much as possible. Engineers are convinced they can avoid unanticipated behavior if they have enough knowledge about the systems they design and make, that is, given enough knowledge about the behavior of the systems' parts and how they are related. However, if emergent properties are by definition unexpected or unpredictable, then this is impossible.

The reason these issues are of particular importance in traditional engineering practice is that they pose a serious threat to what is called the "control paradigm": under the conditions of operation and use laid down in the design specifications, the behavior of a technical system can be fully controlled by controlling the behavior of its constituent parts. Emergent features endanger the control of engineering systems. Apparently, emergent system features with causal powers of their own cannot be controlled through the causal powers of the systems' constituent parts. Functional decomposition is the tool engineers use to construct the behavior of the overall system starting from the behavior of its parts, and it allows them to control the behavior of the whole system by controlling the behavior of these parts. It would be pointless in the context of designing emergent properties. In regard to unanticipated behavior, if there is one thing that does not fit well with the mind-set of engineers, it is that the things they create display unexpected and unpredictable features (certainly when the thing is used or operated within the design specifications). It does not matter so much whether the emergent features are desirable or undesirable; it is the lack of control implied by unexpectedness and unpredictability that unsettles engineers. So it seems that emergence and control are uncomfortable bedfellows. According to Buchli and Santini (2005: 3), "there is a tradeoff between self-organization [and emergence] on one hand and specification or controllability on the other: If you increase the control over your system you will suppress self-organization capabilities."

This chapter does not address the issue of tensions between emergence and control in complex engineering systems in a direct way. That would not only involve a comparison of the various engineering notions of "emergence" that are in use but also require a detailed account of the various kinds of complex engineering systems that are said to show emergent properties (ranging from complex physical and technical systems to complex software systems—particularly multi-agent-software systems—to sociotechnical systems). From the point of view of engineering applications, many discussions about the beneficial exploitation of emergent properties are still highly speculative. Instead we deal here with the question whether, and in what sense, functions of simple, stand-alone technical arti-facts, such as everyday household appliances, may be regarded as emergent, and we explore the implications of various forms of emergence for the control paradigm. A deeper insight into these issues may be of help in understanding emergence in more complex technical systems.

Before we continue, let us heed Van Gulick's advice (2001: 27) and pay attention to our "key." The following general characterization of emergence is our point of departure: emergent features in (complex) systems are 1) *novel, qualitatively different* features in comparison to the features of the system's parts, which 2) cannot be *reduced* to the features of those parts and their relations.[6] Of course we must explicate in more precise terms the meaning of the notions "novel," "qualitatively different," and "reduced." Following Van Gulick (2001: 16 sq), we distinguish between a metaphysical/ontological and epistemic reading of this characterization of emergence; the former concerns emergence with regard to real-world items, the latter with regard to our representations of the world. Our discus-sion of ontological emergence (section 16.2) starts with the question of whether the func-tion of a technical artifact may be regarded as an ontologically emergent property with respect to its physical structure. We argue that this is the case. We also consider the pos-sibility of emergent phenomena in technical artifacts with causal powers of their own and we analyze in more detail the challenge this would create for the control paradigm within traditional engineering practice. Epistemic interpretations of emergence focus on cognitive relations between our knowledge and representations of emergent features and properties of the entities from the emergence base. Epistemically emergent properties have a direct impact on engineering practice since they signify the limits of predictability and explana-tion and therefore the limitations of the control of technical systems. Our discussion of epistemic emergence (section 16.3) first addresses the question of whether the function of a technical artifact is an epistemically emergent property. We argue that for simple techni-cal artifacts, the function is emergent relative to a physical knowledge base, but that a knowledge base may be chosen such that the function is not emergent. Finally, we explore the possible impact of weak and strong forms of epistemic emergence for the control paradigm.

16.2   Technical Functions and Ontological Emergence

Our ontological reading of the general characterization of emergence takes into consider-ation two interpretations of the first condition for emergence (*novel, qualitatively different features*), namely (a) properties that can be attributed sensibly to the system as a whole, but not to the parts of which it is made up, and (b) new causal powers that go beyond the causal powers of its parts.[7]

The first interpretation is rather weak. Taken by itself, it would make almost any macroscopic property (including functional properties) an emergent property. So in that case the ontological interpretation of the second condition (*reduction*) *for* emergence has to discriminate between ontologically emergent and nonemergent properties. The problem is how to interpret the reduction relation in an ontological sense. Van Gulick (2001: 4) discusses at least five different ways this ontological link has been interpreted in the context of the mind-body problem: elimination, identity, composition, supervenience, and realization. Many macroscopic properties will lose their ontologically emergent status under suitable interpretations of, for instance, the composition or realization relation.

We do not enter into a discussion of these various ontological reduction relations here. We are interested in the ontological relation between the function of a technical artifact and its underlying physical structure and, since it is quite common to claim that technical functions are realized by physical structures, we focus on the realization-reduction relation. Given that technical functions satisfy the first condition (in general it makes no sense to attribute the function of an artifact to its components) but fail on the second condition (they are claimed to be realized by physical structures), the conclusion appears to be straightforward that they are not ontologically emergent properties relative to their physical emergence base. Take a relatively simple example of a material technical artifact, for example, a mechanical clock. Its function is to measure time and this function is realized through its physical structure, which consists of many parts.[8] The physical construction as a whole has the functional property of being a clock, a property that cannot be attributed to any of its physical parts separately. These parts, however, in their specific configuration are considered to realize the function of measuring time. This conclusion therefore seems warranted: the function of measuring time is not ontologically emergent on the physical structure of the clock, because it is ontologically reducible to that physical structure through the relation of realization.

This line of reasoning, however, is highly problematic, since it would inevitably lead to the conclusion that the functional property of, for example, being a clock is a mind-independent, intrinsic feature of the physical construction. But technical functions are generally considered to be mind-dependent features of the world; physical objects can acquire a technical function only by being embedded in a context of intentional action. On the one hand, there is a close relationship between the function of a technical artifact and its physical structure (not any physical structure will realize the function); on the other hand, this function is intimately tied to intentional features of the world (practices of human action).[9]

The problem with this line of reasoning concerns the interpretation of the phrase "this function is realized through its physical structure." This phrase is misleading in that it conflates a function with the physical behavior (or capacity) corresponding to that function. What is realized through the physical construction is the physical behavior (capacity) that is necessary for physical constructions of this type[10] to have the function of measuring time.[11] But this behavior is not sufficient for realizing the function. Note that this behavior is not an emergent property of the physical construction; it is generally considered to be realized by the physical construction and therefore to be ontologically reducible to the behavior of its parts without any problems. However, that physical behavior by itself does not endow the physical construction with the function of measuring time—with the property of being a clock. For that, another necessary condition has to be fulfilled, namely that the physical construction is embedded in intentional (social) practices (in which it was designed as a clock and used as a clock, etc.).[12] The foregoing means that the functional property of being a clock is ontologically a relational property, which involves as relata a physical object, with the right physical capacities or behavior, and intentional actions or intentional states of a certain kind.[13]

So the physical structure by itself does not realize the technical function, which means that the second condition for functions to be ontologically emergent properties is also satisfied. Therefore, relative to its physical emergence base, a technical function is ontologically emergent. One might object to this conclusion on the grounds that it is based on an inappropriate choice of the emergence base. Suppose that the emergence base is enlarged to consist not only of a physical construction but also of actions that occur in intentional practices in which this physical construction is designed as, used as, and so on, a clock. A strong case could be made for the claim that relative to such an enlarged emergence base, technical functions are not ontologically emergent. However, the strategy of enlarging the emergence base runs the risk of trivializing the whole notion of "emergence" (it may be possible to turn any property into an ontologically nonemergent one by an appropriate choice of emergence base).

These considerations make clear that the question of whether functions of relatively simple technical artifacts are ontologically emergent properties or not, in the first interpretation described earlier, is not a straightforward matter. Philosophers may worry about this form of ontological emergence of functional properties; engineers usually do not care very much about it. From the point of view of the control paradigm of engineering practice, the second interpretation, (b), which relates emergence to new causal powers, appears to be much more relevant. So let us now turn our attention to that possibility.

To begin with, note that the interpretation of the first condition for emergence (*novel, qualitatively different features*) as new causal powers that go beyond the causal powers of the system's parts already takes care of the second (*nonreduction*) condition if we take the meaning of "going beyond" to be the same as "not reducible to" (which we do from now on). In contrast to the first interpretation of the novel, qualitatively different condition, this interpretation is rather strong: "emergence" in this sense implies that technical artifacts as a whole have new causal powers in comparison to the causal powers of the parts of which they are constituted.

We have already observed that systemic properties of technical artifacts without causal powers of their own are not of much interest for engineers. They do not offer new opportunities to causally transform the physical environment, that is, opportunities that "go beyond" those presented by the (combination of the) causal powers of the parts of the technical artifacts. Therefore the existence of such systemic properties does not undermine the control paradigm. The functional properties of ordinary technical artifacts all fall into this category: they are systemic properties without causal powers of their own, which makes them according to the second interpretation ontologically nonemergent properties (in contrast to the conclusion we reach starting from the first interpretation of ontological emergence). The causal powers or capacities corresponding to the functions of ordinary technical artifacts can all be reduced (or are assumed to be reducible) to the causal powers or capacities of their parts. If that were not the case, the use of techniques like functional and physical decomposition would not make much sense.[14]

Is the possibility of emergent features with causal powers of their own to be taken seriously within engineering practice, or at all? In discussions of the mind-body problem, emergent causal powers play an important role because of mental-to-physical causation; emergent mental states appear to have a causal impact on their own emergence base, namely brain states. This kind of emergent causal power involves what is called "reflexive downward causation": the system as a whole causally influences the state of its own constituents, which in turn determine the causal powers of the whole system (Kim 1999). Whether the notion of "reflexive downward causation" is coherent is much disputed, since it seems to involve the notion of "self-causation" or "self-determination." Arguments that take their cue from analogies with biological systems and advocate incorporating emergent features with their own causal properties into technical systems often hinge precisely on this possibility of reflexive downward causation. Desirable emergent properties such as self-repair, self-optimization, self-learning, and adaptability all imply that systems with these properties change their own emergence base one way or another. Again the question may be raised whether this is a coherent conceptual possibility. Another, equally disputable, conceptual possibility would be the emergence of causal powers in technical systems that allow changing the environment of those systems.[15] Such causal powers would be neither reflexive nor necessarily downward.[16]

If we assume, for the sake of argument, that emergent causal powers, whether reflexive downward or not, are a conceptual possibility, then their actual occurrence in technical systems would pose serious problems for traditional engineering practice. Assuming that ontological nonreducibility implies epistemic nonreducibility,[17] it would not be possible to predict the occurrence of such emergent causal powers in a particular technical system on the basis of considerations that start from the causal powers of its parts. Such phenomena could be predicted only empirically (inductively; see the next section). That by itself may not be a serious drawback for engineering practice, in which direct experience with regard to the functioning of technical artifacts (e.g., the actual testing of prototypes) is more or less standard procedure. But the existence of emergent causal powers would immediately raise the question whether it is possible to harness them and to design functions based on or realized by those emergent causal powers for use in technical systems. If inductive reasoning leads to the conclusion that certain types of complex systems show some kind of emergent causal power, it will be possible of course to introduce functions based on this behavior into the design of a new system by designing that new system such that it is a member of that type. However, this is more like imitation than true design.[18] What is missing is an insight into how the overall function of the system is realized through the subfunctions of its parts; in other words, what is missing is a functional decomposition of the system that (deductively) explains how the overall function results from a combination of all the subfunctions. But if the overall function is realized by an emergent property, it is by definition impossible to come up with a functional decomposition, since the emergent behavior (corresponding to the overall function) cannot be reduced to the behavior of its parts (corresponding to their subfunctions) (Johnson n.d.; Pavard and Dugdale 2000; Buchli and Santini 2005).

From an engineering point of view, the control of emergent causal powers also raises serious problems. If emergent causal powers cannot be ontologically reduced to the causal powers of the parts of a technical system, then it is questionable whether these causal powers can be controlled in the same way that the nonemergent causal powers of the technical system as a whole are controlled. The latter are controlled by means of "local" control parameters, that is, control parameters that affect the causal powers of the parts of the system and the way these causal powers are combined. All the local control variables together exhaust the possibilities for intervening, and thus controlling, a technical system. So it seems that insofar as emergent causal properties can be controlled at all, they must be controlled by local control variables. One form of control seems less problematic, namely to switch on or off emergent causal powers by effectively *changing* the kind of structure of the system (which is conceptually different from controlling the original system) such that the resulting system no longer exhibits the emergent causal powers.

Recalling the three issues described in the introduction, we observe that the first issue involves emergence in an ontological sense. This form of emergence also touches upon the second issue, as it is connected to ideas underlying the technique of functional decomposition. Emergent causal powers prove to be incompatible with the control paradigm of traditional engineering. The occurrence of emergent causal powers in technical systems, however, is, if conceptually coherent at all, highly speculative.

16.3   Technical Functions and Epistemic Emergence

In an epistemic reading, notions such as "novel," "qualitatively different," and "reduced" are to be interpreted in terms of relations between knowledge of emergent features and knowledge of features of the emergence base. This implies that relations of epistemic emergence "turn crucially on our abilities or inabilities to comprehend or explicate the

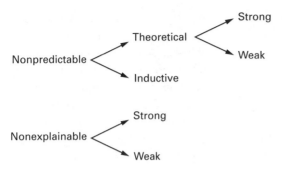

Figure 16.1
Various notions of "nonpredictable" and "nonexplainable."

nature of the links or dependencies among real-world items [the emergent features and the features of the emergence base]" (Van Gulick 2001: 16). According to Van Gulick, epistemic emergence relations are therefore "in a sense subjective." They are subjective in the sense that whether a property is emergent or not depends on a knowing subject or the knowledge base of a cognitive practice. We consider also limiting cases in which issues of epistemic emergence become decoupled from a knowing subject or a knowledge base.

Many different epistemic readings of the the notions "novel," "qualitatively different," and "reduced" may be considered, each leading to its own criteria for deciding whether a feature is epistemically emergent.[19] Here we pick out two such criteria, namely nonpredictability and nonexplainability. The reason for these choices is that prediction and explanation are two cognitive "tools," which are of paramount importance for the engineering control paradigm.

Both criteria come in different forms (see figure 16.1). The first form of nonpredictability, to be called "theoretical" (or *a componentibus*) nonpredictability, implies that a systemic feature of a technical system cannot be predicted on the basis of knowledge about its components. The second form is "inductive" nonpredictability, which implies that a systemic feature cannot be predicted using generalizations based on observed cases. For the moment, inductive nonpredictability does not concern us, since we are interested in predictability on the basis of knowledge of elements from the emergence base. Theoretical nonpredictability comes in a strong and a weak form; in its strong form, a feature is in principle not predictable even on the basis of complete knowledge about the behavior of its components.[20] This kind of nonpredictability has been a topic of intensive research in chaos theory (see, e.g., Bertuglia and Vaio 2005). In the weak sense, theoretical nonpredictability is conditional on the existing state of knowledge (about the emergent features themselves, the features of the emergence base, and the laws of nature). Nonexplainability also comes in a strong and a weak sense: a feature is strongly nonexplainable when it cannot be explained in principle on the basis of knowledge of features of its emergence base, and weakly nonexplainable when it cannot be explained relative to a given state of knowledge. We do not enter into a discussion about whether the two criteria for epistemic emergence are independent of each other. That would require an in-depth analysis of the (logical) relations among the various models of prediction and explanation. The strong form of epistemic emergence occurs when a feature of a system is in principle neither predictable nor explainable. This is a "timeless" form of epistemic emergence. All other forms are weak in the sense that they depend on a given knowledge base, which implies that a feature may acquire or lose its status of being epistemically emergent in the course of time.

With these distinctions in mind, what can we say about the epistemic emergence of functions of technical artifacts? We take knowledge of the physical parts of which they are made as our knowledge base. Since it is easy to come up with weak forms of epistemic emergence by restricting the actual physical knowledge about the components of technical artifacts in the knowledge base, we assume here that there are no limits on this knowledge base; it contains all possible physical knowledge about the components. So we are interested in the strong form of epistemic emergence.

Take again the example of our mechanical clock. In what sense, if any, can its function, measuring time, be claimed to be epistemically emergent on the knowledge of its physical structure? In this case, too, we have to distinguish carefully between the function of the clock and the physical behavior (capacity) of the clock corresponding to that function. The prediction and explanation of the physical behavior of the hands of the clock, starting from the physical properties and the arrangement of the various parts of the clock, do not seem to pose any problems. There is no reason to assume that this physical behavior is an epistemically emergent property of this physical system.

Whether the function of the clock is an epistemic emergent property is another question. Can its function be predicted on the basis of knowledge of its physical structure?[21] Of course that would not be a problem for someone who is familiar with mechanical clocks; on inductive grounds its function can be reliably predicted. But theoretical, not inductive, prediction is the issue here. Suppose some archaeologists who also happen to be mechanical engineers, but who are totally unfamiliar with mechanical clocks, dig up an object that was used by an extinct civilization as a (mechanical) clock and try to "retrodict" its function. By clever reverse engineering they may come up with the claim that its function is to produce a regular motion of the hands (probably they would make use of a functional decomposition of the object that explains how all the parts with their subfunctions together realize a regular motion of the hands). But their engineering and physical knowledge would not enable them to retrodict that its function was to measure time.[22] The function is also not explainable from the given knowledge base.[23] The explanandum is the function of the technical artifact. What can be explained are the regular movements of the hands. But that still does not explain that the function of the physical device is to measure time.

It is one thing to explain a physical feature; it is another to explain how this feature corresponds to a particular function. For someone with only physical knowledge about the world, the function of the clock cannot be explained on the basis of that person's knowledge base. So we may conclude that relative to a knowledge base that consists of only physical knowledge, the function is an emergent property of the physical object.

Suppose we enlarge the physical knowledge base with technical knowledge and all there is to know about the intentions and actions of designers, producers, users, and so forth with regard to the object; then the situation becomes different. This expanded knowledge base does seem to be sufficient for predicting and explaining that the function of the object is to measure time. The knowledge about how an object is embedded within the practices of intentional human action is precisely the kind of knowledge that archaeologists try to reconstruct to determine what the function of an object might be. With respect to this knowledge base, therefore, the function of the clock is no longer an epistemically emergent property. Note that this conclusion is based on a very strong assumption about the knowledge base, namely that it includes knowledge about how the object is going to be embedded in practices of intentional human action. It is, however, not always easy (or possible) to predict user behavior. Occasionally users attribute functions to technical artifacts that were never intended by their designers and that were also not predictable in terms of their knowledge base. These functions are therefore epistemically emergent from the designer's knowledge base (assuming that they are also not explainable on their limited knowledge base).

From an engineering point of view, the occurrence of emergent functional features in user contexts is a familiar issue. However, it is not this kind of emergence that has caused such a stir recently in engineering circles. Current interest focuses on the ways complex systems display emergent behavior (upon which functions may be based). If such complex technical systems show systemic physical capacities that are epistemically emergent, whether in the strong sense or in one of the weak senses, then the design and control of these unpredictable and unexplainable physical capacities would seem to be problematic and thus put the control paradigm in jeopardy. The fear that the control paradigm is in danger is furthermore fueled by the association of emergence with unexpectedness and surprise in discussions on emergence (Potgieter 2004: ch. 2; Deguet, Demazeau, and Magnin 2006; Johnson n.d.).

It is not evident, however, that the occurrence of epistemically emergent physical features in (complex) systems means a break with traditional engineering and therefore poses a real threat for its control paradigm. Let us for the moment discard the possibility of the strong form of epistemic emergence and restrict ourselves to its weak forms. First, note that in the history of technology, countless technical artifacts have been designed and constructed on the basis of physical phenomena that could not be theoretically predicted or explained at the time. Relative to that contemporary knowledge base, these physical phenomena must be qualified as epistemically emergent. The epistemically emergent nature of these phenomena is usually, however, of limited duration, since progress in the knowledge base, mainly driven by a desire for better control, turns them into nonemergent phenomena. For such phenomena, control may temporarily be an issue, but they rarely are considered to constitute a principal threat to the control paradigm. So the history of engineering might be seen as a continuous attempt to turn epistemically emergent technical phenomena into epistemically nonemergent phenomena.[24]

A second point to be noted is the association of unexpectedness with emergence. If we take due account of the distinction between theoretical and inductive (non)predictability, this association turns out to be unfounded. Let us call the occurrence of a phenomenon unexpected if it is at the moment of its occurrence theoretically and inductively unpredictable. On this interpretation of *unexpected,* the occurrence of epistemically emergent features in technical artifacts is not necessarily unexpected. It may still be possible to predict their occurrence inductively. So technical systems may exhibit expected or unexpected emergent features.[25] From the point of view of control, these are very different categories of emergent features. Unexpected emergent features are the most problematic because they take us by surprise; there is no way to control them in advance. They differ from expected emergent features in terms of control. Expected emergent features are theoretically unpredictable and cannot be explained *at a given point in time* (remember our restriction to weak forms of epistemic emergence), but that does not imply that they cannot be controlled to a high degree. In fact the control of expected emergent phenomena is part and parcel of engineering practice; it is based on inductively established rules of practice and correlations between properties of the emergent feature and properties of components of the system (which, on the assumption that these properties may be controlled, may be used as control variables). Of course if it would be possible to come up in these cases with a functional decomposition of the whole system, which would be tantamount to an explanation of the emergent feature, then that would probably greatly enhance the possibilities of control of the system.

Whether "inductive control" is also a viable possibility for features that are epistemically emergent in the strong sense depends on whether the possibility of regularities (correlations) between properties of the emergent feature and properties of the components in the emergence base is compatible with the in-principle theoretical nonpredictability and nonexplainability of (properties of) the emergent feature. We leave that issue open, as well as the issue of whether there are any technical systems that show those kinds of epistemically emergent features.

Returning to the issues mentioned in the introduction, we observe with regard to the second, about emergence and functional decomposition, that weak forms of epistemic emergence pose much less of a threat to the use of techniques such as functional decomposition than often suggested. Engineers have always had to deal with weakly emergent phenomena, and their success in turning them into ordinary, nonemergent phenomena testifies to the viability of techniques such as functional decomposition as well as the strength of the control paradigm. With regard to the third issue, concerning the unexpectedness and/or unpredictability of emergent features, the association of emergent phenomena with unexpected phenomena is based on a confusion of theoretical and inductive (non)predictability; emergent features, even of the strong type, are not necessarily unexpected.

16.4   Conclusion

Even for simple technical artifacts, the issue of emergence is not a straightforward matter. The function of a technical artifact such as a mechanical clock is ontologically emergent on its physical structure in the sense that its function is not a property that may be attributed to its parts and is not realized by the physical structure of the clock. Although ontologically emergent in this sense, the functions of simple technical artifacts have no causal powers of their own. Therefore they pose no threat to the control paradigm of traditional engineering. Ontologically emergent functions (features) with causal powers of their own would seriously undermine that paradigm; such a strong form of ontological emergence, however, does not appear to be very likely. As far as weak forms of epistemic emergence are concerned, they are part and parcel of routine engineering practice and constitute no significant threat to the control paradigm. The functions of technical artifacts as well as the physical phenomena upon which they are based may be weakly epistemically emergent. I argue that it is a mistake to assume that weak epistemic emergence implies unexpectedness and on that ground poses a threat to the control paradigm. Strong epistemic emergence endangers the control paradigm on the grounds that it is incompatible with techniques such as functional decomposition. Whether the extreme complexity of some of the modern technical systems implies a strong kind of epistemic emergence remains to be seen. Without a doubt, the complexity of these systems stretches to the very limit the capabilities of traditional methods of designing and controlling technical systems, stretching them sometimes so far beyond that these methods are no longer applicable. The search by engineers for new principles of design and control appears warranted. However, complexity within systems is not necessarily proof that such systems will display features that are strongly epistemically emergent. For the time being, with respect to the impact of emergence on engineering practice, we may have to revert to the age-old saying *Nihil nove sub sole*.[26]

Acknowledgments

I am grateful to the Netherlands Institute for Advanced Study (NIAS) for providing me with the opportunity, as a fellow-in-residence, to complete this chapter.

Notes

1. For a brief history of the notion of emergence, see O'Connor and Wong (2005). In the following I use the expression "emergent properties (features)" as shorthand to refer to emergent behavior and emergent capacities.

2. See, e.g., the pre-proceedings of the Paris conference (November 14–18, 2005) of the European Complex Systems Society, ECCS'05 (http://complexite.free.fr/ECCS/); this conference hosted satellite workshops on topics such as "Engineering with Complexity and Emergence" and "Embracing Complexity in Design."

3. Note that the occurrence of emergent phenomena in technical systems may raise intricate problems for issues regarding the moral responsibilities of engineers, especially when this behavior is not predictable; see also Johnson (n.d.: 2).

4. Kasser and Palmer (2005) distinguish three types of emergent properties, namely undesired, serendipitous, and desired; serendipitous features are described as "beneficial and desired once discovered, but not part of the original specifications."

5. See Deguet, Demazeau, and Magnin (2006) and Johnson (n.d.).

6. For a discussion of basic ideas associated with the notion of "emergence," see, e.g., Humphreys (1997: sect. 3), Rueger (2000), Chalmers (2002), Van Gulick (2001), and Kim (1999). Note that our "key" does not fit into the overview of Van Gulick.

7. See Van Gulick (2001: 17); as he remarks, these two interpretations may overlap in case the identity criteria for properties are based on causality profiles.

8. For the purpose of this chapter, I treat the functional property of measuring time and the property of being a clock as equivalent.

9. Technical artifacts may be said to have a dual nature: they are physical objects with functional properties that are grounded in physical as well as intentional phenomena; see Kroes and Meijers (2006).

10. The inclusion of the phrase "for physical constructions of this type" is necessary because other kinds of physical systems may measure time in completely different ways.

11. I leave aside here issues regarding malfunctioning technical artifacts.

12. Cf. also Searle's remark (1995: 10) that being a screwdriver presupposes being thought of as, designed as, and used as a screwdriver, etc.

13. See also Meijers (2000).

14. Whether this is also true for the causal powers of status (social) functions of technical artifacts remains to be seen (I thank Jeroen de Ridder for drawing my attention to this point).

15. An example of this kind of emergent causal power in the mind-body situation would be telekinesis, a controversial phenomenon to say the least. But then again, mustn't there be something similar at work in mind-to-matter causation of the reflexive-downward type? Note that engineering attempts to design man-machine interfaces that bypass any (observable) physical human action do not assume causal efficacy of emergent features (mental states) on the human environment; they operate on the basis of the detection of brain states (and thus are based on traditional physical-to-physical causation).

16. It is quite common in discussions about emergence to use notions such as "high-level" (emergent) and "low-level properties"; the notion of "downward causation" fits very well with this level talk. Emergence then results in a view of the world with a multilevel (physical, chemical, biological, psychological, and social) ontology. It is questionable, however, whether it is necessary to couple the notion of "emergence" to a multilevel view of the world. Humphreys (1997) argues that it is not, just as the notion of "supervenience" does not require such a multilevel view of the world. With regard to technical artifacts, the coupling of emergent properties to different levels of reality also is not obvious. If indeed it is the case that a physical construction without a function is not a technical artifact, and that in reverse a function not realized in a physical structure also is not a technical artifact, then it is not clear why the function should be classified as a high-level property of a technical artifact. For these reasons, I try to avoid as much as possible the use of level talk in this chapter.

17. If that would not be the case, then engineers would be very much inclined to apply Occam's razor. Why posit the existence of new entities (emergent causal powers) when complete epistemic access to these new entities is possible through other entities (causal powers of parts)? What is the conceptual gain in

combining the view of complete epistemic access to emergent causal powers with the idea of ontological emergence?

18. Kim (1999) makes a similar remark about designing systems with phenomenal experiences: "But it is difficult to imagine our designing novel devices and structures that will have phenomenal experiences; I don't think we have any idea were to begin. The only way we can hope to manufacture a mechanism with phenomenal consciousness is to produce an appropriate physical duplicate of a system that is known to be conscious. Notice that this involves inductive prediction, whereas theoretical prediction is what is needed to design new physical devices with consciousness."

19. For the epistemic reduction relation alone, Van Gulick (2001, p. 15) mentions five possibilities.

20. Note that the "in principle" clause decouples this type of nonpredictability from a knowing subject or a knowledge base. For an interesting discussion of various forms of in-principle nonpredictability, see Stephan (2002).

21. Can the function of an object be predicted? Here we take the prediction of the function to mean the prediction of the use of that object corresponding to that function.

22. This example, by the way, makes clear that the two criteria for epistemic emergence, nonpredictability and nonexplainability, may become intertwined.

23. There is no generally accepted model for an explanation of the technical function of an object; the notion of an explanation of a technical function used here is of a rather intuitive kind.

24. This observation also holds if the following definition of *weak emergence*, one proposed by Chalmers (2002), is adopted: "Emergence is the phenomenon wherein a system is designed according to certain principles, but interesting properties arise that are not included in the goals of the designer."

25. These two forms of emergent features derive from an ambiguity in the notion of "novel" in the general characterization of emergence; "novel" may be taken to mean novel in time, or novel with respect to the properties of the emergence base.

26. The only exception may be the engineering of sociotechnical systems. The expression "sociotechnical system" refers to complex, large-scale systems such as air transport systems or electric energy supply infrastructures. The behavior of these systems is driven in a significant way by their technical elements, but the functioning of the whole system depends as much on the functioning of these technical components as on the functioning of the social infrastructure and the behavior of human actors. Sociotechnical systems are hybrid systems consisting of elements of various kinds, such as natural objects, technical artifacts, and human actors and social entities (together with the rules and laws governing the behavior of human actors and social entities). The design and control of sociotechnical systems raises fundamental issues for the traditional control paradigm, since the system to be designed and/or controlled contains elements that may change the system from within.

References

Bertuglia, C. S., and Vaio, F. (2005). *Nonlinearity, Chaos, and Complexity: The Dynamics of Natural and Social Systems.* Oxford: Oxford University Press.

Boogerd, F. C., Bruggeman, F. J., Richardson, R. C., Stephan, A., and Westerhoff, H. V. (2005). Emergence and its place in nature: A case study of biochemical networks. *Synthese, 145:* 131–164.

Buchli, J., and Santini, C. C. (2005). Complexity engineering: Harnessing emergent phenomena as opportunities for engineering. In: *Reports of the Santa Fe Institute's Complex Systems Summer School 2005.* Santa Fe: Santa Fe Institute.

Chalmers, D. J. (2002). Varieties of emergence. http://consc.net/papers/granada.html.

Deguet, J., Demazeau, Y., and Magnin, L. (2006). Elements about the emergence issue: A survey of emergence definitions. *Complexus, 3:* 24–31.

Feltz, B., Crommelinck, M., and Goujon, P. (eds.). (2006). Self-organization and emergence in life sciences. *Synthese Library, Vol. 331.* Dordrecht: Springer.

Humphreys, P. (1997). Emergence, not supervenience. *Philosophy of Science, 64 (Proceedings):* S336–S345.

Johnson, C. W. (n.d.). What are emergent properties and how do they affect the engineering of complex systems. http://www.dcs.gla.ac.uk/~johnson/papers/emergence.pdf.

Kasser, J. E., and Palmer, K. D. (2005). Reducing and managing complexity by changing the boundaries of the system. *Proceedings CSER 2005* (March 23–25). Hoboken, N.J.

Kim, J. (1999). Making sense of emergence. *Philosophical Studies, 95:* 3–36.

Kroes, P., and Meijers, A. (guest eds.). (2006). Special issue: The dual nature of technical artefacts. *Studies in History and Philosophy of Science, 37*(1).

Meijers, A. (2000). The relational ontology of technical artifacts. In: *The Empirical Turn in the Philosophy of Technology. Research in Philosophy and Technology, Vol. 20* (Kroes, P., and Meijers, A., eds.), 81–96. Amsterdam: JAI/Elsevier.

O'Connor, T., and Wong, H. Y. (2005). Emergent properties. In: *The Stanford Encyclopedia of Philosophy, Summer 2005 Edition* (Zalta, E. N., ed.), http://plato.stanford.edu/archives/sum2005/entries/properties-emergent.

Pavard, B., and Dugdale, J. (2000). The contribution of complexity theory to the study of socio-technical systems. *InterJournal Complex Systems, 335.* Cambridge, Mass.: New England Complex Systems Institute.

Potgieter, A. E. G. (2004). The engineering of emergence in complex adaptive sytems. Thesis. Pretoria: University of Pretoria.

Rueger, A. (2000). Robust supervenience and emergence. *Philosophy of Science, 67:* 466–489.

Searle, J. R. (1995). *The Construction of Social Reality.* London: Penguin.

Stephan, A. (1998). Varieties of emergence in artificial and natural systems. *Zeitschrift für Naturforschung, 53C:* 639–656.

Stephan, A. (2002). Emergentism, irreducibility, and downward causation. *Grazer Philosophische Studien, 65:* 77–93.

Van Gulick, R. (2001). Reduction, emergence and other recent options in the mind/body problem. *Journal of Consciousness Studies, 8:* 1–34.

Contributors

Paul Sheldon Davies
Department of Philosophy
College of William and Mary
Williamsburg, Virginia
USA

Maarten Franssen
Section of Philosophy
Delft University of Technology
Delft
The Netherlands

Wybo Houkes
Section of Philosophy and Ethics of
Technology
Department of Technology Management
Eindhoven University of Technology
Eindhoven
The Netherlands

Yoshinobu Kitamura
The Institute of Scientific and Industrial
Research
Osaka University
Ibaraki
Osaka
Japan

Peter Kroes
Section of Philosophy
Delft University of Technology
Delft
The Netherlands

Ulrich Krohs
Department of Philosophy
University of Hamburg
Hamburg
Germany

Tim Lewens
Department of History and Philosophy of
Science
University of Cambridge
Cambridge
United Kingdom

Andrew Light
Department of Philosophy
George Mason University
Fairfax, Virginia
USA

Françoise Longy
Institute for the History and Philosophy of
Science and Technology
University of Paris 1
Paris
France

Peter McLaughlin
Department of Philosophy
University of Heidelberg
Heidelberg
Germany

Riichiro Mizoguchi
The Institute of Scientific and Industrial
Research
Osaka University
Ibaraki
Osaka
Japan

Mark Perlman
Department of Philosophy and Religious
Studies
Western Oregon University
Monmouth, Oregon
USA

Beth Preston
Department of Philosophy
University of Georgia
Athens, Georgia
USA

Giacomo Romano
Department of Philosophy and
Social Sciences
University of Siena
Siena
Italy

Marzia Soavi
Department of Philosophy
University of Padova
Padova
Italy

Pieter E. Vermaas
Section of Philosophy
Delft University of Technology
Delft
The Netherlands

Index

Accidental function. *See* Function, accidental
Agency, 173–175, 178, 180
Agent, 27, 43, 46, 48, 56, 69–85, 94–98, 131,
    134–138, 165, 174–178, 204, 210, 212, 214,
    217, 219. *See also* Quasi-agent; System,
    multi-agent
  beliefs of, 14–15, 73–83, 219
  external, 10
  intentions and goals of, 94, 96–98, 100, 134, 214
  mental states of, 69–72, 76, 84–85, 134, 136
Aitia, 5
Alberts, B., 274n21
Alexander, C., 259
Allen, C., 18–19, 31, 142n2
Alston, W., 33n7
Altenberg, L., 263
Amundson, R., 127, 142n1, 246
Analyzing inward/synthesizing laterally, 135–136,
    141
Anti-intentionalism, 240n14
Appel, T., 142n5
Archaeology, 74, 286–287
  evolutionary, 223, 229, 233–237
Archetype, 131, 245–246
Ariew, A., 19, 32, 142n2
Aristotle, 5–6, 17, 19, 45, 139, 143n12, 168, 187
Armstrong, D., 33n7
Arthur, W., 247
Artifact
  biological, 24, 55–61
  geological, 24
Artifact function. *See* Function, artifact
Artificial selection. *See* Selection, artificial
Asymmetric dependence, 34n9
Audi, R., 33n7
Authenticity, 150–152
Ayala, F., 142n2

Bahn, P. G., 240n9
Baird, J. A., 182n1
Baker, L., 168, 182n4

Baldwin, C. Y., 263
Barrett, C., 179
Basalla, G., 5, 243
Bechtel, W., 143n13, 261
Beckner, M., 30
Bedau, M. A., 18–19
Behavior,
  of animals/organisms, 6, 25, 249
  of devices, 83, 104–117, 121, 132–133, 164–165,
    169–170, 203–218, 224, 277–287
  human/of people, 24, 28–29, 46, 57
Beitz, W., 204, 211, 215
Bekoff, M., 18–19, 31, 142n2
Bennett, F. H., III, 273n9
Bentley, P., 231–232, 240nn7–8
Bertuglia, C. S., 285
Beta-oxidation, 261, 265
Bigelow, J., 18, 19, 31, 66n2
Biological artifact/Bioartifact. *See* Artifact, biological
Biological function. *See* Function, biological
Biston betularia, 116
Bloom, P., 172–173, 179
Blumenbach, Johann Friedrich, 131–135
BonJour, L., 33n7
Boogerd, F. C., 225, 269, 277
Boorse, C., 17–19, 31, 85, 105, 115, 122, 208–209,
    218
Boyd, R. 143n18, 224, 233, 245, 248–253,
    255–256
Brandon, R., 142n2, 273n6, 274n20
Breeding, 9, 21–24, 99, 228–229
Brown, J. S., 204
Buchli, J., 277–279, 284
Buller, D., 32, 39, 43–44, 142n2

Cain, J. A., 189–190
Calabretta, R., 264–265, 271, 273n8
Callebaut, W., 261–263, 265, 271, 273nn2,5
Callicott, J. B., 148
Cameron, R., 18–19
Canfield, J., 30